Quantum Science and Technology

Series editors

Raymond Laflamme, Waterloo, ON, Canada
Gaby Lenhart, Sophia Antipolis, France
Daniel Lidar, Los Angeles, CA, USA
Arno Rauschenbeutel, Vienna University of Technology, Vienna, Austria
Renato Renner, Institut für Theoretische Physik, ETH Zürich, Zürich, Switzerland
Maximilian Schlosshauer, Department of Physics, University of Portland, Portland, OR, USA
Yaakov S. Weinstein, Quantum Information Science Group, The MITRE Corporation, Princeton, NJ, USA
H. M. Wiseman, Brisbane, QLD, Australia

Aims and Scope

The book series Quantum Science and Technology is dedicated to one of today's most active and rapidly expanding fields of research and development. In particular, the series will be a showcase for the growing number of experimental implementations and practical applications of quantum systems. These will include, but are not restricted to: quantum information processing, quantum computing, and quantum simulation; quantum communication and quantum cryptography; entanglement and other quantum resources; quantum interfaces and hybrid quantum systems; quantum memories and quantum repeaters; measurement-based quantum control and quantum feedback; quantum nanomechanics, quantum optomechanics and quantum transducers; quantum sensing and quantum metrology; as well as quantum effects in biology. Last but not least, the series will include books on the theoretical and mathematical questions relevant to designing and understanding these systems and devices, as well as foundational issues concerning the quantum phenomena themselves. Written and edited by leading experts, the treatments will be designed for graduate students and other researchers already working in, or intending to enter the field of quantum science and technology.

More information about this series at http://www.springer.com/series/10039

Maria Schuld · Francesco Petruccione

Supervised Learning with Quantum Computers

 Springer

Maria Schuld
School of Chemistry and Physics,
 Quantum Research Group
University of KwaZulu-Natal
Durban, South Africa

and

National Institute for Theoretical
 Physics (NITheP)
KwaZulu-Natal, South Africa

and

Xanadu Quantum Computing Inc
Toronto, Canada

Francesco Petruccione
School of Chemistry and Physics
University of KwaZulu-Natal
Durban, South Africa

and

National Institute for Theoretical
 Physics (NITheP)
KwaZulu-Natal, South Africa

and

School of Electrical Engineering
Korea Advanced Institute of Science
 and Technology (KAIST)
Daejeon, Republic of Korea

ISSN 2364-9054   ISSN 2364-9062 (electronic)
Quantum Science and Technology
ISBN 978-3-319-96423-2   ISBN 978-3-319-96424-9 (eBook)
https://doi.org/10.1007/978-3-319-96424-9

Library of Congress Control Number: 2018950807

© Springer Nature Switzerland AG 2018
This work is subject to copyright. All rights are reserved by the Publisher, whether the whole or part of the material is concerned, specifically the rights of translation, reprinting, reuse of illustrations, recitation, broadcasting, reproduction on microfilms or in any other physical way, and transmission or information storage and retrieval, electronic adaptation, computer software, or by similar or dissimilar methodology now known or hereafter developed.
The use of general descriptive names, registered names, trademarks, service marks, etc. in this publication does not imply, even in the absence of a specific statement, that such names are exempt from the relevant protective laws and regulations and therefore free for general use.
The publisher, the authors and the editors are safe to assume that the advice and information in this book are believed to be true and accurate at the date of publication. Neither the publisher nor the authors or the editors give a warranty, express or implied, with respect to the material contained herein or for any errors or omissions that may have been made. The publisher remains neutral with regard to jurisdictional claims in published maps and institutional affiliations.

This Springer imprint is published by the registered company Springer Nature Switzerland AG
The registered company address is: Gewerbestrasse 11, 6330 Cham, Switzerland

For Chris and Monique

Preface

Quantum machine learning is a subject in the making, faced by huge expectations due to its parent disciplines. On the one hand, there is a booming commercial interest in quantum technologies, which are at the critical point of becoming available for the implementation of quantum algorithms, and which have exceeded the realm of a purely academic interest. On the other hand, machine learning along with artificial intelligence is advertised as a central (if not the central) future technology into which companies are bound to invest to avoid being left out. Combining these two worlds invariably leads to an overwhelming interest in quantum machine learning from the IT industry, an interest that is not always matched by the scientific challenges that researchers are only beginning to explore.

To find out what quantum machine learning has to offer, its numerous possible avenues first have to be explored by an interdisciplinary community of scientists. We intend this book to be a possible starting point for this journey, as it introduces some key concepts, ideas and algorithms that are the result of the first few years of quantum machine learning research. Given the young nature of the discipline, we expect a lot of new angles to be added to this collection in due time. Our aim is not to provide a comprehensive literature review, but rather to summarise themes that repeatedly appear in quantum machine learning, to put them into context and make them accessible to a broader audience in order to foster future research.

On the highest level, we target readers with a background in either physics or computer science that have a sound understanding of linear algebra and computer algorithms. Having said that, quantum mechanics is a field based on advanced mathematical theory (and it does by no means help with a simple physical intuition either), and these access barriers are difficult to circumvent even with the most well-intended introduction to quantum mechanics. Not every section is therefore easy to understand for readers without experience in quantum computing. However, we hope that the main concepts are within reach and try to give higher level overviews wherever possible.

We thank our editors Aldo Rampioni and Kirsten Theunissen for their support and patience. Our thanks also go to a number of colleagues and friends who have helped to discuss, inspire and proofread the book (in alphabetical order): Betony Adams, Marcello Benedetti, Gian Giacomo Guerreschi, Vinayak Jagadish, Nathan Killoran, Camille Lombard Latune, Andrea Skolik, Ryan Sweke, Peter Wittek and Leonard Wossnig.

Durban, South Africa
March 2018

Maria Schuld
Francesco Petruccione

Contents

1 **Introduction** 1
   1.1 Background 2
      1.1.1 Merging Two Disciplines 2
      1.1.2 The Rise of Quantum Machine Learning 4
      1.1.3 Four Approaches 5
      1.1.4 Quantum Computing for Supervised Learning 6
   1.2 How Quantum Computers Can Classify Data 7
      1.2.1 The Squared-Distance Classifier 8
      1.2.2 Interference with the Hadamard Transformation 9
      1.2.3 Quantum Squared-Distance Classifier 13
      1.2.4 Insights from the Toy Example 16
   1.3 Organisation of the Book 17
   References 18

2 **Machine Learning** 21
   2.1 Prediction 22
      2.1.1 Four Examples for Prediction Tasks 23
      2.1.2 Supervised Learning 25
   2.2 Models 28
      2.2.1 How Data Leads to a Predictive Model 30
      2.2.2 Estimating the Quality of a Model 32
      2.2.3 Bayesian Learning 34
      2.2.4 Kernels and Feature Maps 35
   2.3 Training 39
      2.3.1 Cost Functions 40
      2.3.2 Stochastic Gradient Descent 42
   2.4 Methods in Machine Learning 44
      2.4.1 Data Fitting 45
      2.4.2 Artificial Neural Networks 48

		2.4.3 Graphical Models	60
		2.4.4 Kernel Methods	64
	References		71
3	**Quantum Information**		75
	3.1	Introduction to Quantum Theory	76
		3.1.1 What Is Quantum Theory?	76
		3.1.2 A First Taste	78
		3.1.3 The Postulates of Quantum Mechanics	84
	3.2	Introduction to Quantum Computing	91
		3.2.1 What Is Quantum Computing?	91
		3.2.2 Bits and Qubits	93
		3.2.3 Quantum Gates	97
		3.2.4 Quantum Parallelism and Function Evaluation	101
	3.3	An Example: The Deutsch-Josza Algorithm	103
		3.3.1 The Deutsch Algorithm	103
		3.3.2 The Deutsch-Josza Algorithm	104
		3.3.3 Quantum Annealing and Other Computational Models	106
	3.4	Strategies of Information Encoding	108
		3.4.1 Basis Encoding	109
		3.4.2 Amplitude Encoding	110
		3.4.3 Qsample Encoding	112
		3.4.4 Dynamic Encoding	113
	3.5	Important Quantum Routines	114
		3.5.1 Grover Search	114
		3.5.2 Quantum Phase Estimation	117
		3.5.3 Matrix Multiplication and Inversion	119
	References		123
4	**Quantum Advantages**		127
	4.1	Computational Complexity of Learning	127
	4.2	Sample Complexity	131
		4.2.1 Exact Learning from Membership Queries	133
		4.2.2 PAC Learning from Examples	134
		4.2.3 Introducing Noise	135
	4.3	Model Complexity	135
	References		137
5	**Information Encoding**		139
	5.1	Basis Encoding	141
		5.1.1 Preparing Superpositions of Inputs	142
		5.1.2 Computing in Basis Encoding	145
		5.1.3 Sampling from a Qubit	146

	5.2	Amplitude Encoding....................................	148
		5.2.1 State Preparation in Linear Time	150
		5.2.2 Qubit-Efficient State Preparation	154
		5.2.3 Computing with Amplitudes	159
	5.3	Qsample Encoding	159
		5.3.1 Joining Distributions	160
		5.3.2 Marginalisation	160
		5.3.3 Rejection Sampling	162
	5.4	Hamiltonian Encoding	163
		5.4.1 Polynomial Time Hamiltonian Simulation	164
		5.4.2 Qubit-Efficient Simulation of Hamiltonians ..	166
		5.4.3 Density Matrix Exponentiation	167
	References	169

6	**Quantum Computing for Inference**...........................	173
	6.1 Linear Models	173
	6.1.1 Inner Products with Interference Circuits...	174
	6.1.2 A Quantum Circuit as a Linear Model	179
	6.1.3 Linear Models in Basis Encoding	182
	6.1.4 Nonlinear Activations	183
	6.2 Kernel Methods	188
	6.2.1 Kernels and Feature Maps	189
	6.2.2 The Representer Theorem	191
	6.2.3 Quantum Kernels	193
	6.2.4 Distance-Based Classifiers	196
	6.2.5 Density Gram Matrices	201
	6.3 Probabilistic Models	204
	6.3.1 Qsamples as Probabilistic Models	204
	6.3.2 Qsamples with Conditional Independence Relations	205
	6.3.3 Qsamples of Mean-Field Approximations	207
	References	209

7	**Quantum Computing for Training**	211
	7.1 Quantum Blas	212
	7.1.1 Basic Idea	212
	7.1.2 Matrix Inversion for Training	213
	7.1.3 Speedups and Further Applications	218
	7.2 Search and Amplitude Amplification	219
	7.2.1 Finding Closest Neighbours	220
	7.2.2 Adapting Grover's Search to Data Superpositions	221
	7.2.3 Amplitude Amplification for Perceptron Training	223
	7.3 Hybrid Training for Variational Algorithms	224
	7.3.1 Variational Algorithms	226
	7.3.2 Variational Quantum Machine Learning Algorithms	230

		7.3.3	Numerical Optimisation Methods	233
		7.3.4	Analytical Gradients of a Variational Classifier	236
	7.4	Quantum Adiabatic Machine Learning		238
		7.4.1	Quadratic Unconstrained Optimisation	239
		7.4.2	Annealing Devices as Samplers	240
		7.4.3	Beyond Annealing	242
	References			243
8	**Learning with Quantum Models**			247
	8.1	Quantum Extensions of Ising-Type Models		248
		8.1.1	The Quantum Ising Model	249
		8.1.2	Training Quantum Boltzmann Machines	251
		8.1.3	Quantum Hopfield Models	253
		8.1.4	Other Probabilistic Models	255
	8.2	Variational Classifiers and Neural Networks		256
		8.2.1	Gates as Linear Layers	257
		8.2.2	Considering the Model Parameter Count	259
		8.2.3	Circuits with a Linear Number of Parameters	261
	8.3	Other Approaches to Build Quantum Models		263
		8.3.1	Quantum Walk Models	263
		8.3.2	Superposition and Quantum Ensembles	266
		8.3.3	QBoost	269
	References			271
9	**Prospects for Near-Term Quantum Machine Learning**			273
	9.1	Small Versus Big Data		274
	9.2	Hybrid Versus Fully Coherent Approaches		276
	9.3	Qualitative Versus Quantitative Advantages		277
	9.4	What Machine Learning Can Do for Quantum Computing		278
	Reference			279
Index				281

Acronyms

x^m	Training input/feature vector
y^m	Training output/target label
$\tilde{x}$	New input/feature vector
$\tilde{y}$	Predicted label of a new input
γ	Lagrangian parameter
α	Amplitude vector
α_i	Amplitude, ith entry of the amplitude vector
χ	Input domain
$\mathcal{Y}$	Output domain
$\mathcal{D}$	Data set, Training set
v_i	Visible unit of a probabilistic model
h_j	Hidden unit of a probabilistic model
s_i	Visible or hidden unit
b_i	Binary variable
$|i\rangle$	ith computational basis state of a qubit system
D	Number of classes in a classification problem
w	Parameter or weight vector of a machine learning model
w_0	Bias parameter of a machine learning model
θ	Set of parameters of a machine learning model
C	Cost function for training a machine learning model
L	Loss function for training a machine learning model
$\mathcal{L}$	Lagrangian objective function
R	Regulariser for training a machine learning model
ϕ	Feature map from a data input space to a feature space
φ	Nonlinear activation function for neural networks
κ	Kernel function
η	Learning rate
X	Data matrix, Design matrix
U	Unitary matrix, operator, quantum circuit
$U(\theta)$	Parametrised quantum circuit/Variational circuit

Chapter 1
Introduction

Machine learning, on the one hand, is the art and science of making computers learn from data how to solve problems instead of being explicitly programmed. Quantum computing, on the other hand, describes information processing with devices based on the laws of quantum theory. Both machine learning and quantum computing are expected to play a role in how society deals with information in the future and it is therefore only natural to ask how they could be combined. This question is explored in the emerging discipline of *quantum machine learning* and is the subject of this book.

In its broadest definition, quantum machine learning summarises approaches that use synergies between machine learning and quantum information. For example, researchers investigate how mathematical techniques from quantum theory can help to develop new methods in machine learning, or how we can use machine learning to analyse measurement data of quantum experiments. Here we will use a much more narrow definition of quantum machine learning and understand it as *machine learning with quantum computers* or *quantum-assisted machine learning*. Quantum machine learning in this narrow sense looks at the opportunities that the current development of quantum computers open up in the context of intelligent data mining. Does quantum information add something new to how machines recognise patterns in data? Can quantum computers help to solve problems faster, can they learn from fewer data samples or are they able to deal with higher levels of noise? How can we develop new machine learning techniques from the language in which quantum computing is formulated? What are the ingredients of a quantum machine learning algorithm, and where lie the bottlenecks? In the course of this book we will investigate these questions and present different approaches to quantum machine learning research, together with the concepts, language and tricks that are commonly used.

To set the stage, the following section introduces the background of quantum machine learning. We then work through a toy example of how quantum computers can learn from data, which will already display a number of issues discussed in the course of this book.

1.1 Background

1.1.1 Merging Two Disciplines

Computers are physical devices based on electronic circuits which process information. Algorithms (the computer programs or 'software') are the recipes of how to manipulate the current in these circuits in order to execute computations. Although the physical processes involve microscopic particles like electrons, atoms and molecules, we can for all practical purposes describe them with a macroscopic, classical theory of the electric properties of the circuits. But if microscopic systems such as photons, electrons and atoms are directly used to process information, they require another mathematical description to capture the fact that on small scales, nature behaves radically different from what our intuition teaches us. This mathematical framework is called *quantum theory* and since its development at the beginning of the 20th century it has generally been considered to be the most comprehensive description of microscopic physics that we know of. A computer whose computations can only be described with the laws of quantum theory is called a *quantum computer*.

Since the 1990s quantum physicists and computer scientists have been analysing how quantum computers can be built and what they could potentially be used for. They developed several languages to describe computations executed by a quantum system, languages that allow us to investigate these devices from a theoretical perspective. An entire 'zoo' of *quantum algorithms* has been proposed and is waiting to be used on physical hardware. The most famous language in which quantum algorithms are formulated is the circuit model. The central concept is that of a *qubit*, which takes the place of a classical bit, as well as *quantum gates* to perform computations on qubits [1].

Building a quantum computer in the laboratory is not an easy task, as it requires the accurate control of very small systems. At the same time, it is crucial not to disturb the fragile *quantum coherence* of these systems, which would destroy the quantum effects that we want to harvest. In order to preserve quantum coherence throughout thousands of computational operations, error correction becomes crucial. But error correction for quantum systems turns out to be much more difficult than for classical ones, and becomes one of the major engineering challenges in developing a full-scale quantum computer. Implementations of most of the existing quantum algorithms will therefore have to wait a little longer.

However, full-scale quantum computers are widely believed to become available in the future. The research field has left the purely academic sphere and is on the agenda of the research labs of some of the largest IT companies. More and more computer scientists and engineers come on board to add their skills to the quantum computing community. Software toolboxes and quantum programming languages based on most major classical computational languages are available, and more are being developed every year. In summary, the realisation of quantum technology became an international and interdisciplinary effort.

1.1 Background

While targeting full-scale devices, a lot of progress has been made in the development of so called *intermediate-term* or *small-scale devices*. These devices have no error correction, and count around 50–100 qubits that do not necessarily all speak to one another due to limited connectivity. Small-scale quantum devices do in principle have the power to test the advantages of quantum computing, and gave a new incentive to theory-driven research in quantum algorithmic design. The holy grail is currently to find a useful computational problem that can be solved by a small-scale device, and with a (preferably exponential) speed-up in runtime to the best known classical algorithm. In other words, the quest to find a 'killer-app', a compact but powerful algorithm tailor made for early quantum technologies, is on. Machine learning and its core mathematical problem, optimisation, are often mentioned as two promising candidates, a circumstance that has given huge momentum to quantum machine learning research in the last couple of years.

This brings us to the other parent discipline, machine learning. Machine learning lies at the intersection of statistics, mathematics and computer science. It analyses how computers can learn from prior examples - usually large datasets based on highly complex and nonlinear relationships - how to make predictions or solve unseen problem instances. Machine learning was born as the data-driven side of artificial intelligence research and tried to give machines human-like abilities such as image recognition, language processing and decision making. While such tasks come naturally to humans, we do not know in general how to make machines acquire similar skills. For example, looking at an image of a mountain panorama it is unclear how to relate the information that pixel (543,1352) is dark blue to the concept of a mountain. Machine learning approaches this problem by making the computer recover patterns from data, patterns that inherently contain these concepts.

Machine learning is also a discipline causing a lot of excitement in the academic world as well as the IT sector (and certainly on a much larger scale than quantum computing). It is predicted to change the way a large share of the world's population interacts with technology, a trend that has already started. As data is becoming increasingly accessible, machine learning systems mature from research to business solutions and are integrated into PCs, cell phones and household devices. They scan through huge numbers of emails every day in order to pick out spam mail, or through masses of images on social platforms to identify offensive contents. They are used in forecasting of macroeconomic variables, risk analysis as well as fraud detection in financial institutions, as well as medical diagnosis.

What has been celebrated as 'breakthroughs' and innovation is thereby often based on the growing sizes of datasets as well as computational power, rather than on fundamentally new ideas. Methods such as neural networks, support vector machines or AdaBoost, as well as the latest trend towards deep learning were basically invented in the 1990s and earlier. Finding genuinely new approaches is difficult as many tasks translate into hard optimisation problems. To solve them, computers have to search more or less blindly through a vast landscape of solutions to find the best candidate. A lot of research therefore focuses on finding variations and approximations of methods that work well in practice, and machine learning is known to contain a fair share of "black art" [2]. This is an interesting point of leverage for quantum computing, which has the potential of contributing fundamentally new approaches to machine learning.

1.1.2 The Rise of Quantum Machine Learning

In recent years, there has been a growing body of literature with the objective of combining the disciplines of quantum information processing and machine learning. Proposals that merge the two fields have been sporadically put forward since quantum computing established itself as an independent discipline. Perhaps the earliest notions were investigations into quantum models of neural networks starting in 1995 [3]. These were mostly biologically inspired, hoping to find explanations within quantum theory for how the brain works (an interesting quest which is still controversially disputed for lack of evidence). In the early 2000s the question of statistical learning theory in a quantum setting was discussed, but received only limited attention. A series of workshops on 'Quantum Computation and Learning' were organised, and in the proceedings of the third event, Bonner and Freivals mention that "[q]uantum learning is a theory in the making and its scientific production is rather fragmented" [4]. Sporadic publications on quantum machine learning algorithms also appeared during that time, such as Ventura and Martinez' quantum associative memory [5] or Hartmut Neven's 'QBoost' algorithm, which was implemented on the first commercial quantum annealer, the D-Wave device, around 2009 [6].

The term 'quantum machine learning' came into use around 2013. Lloyd, Mohseni and Rebentrost [7] mention the expression in their manuscript of 2013, and in 2014, Peter Wittek published an early monograph with the title *Quantum Machine Learning—What quantum computing means to data mining* [8], which summarises some of the early papers. From 2013 onwards, interest in the topic increased significantly [9] and produced a rapidly growing body of literature that covers all sorts of topics related to joining the two disciplines. Various international workshops and conferences[1] have been organised and their number grew with every year. Numerous groups, most of them still rooted in quantum information science, started research projects and collaborations. Combining a dynamic multi-billion dollar market with the still 'mysterious' and potentially profitable technology of quantum computing has also sparked a lot of interest in industry.[2]

---

[1]Some early events include a workshop at the *Neural Information Processing Systems (NIPS)* conference in Montreal, Canada in December 2015, the Quantum Machine Learning Workshop in South Africa in July 2016 as well as a Quantum Machine Learning conference at the Perimeter Institute in Waterloo, Canada, in August 2016.

[2]Illustrative examples are Google's *Quantum Artificial Intelligence Lab* established in 2013, Microsoft's *Quantum Architectures and Computation* group and IBM's *IBM Q* initiative.

1.1 Background

1.1.3 Four Approaches

As mentioned before, there are several definitions of the term *quantum machine learning*, and in order to clarify the scope of this book it is useful to locate our definition in the wider research landscape. For this we use a typology introduced by Aimeur, Brassard and Gambs [10]. It distinguishes four approaches of how to combine quantum computing and machine learning, depending on whether one assumes the data to be generated by a quantum (Q) or classical (C) system, and if the information processing device is quantum (Q) or classical (C) (see Fig. 1.1).

The case *CC* refers to classical data being processed classically. This is of course the conventional approach to machine learning, but in this context it relates to machine learning based on methods borrowed from quantum information research. An example is the application of tensor networks, which have been developed for quantum many-body-systems, to neural network training [11]. There are also numerous 'quantum-inspired' machine learning models, with varying degrees of foundation in rigorous quantum theory.

The case *QC* investigates how machine learning can help with quantum computing. For example, when we want to get a comprehensive description of the internal state of a quantum computer from as few measurements as possible we can use machine learning to analyse the measurement data [12]. Another idea is to learn phase transitions in many-body quantum systems, a fundamental physical problem with applications in the development of quantum computers [13]. Machine learning has also been found useful to discriminate between quantum states emitted by a source, or transformations executed by an experimental setup [14–16], and applications are plenty.

In this book we use the term 'quantum machine learning' synonymously with the remaining *CQ* and *QQ* approach on the right of Fig. 1.1. In fact, we focus mainly

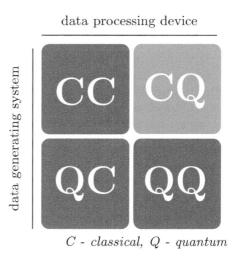

Fig. 1.1 Four approaches that combine quantum computing and machine learning

on the *CQ* setting, which uses quantum computing to process classical datasets. The datasets consist of observations from classical systems, such as text, images or time series of macroeconomic variables, which are fed into a quantum computer for analysis. This requires a quantum-classical interface, which is a challenge we discuss in detail in the course of the book. The central task of the *CQ* approach is to design quantum algorithms for data mining, and there are a number of strategies that have been proposed by the community. They range from translations of classical machine learning models into the language of quantum algorithms, to genuinely new models derived from the working principles of quantum computers.

We will mostly be concerned with *supervised learning*, in which one has access to a dataset with solutions to a problem, and uses these solutions as a supervision or figure of merit when solving a new problem. However, we note that quantum reinforcement learning [17, 18] and unsupervised learning [19–21] are active research fields adding interesting angles to the content of this book.

The last approach, *QQ*, looks at 'quantum data' being processed by a quantum computer. This can have two different meanings. First, the data could be derived from measuring a quantum system in a physical experiment and feeding the values back into a separate quantum processing device. A much more natural setting however arises where a quantum computer is first used to simulate the dynamics of a quantum system (as investigated in the discipline of *quantum simulation* and with fruitful applications to modeling physical and chemical properties that are otherwise computationally intractable), and consequently takes the state of the quantum system as an input to a quantum machine learning algorithm executed on the very same device. The advantage of such an approach is that while measuring all information of a quantum state may require a number of measurements that is exponential in the system size, the quantum computer has immediate access to all this information and can produce the result, for example a yes/no decision, directly—an exponential speedup by design.

The *QQ* approach is doubtless very interesting, but there are presently only few results in this direction (for example [17]). Some authors claim that their quantum machine learning algorithm can easily be fed with quantum data, but the details may be less obvious. Does learning from quantum data (i.e. the wave function of a quantum system) produce different results to classical data? How can we combine the data generation and data analysis unit effectively? Can we design algorithms that answer important questions from experimentalists which would otherwise be intractable? In short, although much of what is presented here can be used in the 'quantum data' setting as well, there are a number of interesting open problems specific to this case that we will not be able to discuss in detail.

1.1.4 Quantum Computing for Supervised Learning

From now on we will focus on the *CQ* case for supervised learning problems. There are two different strategies when designing quantum machine learning algorithms, and of course most researchers are working somewhere between the two extremes. The first strategy aims at *translating* classical models into the language of quantum mechanics in the hope to harvest algorithmic speedups. The sole goal is to reproduce the results of a given model, say a neural net or a Gaussian process, but to 'outsource' the computation or parts of the computation to a quantum device. The *translational approach* requires significant expertise in quantum algorithmic design. The challenge is to assemble quantum routines that imitate the results of the classical algorithm while keeping the computational resources as low as possible. While sometimes extending the toolbox of quantum routines by some new tricks, learning does not pose a genuinely new problem here. On the contrary, the computational tasks to solve resemble rather general mathematical problems such as computing a nonlinear function, matrix inversion or finding the optimum of a non-convex objective function. Consequently, the boundaries of speedups that can be achieved are very much the same as in 'mainstream' quantum computing. Quantum machine learning becomes an application of quantum computing rather than a truly interdisciplinary field of research.

The second strategy, whose many potential directions are still widely unexplored, leaves the boundaries of known classical machine learning models. Instead of starting with a classical algorithm, one starts with a quantum computer and asks what type of machine learning model might fit its physical characteristics and constraints, its formal language and its proposed advantages. This could lead to an entirely new model or optimisation objective—or even an entirely new branch of machine learning—that is derived from a quantum computational paradigm. We will call this the *exploratory approach*. The exploratory approach does not necessarily rely on a digital, universal quantum computer to implement quantum algorithms, but may use any system obeying the laws of quantum mechanics to derive (and then train) a model that is suitable to learn from data. The aim is not only to achieve runtime speedups, but to contribute innovative methods to the machine learning community. For this, a solid understanding—and feeling—for the intricacies of machine learning is needed, in particular because the new model has to be analysed and benchmarked to access its potential. We will investigate both strategies in the course of this book.

1.2 How Quantum Computers Can Classify Data

In order to build a first intuition of what it means to learn from classical data with a quantum computer we want to present a toy example that is supposed to illustrate a range of topics discussed throughout this book, and for which no previous knowledge in either field is required. More precisely, we will look at how to implement a type of *nearest neighbour* method with quantum interference induced by a *Hadamard gate*. The example is a strongly simplified version of a quantum machine learning algorithm proposed in [22], which will be presented in more detail in Sect. 6.2.4.2.

Table 1.1 Mini-dataset for the quantum classifier example

	Raw data		Preprocessed data		Survival
	Price	Cabin	Price	Cabin	
Passenger 1	8,500	0910	0.85	0.36	1 (yes)
Passenger 2	1,200	2105	0.12	0.84	0 (no)
Passenger 3	7,800	1121	0.78	0.45	?

1.2.1 The Squared-Distance Classifier

Machine learning always starts with a dataset. Inspired by the *kaggle*[3] Titanic dataset, let us consider a set of 2-dimensional input vectors $\{x^m = (x_0^m, x_1^m)^T\}, m = 1, \ldots, M$. Each vector represents a passenger who was on the Titanic when the ship sank in the tragic accident of 1912, and specifies two features of the passenger: The *price* in dollars which she or he paid for the ticket (feature 0) and the passenger's *cabin number* (feature 1). Assume the ticket price is between $0 and $10,000, and the cabin numbers range from 1 to 2,500. Each input vector x^m is also assigned a label $\tilde{y}^m$ that indicates if the passenger survived ($y^m = 1$) or died ($y^m = 0$).

To reduce the complexity even more (possibly to an absurd extent), we consider a dataset of only 2 passengers, one who died and one who survived the event (see Table 1.1). The task is to find the probability of a third passenger of features $\tilde{x} = (\tilde{x}_0, \tilde{x}_1)^T$ and for whom no label is given, to survive or die. As is common in machine learning, we preprocess the data in order to project it onto roughly the same scales. Oftentimes, this is done by imposing zero mean and unit variance, but here we will simply rescale the range of possible ticket prices and cabin numbers to the interval [0, 1] and round the values to two decimal digits.

Possibly the simplest supervised machine learning method, which is still surprisingly successful in many cases, is known as *nearest neighbour*. A new input is given the same label as the data point closest to it (or, in a more popular version, the majority of its k nearest neighbours). Closeness has to be defined by a distance measure, for example the Euclidean distance between data points. A less frequent strategy which we will consider here is to include all data points $m = 1, \ldots, M$, but weigh each one's influence towards the decision by a weight

$$\gamma_m = 1 - \frac{1}{c}|\tilde{x} - x^m|^2, \tag{1.1}$$

where c is some constant. The weight γ_m measures the squared distance between x^m and the new input $\tilde{x}$, and by subtracting the distance from one we get a higher weight for closer data points. We define the probability of assigning label $\tilde{y}$ to the

---

[3] Kaggle (www.kaggle.com) is an open data portal that became famous for hosting competitions which anyone can enter to put her or his machine learning software to the test.

1.2 How Quantum Computers Can Classify Data

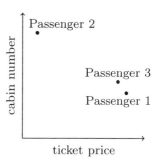

Fig. 1.2 The mini-dataset displayed in a graph. The similarity (Euclidean distance) between Passengers 1 and 3 is closer than between Passengers 2 and 3

new input $\tilde{x}$ as the sum over the weights of all M_1 training inputs which are labeled with $y^m = 1$,

$$p_{\tilde{x}}(\tilde{y} = 1) = \frac{1}{\chi} \frac{1}{M_1} \sum_{m|y^m=1} \left(1 - \frac{1}{c}|\tilde{x} - x^m|^2\right). \quad (1.2)$$

The probability of predicting label 0 for the new input is the same sum, but over the weights of all inputs labeled with 0. The factor $\frac{1}{\chi}$ is included to make sure that $p_{\tilde{x}}(\tilde{y} = 0) + p_{\tilde{x}}(\tilde{y} = 1) = 1$. We will call this model the *squared-distance classifier*.

A nearest neighbour method is based on the assumption that similar inputs should have a similar output, which seems reasonable for the data at hand. People from a similar income class and placed at a similar area on the ship might have similar fates during the tragedy. If we had another feature without expressive power to explain death or survival of a person, for example a ticket number that was assigned randomly to the tickets, this method would obviously be less successful because it tries to consider the similarity of ticket numbers. Applying the squared-distance classifier to the mini-dataset, we see in Fig. 1.2 that Passenger 3 is closer to Passenger 1 than to Passenger 2, and our classifier would predict 'survival'.

1.2.2 Interference with the Hadamard Transformation

Now we want to discuss how to use a quantum computer in a trivial way to compute the result of the squared-distance classifier. Most quantum computers are based on a mathematical model called a qubit, which can be understood as a random bit (a Bernoulli random variable) whose description is not governed by classical probability theory but by quantum mechanics. The quantum machine learning algorithm requires us to understand only one 'single-qubit operation' that acts on qubits, the so called *Hadamard transformation*. We will illustrate what a Hadamard gate does to two qubits by comparing it with an equivalent operation on two random bits. To rely even more on intuition, we will refer to the two random bits as two coins that can be tossed, and the quantum bits can be imagined as quantum versions of these coins.

Table 1.2 Probability distribution over possible outcomes of the coin toss experiment, and its equivalent with qubits

State	Classical coin			State	Qubit		
	Step 1	Step 2	Step 3		Step 1	Step 2	Step 3
(heads, heads)	1	0.5	0.5	|heads⟩|heads⟩	1	0.5	1
(heads, tails)	0	0	0	|heads⟩|tails⟩	0	0	0
(tails, heads)	0	0.5	0.5	|tails⟩|heads⟩	0	0.5	0
(tails, tails)	0	0	0	|tails⟩|tails⟩	0	0	0

Imagine two fair coins c_1 and c_2 that can each be in state *heads* or *tails* with equal probability. The space of possible states after tossing the coins (c_1, c_2) consists of (heads, heads), (heads, tails), (tails, heads) and (heads, heads). As a preparation Step 1, turn both coins to 'heads'. In Step 2 toss the first coin only and check the result. In Step 3 toss the first coin a second time and check the result again. Consider repeating this experiment from scratch a sufficiently large number of times to count the statistics, which in the limiting case of infinite repetitions can be interpreted as probabilities.[4] The first three columns of Table 1.2 show these probabilities for our little experiment. After the preparation step 1 the state is by definition (heads, heads). After the first toss in step 2 we observe the states (heads, heads) and (tails, heads) with equal probability. After the second toss in step 3, we observe the same two states with equal probability, and the probability distribution hence does not change between step 2 and 3. Multiple coin tosses maintain the state of maximum uncertainty for the observer regarding the first coin.

Compare this with two qubits q_1 and q_2. Again, performing a measurement called a projective z-measurement (we will come to that later) a qubit can be found to be in two different states (let us stick with calling them |heads⟩ and |tails⟩, but later it will be |0⟩ and |1⟩). Start again with both qubits being in state |heads⟩|heads⟩. This means that repeated measurements would always return the result |heads⟩|heads⟩, just as in the classical case. Now we apply an operation called the Hadamard transform on the first qubit, which is sometimes considered as the quantum equivalent of a fair coin toss. Measuring the qubits after this operation will reveal the same probability distribution as in the classical case, namely that the probability of |heads⟩|heads⟩ and |tails⟩|heads⟩ is both 0.5. However, if we apply the 'Hadamard coin toss' twice without intermediate observation of the state, one will measure the qubits always in state (heads, heads), no matter how often one repeats the experiment. This transition from high uncertainty to a state of lower uncertainty is counterintuitive for classical stochastic operations. As a side note beyond the scope of this chapter, it is crucial that we do not measure the state of the qubits after Step 2 since this would return a different distribution for Step 3—another interesting characteristic of quantum mechanics.

---

[4] We are assuming a frequentist's viewpoint for the moment.

1.2 How Quantum Computers Can Classify Data

Let us have a closer look at the mathematical description of the Hadamard operation (and have a first venture into the world of quantum computing). In the classical case, the first coin toss imposes a transformation

$$p = \begin{pmatrix} 1 \\ 0 \\ 0 \\ 0 \end{pmatrix} \to p' = \begin{pmatrix} 0.5 \\ 0 \\ 0.5 \\ 0 \end{pmatrix},$$

where we have now written the four probabilities into a *probability vector*. The first entry of that vector gives us the probability to observe state (heads, heads), the second (heads, tails) and so forth. In linear algebra, a transformation between probability vectors can always be described by a special matrix called a *stochastic matrix*, in which rows add up to 1. Performing a coin toss on the first coin corresponds to a stochastic matrix of the form

$$S = \frac{1}{2} \begin{pmatrix} 1 & 0 & 1 & 0 \\ 0 & 1 & 0 & 1 \\ 1 & 0 & 1 & 0 \\ 0 & 1 & 0 & 1 \end{pmatrix}.$$

Applying this matrix to p' leads to a new state $p'' = Sp'$, which is in this case equal to p'.

This description works fundamentally differently when it comes to qubits governed by the probabilistic laws of quantum theory. Instead of stochastic matrices acting on probability vectors, quantum objects can be described by unitary (and complex) matrices acting on complex *amplitude vectors*. There is a close relationship between probabilities and amplitudes: The probability of the two qubits to be measured in a certain state is the *absolute square* of the corresponding amplitude. The amplitude vector α describing the two qubits after preparing them in |heads⟩|heads⟩ would be

$$\alpha = \begin{pmatrix} 1 \\ 0 \\ 0 \\ 0 \end{pmatrix},$$

which makes the probability of |heads⟩|heads⟩ equal to $|1|^2 = 1$. In this case the amplitude vector is identical to the probability vector of the quantum system. In the quantum case, the stochastic matrix is replaced by a Hadamard transform acting on the first qubit, which can be written as

$$H = \frac{1}{\sqrt{2}} \begin{pmatrix} 1 & 0 & 1 & 0 \\ 0 & 1 & 0 & 1 \\ 1 & 0 & -1 & 0 \\ 0 & 1 & 0 & -1 \end{pmatrix}.$$

applied to the amplitude vector. Although H does not have complex entries, there are negative entries, which is not possible for stochastic matrices and the laws of classical probability theory. Multiplying this matrix with a results in

$$\alpha' = \frac{1}{\sqrt{2}} \begin{pmatrix} 1 \\ 0 \\ 1 \\ 0 \end{pmatrix}.$$

The probability of the outcomes $|\text{heads}\rangle|\text{heads}\rangle$ and $|\text{tails}\rangle|\text{heads}\rangle$ is equally given by $|\sqrt{0.5}|^2 = 0.5$ while the other states are never observed, as claimed in Table 1.2. If we apply the Hadamard matrix altogether twice, something interesting happens. The negative sign 'interferes amplitudes' to produce again the initial state,

$$\alpha'' = \begin{pmatrix} 1 \\ 0 \\ 0 \\ 0 \end{pmatrix}.$$

This is exactly what we claimed in Table 1.2.

More generally, if we apply the Hadamard to the first of n qubits in total, the transformation matrix looks like

$$H_n^{(q_1)} = \frac{1}{\sqrt{2}} \begin{pmatrix} \mathbb{1} & \mathbb{1} \\ \mathbb{1} & -\mathbb{1} \end{pmatrix}, \tag{1.3}$$

where $\mathbb{1}$ is the identity matrix of dimension $\frac{N}{2} \times \frac{N}{2}$, and $N = 2^n$. Applied to a general amplitude vector that describes the state of n qubits we get

$$\begin{pmatrix} \alpha_1 \\ \vdots \\ \alpha_{\frac{N}{2}} \\ \alpha_{\frac{N}{2}+1} \\ \vdots \\ \alpha_N \end{pmatrix} \rightarrow \frac{1}{\sqrt{2}} \begin{pmatrix} \alpha_1 + \alpha_{\frac{N}{2}+1} \\ \vdots \\ \alpha_{\frac{N}{2}} + \alpha_N \\ \alpha_1 - \alpha_{\frac{N}{2}+1} \\ \vdots \\ \alpha_{\frac{N}{2}} - \alpha_N \end{pmatrix}.$$

If we summarise the first half of the original amplitude vector's entries as a and the second half as b, the Hadamard transform produces a new vector of the form $(a+b, a-b)^T$.

Note that the Hadamard transformation was applied on one qubit only, but acts on all 2^n amplitudes. This 'parallelism' is an important source of the power of quantum computation, and with 100 qubits we can apply the transformation to 2^{100} amplitudes.

1.2 How Quantum Computers Can Classify Data

Of course, parallelism is a consequence of the probabilistic description and likewise true for classical statistics. However, together with the effects of interference (i.e., the negative signs in the matrix), quantum computing researchers hope to gain a significant advantage over classical computation.

1.2.3 Quantum Squared-Distance Classifier

Let us get back to our toy quantum machine learning algorithm. We can use the Hadamard operation to compute the prediction of the squared-distance classifier by following these four steps:

Step A—Some more data preprocessing

To begin with we need another round of data preprocessing in which the length of each input vector (i.e., the ticket price and cabin number for each passenger) gets normalised to one. This requirement projects the data onto a unit circle, so that only information about the angles between data vectors remains. For some datasets this is a desired effect because the length of data vectors has no expressive power, while for others the loss of information is a problem. In the latter case one can use tricks which we will discuss in Chap. 5. Luckily, for the data points chosen in this example the normalisation does not change the outcome of a distance-based classifier (see Fig. 1.3).

Step B—Data encoding

The dataset has to be encoded in a quantum system in order to use the Hadamard transform. We will discuss different ways of doing so in Chap. 5. In this example the data is represented by an amplitude vector (in a method we will later call *amplitude encoding*). Table 1.3 shows that we have six features to encode, plus two class labels. Let us have a look at the features first. We need three qubits or 'quantum coins' (q_1, q_2, q_3) with values $q_1, q_2, q_3 = 0, 1$ to have 8 different measurement results. (Only two qubits would not be sufficient, because we would only have four possible

	price	room	survival
Passenger 1	0.921	0.390	yes (1)
Passenger 2	0.141	0.990	no (0)
Passenger 3	0.866	0.500	?

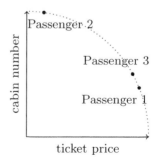

Fig. 1.3 Left: Additional preprocessing of the data. Each feature vector gets normalised to unit length. Right: Preprocessed data displayed in a graph. The points now lie on a unit circle. The Euclidean distance between Passengers 1 and 3 is still smaller than between Passengers 2 and 3

Table 1.3 The transformation of the amplitude vector in the quantum machine learning algorithm

Qubit state				Transformation of amplitude vector		
q_1	q_2	q_3	q_4	Step B $\alpha_{\text{init}} \to$	Step C $\alpha_{\text{inter}} \to$	Step D α_{final}
0	0	0	0	0	0	0
0	0	0	1	$\frac{1}{\sqrt{4}}0.921$	$\frac{1}{\sqrt{4}}(0.921+0.866)$	$\frac{1}{\sqrt{4\chi}}(0.921+0.866)$
0	0	1	0	0	0	0
0	0	1	1	$\frac{1}{\sqrt{4}}0.390$	$\frac{1}{\sqrt{4}}(0.390+0.500)$	$\frac{1}{\sqrt{4\chi}}(0.390+0.500)$
0	1	0	0	$\frac{1}{\sqrt{4}}0.141$	$\frac{1}{\sqrt{4}}(0.141+0.866)$	$\frac{1}{\sqrt{4\chi}}(0.141+0.866)$
0	1	0	1	0	0	0
0	1	1	0	$\frac{1}{\sqrt{4}}0.990$	$\frac{1}{\sqrt{4}}(0.990+0.500)$	$\frac{1}{\sqrt{4\chi}}(0.990+0.500)$
0	1	1	1	0	0	0
1	0	0	0	0	0	0
1	0	0	1	$\frac{1}{\sqrt{4}}0.866$	$\frac{1}{\sqrt{4}}(0.921-0.866)$	0
1	0	1	0	0	0	0
1	0	1	1	$\frac{1}{\sqrt{4}}0.500$	$\frac{1}{\sqrt{4}}(0.390-0.500)$	0
1	1	0	0	$\frac{1}{\sqrt{4}}0.866$	$\frac{1}{\sqrt{4}}(0.141-0.866)$	0
1	1	0	1	0	0	0
1	1	1	0	$\frac{1}{\sqrt{4}}0.500$	$\frac{1}{\sqrt{4}}(0.990-0.500)$	0
1	1	1	1	0	0	0

Data encoding starts with a quantum system whose amplitude vector contains the features as well as some zeros (Step 2). The Hadamard transformation "interferes" blocks of amplitudes (Step 3). Measuring the first qubit in state 0 (and aborting/repeating the entire routine if this observation did not happen) effectively turns all amplitudes of the second block to zero and renormalises the first block (Step 4). The renormalisation factor is given by $\chi = \frac{1}{4}(|0.921+0.866|^2 + |0.390+0.500|^2 + |0.141+0.866|^2 + |0.990+0.500|^2)$

outcomes as in the example above). Each measurement result is associated with an amplitude whose absolute square gives us the probability of this result being observed. Amplitude encoding 'writes' the values of features into amplitudes and uses operations such as the Hadamard transform to perform computations on the features, for example additions and subtractions.

The amplitude vector we need to prepare is equivalent to the vector constructed by concatenating the features of Passenger 1 and 2, as well as two copies of the features of Passenger 3,

$$\alpha = \frac{1}{\sqrt{4}}(0.921, 0.39, 0.141, 0.99, 0.866, 0.5, 0.866, 0.5)^T.$$

The absolute square of all amplitudes has to sum up to 1, which is why we had to include another scaling or normalisation factor of $1/\sqrt{4}$ for the 4 data points. We

1.2 How Quantum Computers Can Classify Data

now extend the state by a fourth qubit. For each feature encoded in an amplitude, the fourth qubit is in the state that corresponds to the label of that feature vector. (Since the new input does not have a target, we associate the first copy with the target of Passenger 1 and the second copy with the target of Passenger 2, but there are other choices that would work too). Table 1.3 illuminates this idea further. Adding the fourth qubit effectively pads the amplitude vector by some intermittent zeros,

$$\alpha_{\text{init}} = \frac{1}{\sqrt{4}} (0, 0.921, 0, 0.39, 0.141, 0, 0.99, 0, 0, 0.866, 0, 0.5, 0.866, 0, 0.5, 0)^T.$$

This way of associating an amplitude vector with data might seem arbitrary at this stage, but we will see that it fulfils its purpose.

Step C—Hadamard transformation

We now 'toss' the first 'quantum coin' q_1, or in other words, we multiply the amplitude vector by the Hadamard matrix from Eq. (1.3). Chapter 3 will give a deeper account of what this means in the framework of quantum computing, but for now this can be understood at a single standard computational operation on a quantum computer, comparable with an AND or OR gate on a classical machine. The result can be found in column α_{inter} of Table 1.3. As stated before, the Hadamard transform computes the sums and differences between blocks of amplitudes, in this case between the copies of the new input to every training input.

Step D—Measure the first qubit

Now measure the first qubit, and only continue the algorithm if it is found in state 0 (otherwise start from scratch). This introduces an 'if' statement into the quantum algorithm, and is similar to rejection sampling. After this operation we know that the first qubit cannot be in state 1 (by sheer common sense). On the level of the amplitude vector, we have to write zero amplitudes for states in which $q_1 = 1$ and renormalise all other amplitudes so that the amplitude vector is again overall normalised (see column α_{final} of Table 1.3).

Step E—Measure the last qubit

Finally, we measure the last qubit. We have to repeat the entire routine for a number of times to resolve the probability $p(q_4)$ (since measurements only take samples from the distribution). The probability $p(q_4 = 0)$ is interpreted as the output of the machine learning model, or the probability that the classifier predicts the label 0 for the new input. We now want to show that this is exactly the result of the squared-distance classifier (1.2).

By the laws of quantum mechanics, the probability of observing $q_4 = 0$ after the data encoding and the Hadamard transformation can be computed by adding the absolute squares of the amplitudes corresponding to $q_4 = 0$ (i.e. the values of even rows in Table 1.3),

$$p(q_4 = 0) = \frac{1}{4\chi} \left(|0.141 + 0.866|^2 + |0.990 + 0.500|^2 \right) \approx 0.448,$$

with $\chi = \frac{1}{4}(|0.921 + 0.866|^2 + |0.390 + 0.500|^2 + |0.141 + 0.866|^2 + |0.990 + 0.500|^2)$. Equivalently, the probability of observing $q_4 = 1$ is given by

$$p(q_4 = 1) = \frac{1}{4\chi}\left(|0.921 + 0.866|^2 + |0.390 + 0.500|^2\right) \approx 0.552,$$

and is obviously equal to $1 - p(q_4 = 0)$. This is the same as

$$p(q_4 = 0) = \frac{1}{\chi}\left(1 - \frac{1}{4}(|0.141 - 0.866|^2 + |0.990 - 0.500|^2)\right) \approx 0.448,$$

$$p(q_4 = 1) = \frac{1}{4\chi}\left(1 - \frac{1}{4}(|0.921 - 0.866|^2 + |0.390 - 0.500|^2)\right) \approx 0.552.$$

Of course, the equivalence is no coincidence, but stems from the normalisation of each feature vector in Step 1 and can be shown to always be true for this algorithm. If we compare these last results to Eq. (1.2), we see that this is in fact exactly the output of the squared-distance classifier, with the constant c now specified as $c = 4$.

The crux of the matter is that *after data encoding*, only one single computational operation and two simple measurements (as well as a couple of repetitions of the entire routine) were needed to get the result, the output of the classifier. This holds true for any size of the input vectors or dataset. For example, if our dataset had 1 billion training vectors of size 1 million, we would still have an algorithm with the same constant runtime of three elementary operations.

1.2.4 Insights from the Toy Example

As much as nearest neighbour is an oversimplification of machine learning, using the Hadamard to calculate differences is a mere glimpse of what quantum computers can do. There are many other approaches to design quantum machine learning algorithms, for example to encode information into the state of the qubits, or to use the quantum computer as a sampler. And although the promise of a data-size-independent algorithm first sounds too good to be true, there are several things to consider. First, the initial state α_{init} encoding the data has to be prepared, and if no shortcuts are available, this requires another algorithm with a number of operations that is linear in the dimension and size of the dataset (and we are back to square one). What is more, the Hadamard transform belongs to the so called *Clifford group* of quantum gates, which means that the simulation of the algorithm is classically tractable. On the other hand, both arguments do not apply to the QQ approach discussed above, in which we process quantum data and where true exponential speedups are to be expected.

1.2 How Quantum Computers Can Classify Data

Putting these considerations aside, the toy example of a quantum machine learning algorithm fulfilled a number of purposes here. First, it gave a basic idea of how machine learning works, and second, it introduced the very basic logic of quantum computing by manipulating amplitude vectors. Third, it actually introduced a toy model quantum machine learning algorithm we can use to predict the survival probability in the Titanic mini-dataset. And finally, it allows us to derive some insights about quantum machine learning algorithms that will come up frequently in the following chapters:

1. Data encoding is a crucial step of quantum machine learning with classical data, and defines the working principle of the algorithm (here: interference with a Hadamard transform). It is often a bottleneck for the runtime.
2. The quantum algorithm imposes certain requirements on preprocessing, for example the unit length of input vectors.
3. The result of a quantum machine learning algorithm is a measurement.
4. Quantum machine learning algorithms are often inspired by a classical model, in this case a special version of nearest neighbour.
5. The way quantum computers work can give rise to variations of models. For example, we used the squared distance here because it suited the quantum formalism.

We will see more of these ideas in the following.

1.3 Organisation of the Book

The next seven chapters of this book can be distinguished into three parts:

- Chapters 2 and 3 are an introduction to the parent disciplines, machine learning and quantum computing. These chapters intend to give a non-expert the concepts, definitions and pointers necessary to grasp most of the content of the rest of the chapters, and introduce a range of machine learning models as well as quantum algorithms that will become important later.
- Chapters 4 and 5 give a background to quantum machine learning. The short Chap. 4 gives an overview of learning theory and its extensions to quantum mechanics. It presents advantages that quantum computing has to offer for machine learning. Chapter 5 discusses different strategies and simple algorithms to encode information into quantum states. The concepts developed here are crucial for our approach to quantum machine learning, and have to our knowledge not been laid out in the literature before.
- Chapters 6, 7 and 8 present the main part of this book, namely different methods of designing quantum machine learning algorithms. As described before, the intention is not a comprehensive literature review, but to give a number of illustrative examples from which the reader can venture into the literature on her own behalf. Chapter 6 looks at tricks to create quantum algorithms for inference, in

other words to compute a prediction if given an input. Chapter 7 is dedicated to training and optimisation with quantum devices. Chapter 8 finally introduces some ideas that leave the realm of what is known in machine learning and think about *quantum extensions* of classical models, or how to use quantum devices as a model black-box.

The conclusion in Chap. 9 is dedicated to the question of what role machine learning plays for applications with intermediate-term quantum devices. The discussion of quantum machine learning in a closer time horizon will also be useful to summarise some problems and solutions which came up as the main themes of the book, while giving an outlook to potential further research.

References

1. Nielsen, M.A., Chuang, I.L.: Quantum Computation and Quantum Information. Cambridge University Press, Cambridge (2010)
2. Domingos, P.: A few useful things to know about machine learning. Commun. ACM **55**(10), 78–87 (2012)
3. Schuld, M., Sinayskiy, I., Petruccione, F.: The quest for a quantum neural network. Quant. Inf. Process. **13**(11), 2567–2586 (2014)
4. Bonner, R., Freivalds, R.: A survey of quantum learning. In: Quantum Computation and Learning, p. 106 (2003)
5. Ventura, D., Martinez, T.: Quantum associative memory. Inf. Sci. **124**(1), 273–296 (2000)
6. Neven, H., Denchev, V.S., Rose, G., Macready, W.G.: Training a large scale classifier with the quantum adiabatic algorithm. arXiv:0912.0779 (2009)
7. Lloyd, S., Mohseni, M., Rebentrost, P.: Quantum algorithms for supervised and unsupervised machine learning. arXiv:1307.0411 (2013)
8. Wittek, P.: Quantum machine learning: what quantum computing means to data mining. Academic Press (2014)
9. Schuld, M., Sinayskiy, I., Petruccione, F.: Introduction to quantum machine learning. Contemp. Phys. **56**(2), 172–185 (2015)
10. Aïmeur, E., Brassard, G., Gambs, S.: Machine learning in a quantum world. In: Advances in Artificial Intelligence, pp. 431–442. Springer (2006)
11. Stoudenmire, E., Schwab, D.J.: Supervised learning with tensor networks. In: Advances in Neural Information Processing Systems, pp. 4799–4807 (2016)
12. Aaronson, S.: The learnability of quantum states. In: Proceedings of the Royal Society of London A: Mathematical, Physical and Engineering Sciences, vol. 463, pp. 3089–3114. The Royal Society (2007)
13. Carrasquilla, J., Melko, R.G.: Machine learning phases of matter. Nature Phys. **13**, 431–434 (2017)
14. Sasaki, M., Carlini, A.: Quantum learning and universal quantum matching machine. Phys. Rev. A **66**(2), 022303 (2002)
15. Bisio, A., Chiribella, G., Mauro, G., Ariano, D., Facchini, S., Perinotti, P.: Optimal quantum learning of a unitary transformation. Phys. Rev. A **81**(3), 032324 (2010)
16. Sentís, G., Bagan, E., Calsamiglia, J., Munoz-Tapia, R.: Multicopy programmable discrimination of general qubit states. Phys. Rev. A **82**(4), 042312 (2010)
17. Dunjko, V., Taylor, J.M., Briegel, H.J.: Quantum-enhanced machine learning. Phys. Rev. Lett. **117**(13), 130501 (2016)
18. Lamata, L.: Basic protocols in quantum reinforcement learning with superconducting circuits. Sci. Reports **7**(1), 1609 (2017)

References

19. Aïmeur, E., Brassard, G., Gambs, S.: Quantum speed-up for unsupervised learning. Mach. Learn. **90**(2), 261–287 (2013)
20. Lloyd, S., Mohseni, M., Rebentrost, P.: Quantum principal component analysis. Nature Phys. **10**, 631–633 (2014)
21. Benedetti, M., Realpe-Gómez, J., Perdomo-Ortiz, A.: Quantum-assisted Helmholtz machines: a quantum-classical deep learning framework for industrial datasets in near-term devices. Quant. Sci. Technol. **3**, 034007 (2018)
22. Schuld, M., Fingerhuth, M., Petruccione, F.: Implementing a distance-based classifier with a quantum interference circuit. EPL (Europhys. Lett.) **119**(6), 60002 (2017)

Chapter 2
Machine Learning

Machine learning originally emerged as a sub-discipline of artificial intelligence research where it extended areas such as computer perception, communication and reasoning [1]. For humans, learning means finding patterns in previous experience which help us to deal with an unknown situation. For example, someone who has lived on a farm for thirty years will be very good at predicting the local weather. Financial analysts pride themselves on being able to predict the immediate stock market trajectory. This expertise is the result of many iterations of observing meaningful indicators such as the clouds, wind and time of the year, or the global political situation and macroeconomic variables. When speaking about machines, observations come in the form of data, while the solution to a new problem may be understood as the output of an algorithm. Machine learning means to automate the process of generalising from experience in order to make a prediction in an unknown situation, and thereby tries to reproduce and enhance a skill typical to humans.

Although still related to artificial intelligence research,[1] by the 1990s machine learning had outgrown its foundations and became an independent discipline that has—with some intermediate recessions—been expanding ever since. Today, machine learning is another word for data-driven decision making or prediction. As in the weather and stock market examples, the patterns from which the predictions have to be derived are usually very complex and we have only little understanding ourselves of the mechanisms of each system's dynamics. In other words, it is very difficult to hand shape a model of dynamic equations that captures the mechanism we expect to produce the weather or stock market data (although such models do of course exist). A machine learning approach starts instead with a very general, agnostic mathematical model, and uses the data to adapt it to the case. When looking at the final model we do not gain information on the physical mechanism, but consider

---

[1] For example, connections between deep neural networks and the visual cortex of our brain are an active topic of research [2].

it as a black box that has learned the patterns in the data in a sense that it produces reliable predictions.

This chapter is an attempt to give a quantum physicist access to major ideas in the vast landscape of machine learning research. There are four main concepts we want to introduce:

1. the task of supervised learning for *prediction* by generalising from labelled datasets (Sect. 2.1),
2. how to do inference with machine learning *models* (Sect. 2.2),
3. how to *train* models through optimisation (Sect. 2.3),
4. how well-established *methods* of machine learning combine specific models with training strategies (Sect. 2.4).

The terminology, logic and in particular the machine learning methods presented in Sect. 2.4 will be referred to heavily when discussing quantum machine learning algorithms in later chapters. Readers familiar with machine learning can skip this chapter and refer to selected sections only when it becomes necessary. Excellent textbooks for further reading have been written by Bishop [3] as well as Hastie et al. [4], but many other good introductions were also used as a basis for this chapter [5–10].

2.1 Prediction

Almost all machine learning algorithms have one thing in common: They are fed with data and produce an answer to a question. In extreme cases the datasets can consist of billions of values while the answer is only one single bit (see Fig. 2.1). The term 'answer' can stand for many different types of outputs. For example, when future values of a time series have to be predicted, it is a *forecast*, and when the content of images is recognised it is a *classification*. In the context of medical or

Fig. 2.1 Machine learning algorithms are like a funnel that turns large datasets into a simple decision

2.1 Prediction

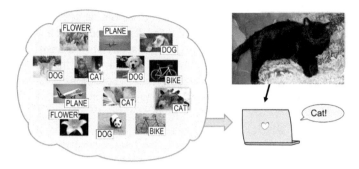

Fig. 2.2 Illustration of image recognition (Example 2.1). While for humans, recognising the content of pictures is a natural task, for computers it is much more difficult to make sense of their numerical representation. Machine learning feeds example images and their content label to the computer, which learns the underlying structure to classify previously unseen images

fault finding applications the computer produces a *diagnosis* and if a robot has to act in an environment one can speak of a *decision*. Since the result of the algorithm is always connected to some uncertainty, another common expression is a *guess*. Here we will mostly use the term *output* or *prediction* of a model, while the data is considered to be the *input*.

2.1.1 Four Examples for Prediction Tasks

Before looking at basic concepts of models, training and generalisation, let us have a look at four typical prediction problems in machine learning.

Example 2.1 (*Image recognition*) While our brain seems to be optimised to recognise concepts such as 'house' or 'mountain panorama' from optical stimuli, it is not obvious how to program a computer to do the same, as the relation between the pixels' Red-Green-Blue (RGB) values and the image's content can be very complex. In machine learning one does not try to explicitly implement such an algorithm, but presents a large number of already labelled images to the computer from which it is supposed to learn the relationship of the digital image representation and its content (see Fig. 2.2). In other words, the complex and unknown input-output function of *pixel matrix* → *content of image* has to be approximated. A 'fruit-fly example' for image recognition is the famous MNIST dataset consisting of black and white images of handwritten digits that have to be identified automatically. Current algorithms guess the correct digit with a success rate of up to 99.65%.[2] An important real-life application for handwritten digit recognition is the processing of postal addresses on mail.

---

[2] See http://yann.lecun.com/exdb/mnist/ as of January 2018.

Example 2.2 (*Time series forecasting*) A time series is a set of data points recorded in consecutive time intervals. An example is the development of the global oil price. Imagine that for every day in the last two years one also records the values of important macroeconomic variables such as the gold price, the DAX index and the Gross Domestic Products of selected nations. These indicators will likely be correlated to the oil price, and there will be many more independent variables that are not recorded. In addition, the past oil price might itself have explanatory power with regards to any consecutive one. The task is to predict on which day in the upcoming month oil will be cheapest. This is an important question for companies who use large amounts of natural resources in their production line.

Example 2.3 (*Hypothesis guessing*) In a notorious assessment test for job interviews, a candidate is given a list of integers between 1 and 100, for example {4, 16, 36, 100} and has to 'complete' the series, i.e. find new instances produced by the same rule. In order to do so, the candidate has to guess the rule or hypothesis with which these numbers were randomly generated. One guess may be the rule 'even numbers out of 100' (H1), but one could also think of 'multiples of 4' (H2), or 'powers to the 2' (H3). One intuitive way of judging different hypotheses that all fit the data is to prefer those that have a smaller amount of options, or in this example, that are true for a smaller amount of numbers. For example, while H1 is true for 50 numbers, H2 is true for only 25 numbers, and H3 only fits to 10 numbers. It would be a much bigger coincidence to pick data with H1 that also fulfills H3 than the other way around. In probabilistic terms, one may prefer the hypothesis for which generating exactly the given dataset has the highest probability. (This example was originally proposed by Josh Tenenbaum [6] and is illustrated in Fig. 2.3.)

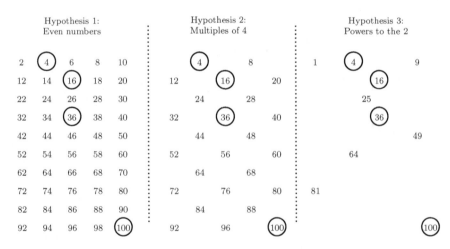

Fig. 2.3 Illustration of hypothesis testing (Example 2.3). The circled numbers are given with the task to find a natural number between 1 and 100 generated by the same rule. There are several hypotheses that match the data and define the space of numbers to pick from

2.1 Prediction

Fig. 2.4 The black and white marbles in the board game Go can be arranged in a vast number of configurations

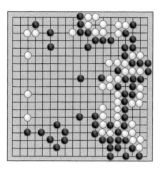

Example 2.4 (*Board games*) A very obvious application for machine learning is to program machines to play games such as chess. The machine—in this context called an agent—learns good strategies through trial games in which it gets rewarded for successful policies and punished for unsuccessful ones. Often the agent does not start its exploration completely clueless, but is pre-trained with data of moves from professional games. While the world champion in chess was beaten by machine learning software as early as the 1990s, the Asian board game Go is more complex (see Fig. 2.4). There are more possible positions of marbles on the board than atoms in the universe, which can make brute force calculations of all possibilities prohibitive for even one single step. However, one of the leading masters of Go had to admit defeat to computer software in 2016, namely Google's AlphaGo [11].

In these four examples, the inputs were given by images, time series, integers and configurations of a board game while the outputs were the content of an image, a price forecast, a number from a set or the next move in the game. If the training data consists of input-output pairs as in Examples 2.1 and 2.2 one speaks of *supervised learning*, while data that does not come with target outputs poses an *unsupervised learning* problem as in Example 2.3. A third area of machine learning, illustrated by Example 2.4 is *reinforcement learning*, in which an agent gets rewarded or punished according to a given rule for its decisions, and the agent learns an optimal strategy by trial and error. This book will focus on the task of supervised learning.

2.1.2 Supervised Learning

The basic structure of a supervised pattern recognition or prediction task can be formally defined as follows.

Definition 2.1 (*Supervised learning problem*) Given an input domain $\mathcal{X}$ and an output domain $\mathcal{Y}$, a training data set $\mathcal{D} = \{(x^1, y^1), \ldots, (x^M, y^M)\}$ of training pairs $(x^m, y^m) \in \mathcal{X} \times \mathcal{Y}$ with $m = 1, \ldots, M$ of training inputs x^m and target outputs y^m, as well as a new unclassified input $\tilde{x} \in \mathcal{X}$, guess or predict the corresponding output $\tilde{y} \in \mathcal{Y}$.

In most applications considered here (and if not stated otherwise) the input domain $\mathcal{X}$ is chosen to be the the space $\mathbb{R}^N$ of real N-dimensional vectors, or for binary variables, the space of N-bit binary strings $\{0, 1\}^N$. The input vectors are also called *feature vectors* as they represent information on carefully selected features of an instance. In cases where the raw data is not from a numerical domain or does not have an obvious distance measure between instances one has to first find a suitable representation that maps the elements to numerical values. For example, in text recognition one often uses so called 'bags of words', where each word in a dictionary is associated with a standard basis vector of the $\mathbb{R}^N$, and N is the number of words in the dictionary. A document is then defined by a vector where each element corresponds to the number of times the corresponding word appears in the text.

Pre-processing the data is central for any machine learning application and in practice often more important than which model is used for prediction [12]. The choice of which features of an instance to consider is called *feature selection*. For example, when preprocessing text data it is useful to exclude stopwords such as 'and' or 'the' which are very unlikely to tell us anything about the topic of the text document. We can therefore ignore the corresponding dimensions in the bag of words vector. *Feature scaling* changes some statistical properties of the data, such as transforming it to have a zero mean and unit variance, which helps to avoid the unwanted effect of vastly different scales (think for example of the yearly income and age of a person). Lastly, *feature engineering* has the goal of crafting powerful features, for example by combining several features in the original dataset to a single one. In some sense, successful feature engineering is "half the job done" of prediction, because we already recognise and extract the important patterns in the data, which can then be further processed by a relatively simple model.

The choice of the output domain determines another important distinction in the type of problem to be solved. If $\mathcal{Y}$ is a set of D discrete class labels $\{l_1, \ldots, l_D\}$ one speaks of a *classification* task. Every D-class classification problem can be converted into $D - 1$ *classification!binary* problems by successively asking whether the input is in class $l_d, d = 1, \ldots, D - 1$ or not (also called the *one-versus-all* scheme). A second strategy used in practice is to construct binary classifiers for pairs of labels, train them on the training data available for the two labels only, and use all classifiers together to make a decision on a new input (the *one-versus-one* scheme). An important concept for the output domain is the so called *one-hot encoding*, which writes the output of a model as a vector in which each dimension corresponds to one of the D possible classes. For example, with $D = 3$ the prediction of the second class would be written as

$$y = \begin{pmatrix} 0 \\ 1 \\ 0 \end{pmatrix}.$$

This is not only handy in the context of categorical classes (which cannot be easily translated to a numerical scale), but has shown advantages in the field of neural networks. The output of the model is usually not a binary vector, but a general real vector in $\mathbb{R}^D$. This vector can be normalised to one and interpreted as a collection of probabilities for the respective classes. A popular normalisation strategy is to add a so called *softmax* layer. For example, a 3-class classification model which produces some raw output

$$f(x; \theta) = \begin{pmatrix} f_1 \\ f_2 \\ f_3 \end{pmatrix}$$

can be mapped to a probability distribution (p_1, p_2, p_3) with $p_1 + p_2 + p_3 = 1$ via the *softmax* function,

$$p_i = \frac{e^{f_i}}{\sum_j e^{f_j}}, \quad i = 1, 2, 3.$$

From there, picking the label with the highest probability, say p_1, produces the prediction $(1, 0, 0)^T$. Using probabilistic outputs has the advantage of revealing the uncertainty associated with a prediction.

Regression refers to problems in which $\mathcal{Y}$ is the space of real numbers $\mathbb{R}$ or an interval therein. Although classification and regression imply two different mathematical structures, most machine learning methods have been formulated for both versions. A classification method can often be generalised to regression by switching to continuous variables and adjusting the functions or distributions accordingly, while the outcome of regression can be discretised (i.e. through interpreting $y > 0 \to 1$ and $y \leq 0 \to 0$).

Examples of inputs and outputs for classification and regression problems can be found in Table 2.1. They might also illustrate why machine learning has gained so much interest from industry and governments: Good solutions to any of these

Table 2.1 Examples of supervised pattern classification tasks in real-life applications

Input	Output
Regression tasks	
Last month's oil price	Tomorrow's oil price
Search history of a user	Chance to click on a car ad
Insurance customer details	Chance of claiming
Multi-label classification tasks	
Images	Cat, dog or plane?
Recording of speech	Words contained in speech
Text segment	Prediction of next word to follow
Binary classification tasks	
Text	Links to terrorism?
Video	Contains a cat?
Email	Is spam?
Spectrum of cancer cell	Malicious?

problems are worth billions of dollars in military, medical, financial, technical or commercial applications.

2.2 Models

The term 'model' and its role in machine learning is comparable with the term of a 'state' in quantum mechanics—it is a central concept with a clear definition for those who use it, but it takes practice to grasp all dimensions of it. Mathematically speaking, a map from inputs to outputs as defined above, $\mathbb{R}^N \to \mathbb{R}$ or $\{0, 1\}^N \to \{0, 1\}$, is a function. From a computational perspective, mapping inputs to outputs is done via an algorithm. On a more abstract level, a model specifies the rule or hypothesis that leads from input to output. One could therefore define models in supervised machine learning as *functions, algorithms or rules that define a relationship between input data and predictions*. For us, the mathematical viewpoint will be the most important one.

We want to distinguish between *deterministic* and *probabilistic* models. The mathematical object defining a deterministic model is the actual function $y = f(x)$ that maps from inputs to outputs. We therefore define a deterministic model as follows:

Definition 2.2 (*Deterministic model*) Let $\mathcal{X}$ be an input domain and $\mathcal{Y}$ be an output domain for a supervised learning problem. A deterministic model is a function

$$y = f(x; \theta), \tag{2.1}$$

with $x \in \mathcal{X}$, $y \in \mathcal{Y}$, and a set of real parameters $\theta = \{\theta_1, \ldots, \theta_D\}$. We also call f the *model function*. For general parameters Eq. (2.1) defines a *model family*.

Note that when we need the parameters to be organised as a vector rather than a set we will often use the notation $w = (w_1, \ldots, w_D)^T \in \mathbb{R}^D$ and refer to it as the *weight vector*.

A model can also depend on a set of hyperparameters which we do not include explicitly. These hyperparameters define a certain model family from an even larger set of possible models and they are usually high-level choices made by the person selecting a model. In the example with the nearest neighbour method discussed in Chap. 1, the hyperparameter was the distance measure that we chose, for example the squared distance. The very same example also shows that the parameter set θ may be empty and the model family only consists of one model.

Machine learning, as does statistics, has to deal with uncertainty, and a lot of methods can be formulated or reformulated in a probabilistic language or as *probabilistic models* (as demonstrated in Ref. [6]). Probabilistic models understand the data inputs and outputs as random variables drawn from an underlying probability distribution $p(x, y)$ or 'ground truth'. One uses the data to construct a model distribution that approximates the ground truth. From there one can compute the probability of a certain label given an input $p(y|x)$ (a procedure called *marginalisation* which is very

2.2 Models

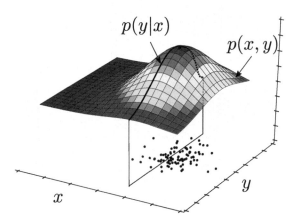

Fig. 2.5 Probabilistic models treat inputs and outputs (black dots on the x-y plane) as random variables drawn from a "ground truth" probability distribution. We use the data to construct a model distribution $p(x, y)$, and derive the marginalised distribution over the outputs given the input $p(y|x)$. The maximum or mean of the marginal (also called class-conditional) distribution can be interpreted as the output of the model. This is of course also possible for discrete inputs and outputs, where the distribution can be displayed in a table listing all possible states of the random variables

similar to tracing out part of density matrices in quantum theory), and translate this to a prediction (see Fig. 2.5).

We formally define a probabilistic model as follows:

Definition 2.3 (*probabilistic model*) Let $\mathcal{X}$ be an input domain and $\mathcal{Y}$ be an output domain for a supervised learning problem. Let X, Y be random variables from which we can draw samples $x \in \mathcal{X}$, $y \in \mathcal{Y}$, and let θ be a set of real parameters. A *probabilistic model* refers to either the *generative model distribution*

$$p(x, y; \theta),$$

or the *discriminative model distribution*

$$p(y|x; \theta),$$

over the data.

Generative probabilistic models derive the full distribution from the dataset, while discriminative models directly try to obtain the slimmer class-conditional distribution $p(y|x)$. Even though generative models contain a lot more information and it may seem that they need more resources to be derived from data, it is by no means clear which type of model is easier to learn [13].

Similarly to the deterministic case, probabilistic models depend on a set of parameters θ. Examples of such distributions in one dimension are the normal or Gaussian distribution with θ consisting of mean μ and variance σ^2,

$$p(x; \mu, \sigma) = \frac{1}{\sqrt{2\pi}\sigma} e^{-\frac{(x-\mu)^2}{2\sigma^2}},$$

or the Bernoulli distribution for binary variables with $\theta = q$, $0 \leq q \leq 1$,

$$p(x; q) = q^x (1-q)^{1-x}.$$

Depending on the context, σ, μ and q can also be understood as hyperparameters of a model, especially if another set of parameters is given.

As mentioned above, the probabilistic model can be used for prediction if we include a rule of deriving outputs from the class-conditional distribution $p(y|x)$ given an input $\tilde{x}$. There are two common practices [14]: The *maximum a posteriori estimate* chooses the output $\tilde{y}$ for which $p(y|\tilde{x})$ is maximised,

$$\tilde{y} = \max_y p(y|\tilde{x}), \qquad (2.2)$$

while an alternative is to take the mean of the distribution,

$$\tilde{y} = \int dy\, p(y|\tilde{x})\, y, \qquad (2.3)$$

which in classification tasks reduces to a sum.

The deterministic and probabilistic approach are highly interlinked. As mentioned above, the continuous outcome of a deterministic model can be turned into a probability to predict a class. Introducing noise into a model makes the outcome probabilistic. Vice versa, probability distributions define a deterministic input-output relation if we interpret the most likely label for a given input as the prediction. However, the logic of model design and training is rather different for the two perspectives.

2.2.1 How Data Leads to a Predictive Model

Figure 2.6 shows four steps of how data leads to a predictive model in a supervised learning problem. Given some preprocessed and carefully featurised data (upper left), a generic model family has to be chosen (upper right). This can for example be a linear function $f(x_1 \ldots x_N) = w_1 x_1 + \cdots + w_N x_N$ with parameters $w_1, \ldots, w_N$, or a sum of Heaveside functions with given thresholds as in the figure. However, we could also choose a probabilistic model that defines a distribution over the input-output space. The model is trained by fitting the parameters and hyperparameters to the data (lower left). This means that a specific model function or distribution is chosen from the family. In the case of a classification task, training defines a decision boundary that separates the input space into regions of different classes. For instance,

2.2 Models

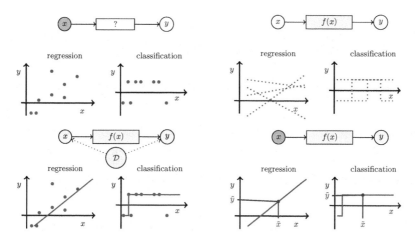

Fig. 2.6 The four basic steps of a generic supervised machine learning algorithm. Upper left: a dataset of one-dimensional inputs x and outputs y is given, produced by an unknown relationship between the two. Upper right: one chooses a model family (in this case a linear model for regression, and a step function for classification). Lower left: training means to fit the model to the data $\mathcal{D}$ using the training set. The test set is used to validate the trained model. Lower right: the model can then be used to predict the output $\tilde{y}$ for a new input $\tilde{x}$

in the classification example of Fig. 2.6 the decision boundary chops the x-axis into intervals in which data is predicted as a positive or negative value respectively. The training step can be of a very different nature for different approaches to machine learning. While sometimes it means to fit thousands of parameters after which the data can be discarded, it can also refer to finding parameters that weigh the data points to compute a prediction. A few models—so called *lazy learners*—derive the prediction directly from the data without generalising from it first, which means that there is no explicit training phase. Once a specific, trained model is derived from the data it can be used for prediction (lower right). Another word for the process of computing a prediction for a model is *inference*.

Machine learning research has developed an entire landscape of models and training strategies to solve supervised pattern recognition tasks and we will introduce some of them below. Each method comes with a distinct language, mathematical background and its own separate scientific community. It is interesting to note that these methods are not only numerous and full of variations, but remarkably interlinked. One method can often be derived from another even though the two are rooted in very different theoretical backgrounds.[3]

---

[3] Different models also have different *inductive biases*. The inductive bias is the set of assumptions that a model uses to generalise from data. For example, linear regression assumes some kind of linear relationship between inputs and outputs, while nearest neighbour methods assume that close neighbours share the same class, and support vector machines assume that a good model maximises the margin between the decision boundary and the samples of each class.

2.2.2 Estimating the Quality of a Model

One of the four steps in the preceding section was to fit the model of choice to the data in order to select a 'good' model from the more general model family. We mentioned several times that a good model generalises from the data. But what does this actually mean? While diving into training only in Sect. 2.3.1, we want to motivate the general concepts of generalisation and overfitting here.

In order to optimise the parameters θ of a model, we need an objective that defines the 'cost' of a certain set of parameters, formalised in terms of a *cost function* C. The cost function can consist of different terms. The most important term is the *loss* which measures how close the model predictions are to the target labels. For example, for a classification task one could consider the *accuracy*,

$$\text{accuracy} = \frac{\text{number of correctly classified examples}}{\text{total number of examples}}.$$

The opposite of the accuracy is the *error*,

$$\text{error} = 1 - \text{accuracy} = \frac{\text{number of incorrectly classified examples}}{\text{total number of examples}}.$$

There are more sophisticated measures to be derived by counting correct predictions, for example those that take *false positives* and *false negatives* into consideration. For many optimisation algorithms the loss has to be continuous, as for example the Eulidean distance between predictions and targets.

But optimising the loss on the training data cannot be the only figure of merit. The goal of machine learning is to predict unseen instances, and a good model is defined by its ability to *generalise* from the given data. In fact, in many cases perfect fits to the training data produce much worse generalisation performance than models that fit less well. This problem known as *overfitting*, and it is a central concept in machine learning.

Overfitting becomes intuitive for a physicist when we consider the following example. Assume you are an experimentalist who wants to recover the law with which an experiment produced some data (see Fig. 2.7). One can always find a high-order polynomial that goes through every data point and thus fits the data (here the training set) perfectly well. However, if we look at an additional data point produced by the same experiment, one might discover that the high-order polynomial (blue dashed line) has a large prediction error, simply because it did not recover the trend in the data. Meanwhile, a less flexible model (red dotted line) is much better suited to reproduce the general structure of the physical law on which the experiment is based.

A lot of machine learning research goes into the prevention of overfitting, or finding the balance between flexible, powerful models and generalisation. Preventing a model from overfitting is called *regularisation*. Regularisation can be achieved in

2.2 Models

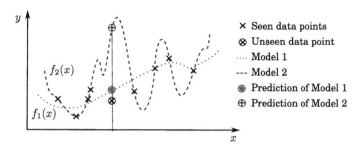

Fig. 2.7 Illustration of the principle of overfitting in linear regression. While the blue model function fits the data points perfectly well, it does not give a very accurate prediction for the new input. Meanwhile, the red model recovers the trend in the data a lot better, even though it does not predict any of them without error

many ways. On the level of model selection one might choose a less flexible model family. One can also consider regularisation as part of the optimisation strategy, for example by stopping iterative algorithms prematurely or *pruning* subsets of parameters to zero. The most common strategy is to write regularisation as a *regulariser* into the cost function,

$$C(\theta) = \text{loss} + \text{regulariser}.$$

Regularisers are penalty terms that impose additional constraints on the parameters selected in training, for example forcing the solution to be sparse or to have a small norm.

The goal of achieving a high generalisation performance has another important consequence for training. If we use the entire labelled data set to train the model, we can compute how well it fits the data, but have no means to estimate the generalisation performance (since new inputs are unlabelled). This is why one never trains with the entire dataset, but divides it into three subsets. The *training set* is used to minimise the cost function, while the *validation* and the *test set* serve as an estimator for the generalisation performance. While the validation set is used to estimate the performance after training in order to adapt hyperparamters (for example, the strength of the regularisation term of the cost function), the test set is only touched once the model is fully specified. This is necessary because while adapting hyperparameters the model is implicitly fitted to the validation set. However, here we will limit ourselves to the distinction between training and test set only, assuming only one 'round' of training so that the test set is neither explicitly nor implicitly used to choose parameters and hyperparameters. Figure 2.8 illustrates the importance of a test set to define the generalisation performance of a model. Both plots show the same training set but two different test sets that suggest two different ways to generalise from the same training data.

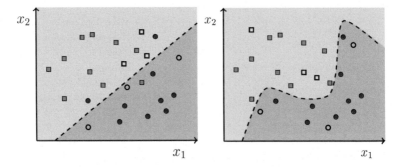

Fig. 2.8 Two different decision boundaries for a binary classification problem in two dimensions and with the two classes 'red circles' and 'blue squares'. Two different test datasets are displayed as black outlines. For the test set displayed in the left case, a linear decision boundary shows a good performance on the test set, while in the right case, the more flexible decision boundary is favourable. Before looking at the test data we do not know which decision boundary is better

2.2.3 Bayesian Learning

What we have presented so far is a very common approach to learning, but there are other approaches with a fundamentally different logic. An important alternative is called *Bayesian learning* [5, 6, 15], and looks at learning from a probabilistic perspective. Not surprisingly, this perspective considers probabilistic models and translates learning into the mathematical or computational problem of integration rather than optimisation.

Bayesian learning is based on Bayes famous rule

$$p(a|b) = \frac{p(b|a)p(a)}{p(b)}, \qquad (2.4)$$

where a, b are values of random variables A, B, and $p(a|b)$ is the conditional probability of a given b defined as

$$p(a|b) = \frac{p(a,b)}{p(b)}.$$

Here, $p(a, b)$ is the probability that both a and b occur. This rule comes with a specific terminology that has influenced probabilistic machine learning significantly. The term $p(b|a)$ in Eq. (2.4) is called the *likelihood*, $p(a)$ is the *prior* and $p(a|b)$ the *posterior*.

How can we use this rule in the context of learning? Given a training dataset $\mathcal{D}$ as before and understanding inputs x and outputs y as random variables, we want to find the probabilistic model $p(x, y|\mathcal{D})$ which is likely to produce the data. Note that we have now made the dependence on the data explicit. Formally one can write

2.2 Models

$$p(x, y|\mathcal{D}) = \int d\theta \; p(x, y|\theta) p(\theta|\mathcal{D}). \tag{2.5}$$

The first part of the integrand, $p(x, y|\theta)$, is the parametrised model distribution $p(x, y; \theta)$ that we chose (see Definition 2.3), but written as a conditional probability. The second part of the integrand, $p(\theta|\mathcal{D})$, is the probability that a certain set of parameters is the 'right one', given the data. In a sense, this is exactly what we want to achieve by training, namely to find an optimal θ given the data.

In order to compute the unknown term $p(\theta|\mathcal{D})$ we can use Bayes formula, which reveals

$$p(\theta|\mathcal{D}) = \frac{p(\mathcal{D}|\theta) p(\theta)}{\int d\theta \; p(\mathcal{D}|\theta) p(\theta)}. \tag{2.6}$$

The prior $p(\theta)$ describes our previous assumption as to which parameters lead to the best model *before* seeing the data. For example, one could assume that the 'correct' parameters have a Gaussian distribution around zero. The prior is a very special characteristic of Bayesian learning which allows us to make an educated guess at the parameters without consulting the data, and under certain circumstances it can be shown to have a close relation to regularisers in an objective function. If nothing is known, the prior can be chosen to be uniform over a certain interval. The likelihood $p(\mathcal{D}|\theta)$ is the probability of seeing the data given certain parameters. Together with the normalisation factor $p(\mathcal{D}) = \int p(\mathcal{D}|\theta) p(\theta) d\theta$ we get the desired posterior $p(\theta|\mathcal{D})$ which is the probability of parameters θ *after* seeing the data.

The equation shifts the problem from finding the probability of parameters given some data to finding how likely it is to observe our dataset if we assume some parameters. This second question can be answered by consulting the model. Under the assumption that the data samples are drawn independently, one can factorise the distribution and write

$$p(\mathcal{D}|\theta) = \prod_m p(x^m, y^m|\theta).$$

Since the expression on the right can be computed, we could in theory integrate over Eq. (2.5) to find the final answer. However, for high-dimensional parameters the exact solution quickly becomes computationally intractable, and may be even difficult to approximate via sampling. It is therefore common to approach the problem with optimisation which leads to the famous *maximum likelihood problem* [15] (see Sect. 2.3.1), and considering the prior in a truly Bayesian fashion leads to the closely related *maximum a posteriori estimation*.

In conclusion, it seems that optimisation and integration are two sides of the same coin, and both are hard problems to solve when no additional structure is given.

2.2.4 Kernels and Feature Maps

It was mentioned above that there are two types of models. Some 'absorb' all information gained from the data in the parameters and we can discard the data after training. However, some model functions depend on all or a subset of data points $\mathcal{D}' \subseteq \mathcal{D}$,

$$f(x; \theta, \mathcal{D}').$$

The idea of data-dependent models is to define a similarity measure for the input space $\mathcal{X}$ and compare training points with the new input. Similar inputs are assumed to have similar outputs. One can weigh training inputs according to their importance for the prediction, which is why the model includes some trainable parameters θ. If many weights are zero, only very few data points are used for inference. A simple example is again the nearest neighbour method from the introduction, where the squared-distance was used to weigh the contribution a training point made to the prediction of a new input.

When the similarity measure can be described as a so-called *kernel*, an entire world of theory opens up which sheds a rather different light on machine learning models, and has interesting parallels to quantum computing that will be explored in Sect. 6.2. We therefore want to give a taste of kernel methods before revisiting them in more detail later. In the context of machine learning, kernels are defined as follows:

Definition 2.4 (*Kernel*) A (positive definite, real-valued) *kernel* is a bivariate function $\kappa : \mathcal{X} \times \mathcal{X} \to \mathbb{R}$ such that for any set $\mathcal{X}_f = \{x^1, \ldots, x^M\} \subset \mathcal{X}$ the matrix K with entries

$$K_{m,m'} = \kappa(x^m, x^{m'}), \quad x^m, x^m \in \mathcal{X}_f \qquad (2.7)$$

called *kernel* or *Gram matrix*, is positive semidefinite. As a consequence, $\kappa(x^m, x^{m'}) \geq 0$ and $\kappa(x^m, x^{m'}) = \kappa(x^m, x^{m'})$.

Examples of popular kernel functions can be found in Table 2.2.

Under certain—but as it turns out fairly general—conditions formalised in the so called *representer theorem* [16, 17] a model $f(x; \theta)$ that does not depend on the data can be rewritten in terms of kernel functions. As an illustration, consider a linear model with weights $w = (w_0, w_1, \ldots, w_N)^T$ and input $x = (1, x_1, \ldots, x_N)^T$,

$$f(x; \theta = w) = w^T x.$$

If we fit this model to data $\{(x^m, y^m)\}, m = 1, \ldots, M$, using a common cost function, the representer theorem guarantees that the optimal weight vector can be written as

Table 2.2 Examples of kernel functions as a distance measure between data points $x, x' \in \mathbb{R}^N$

Name	Kernel	Hyperparameters
Linear	$x^T x'$	–
Polynomial	$(x^T x' + c)^p$	$p \in \mathbb{N}, c \in \mathbb{R}$
Gaussian	$e^{-\gamma \|x-x'\|^2}$	$\gamma \in \mathbb{R}^+$
Exponential	$e^{-\gamma \|x-x'\|}$	$\gamma \in \mathbb{R}^+$
Sigmoid	$\tanh(x^T x' + c)$	$c \in \mathbb{R}$

2.2 Models

a linear combination of the data points,

$$w_{\text{opt}} = \frac{1}{M} \sum_{m=1}^{M} \alpha_m x^m,$$

with real coefficients α_m. If we insert this weight back into the linear model, we get

$$f(x; \theta = \{\alpha_m\}, \mathcal{D}) = \frac{1}{M} \sum_{m=1}^{M} \alpha_m (x^m)^T x = \frac{1}{M} \sum_{m=1}^{M} \alpha_m \kappa(x^m, x).$$

The inner product $(x^m)^T x$ is a linear kernel. We have translated the linear model with $N+1$ parameters $w_0, \ldots, w_N$ into a 'kernelised' model with M parameters $\alpha_1, \ldots, \alpha_M$. This formulation of a model is also called the *dual* form (as opposed to the original *primal* form). The objective of training is now to fit the new parameters which weigh the training data points.

Expressing a model in terms of a kernel function that measures distances to data points allows us to use the so called *kernel trick*. The kernel trick can be described as follows:

> Given an algorithm which is formulated in terms of a positive definite kernel κ, one can construct an alternative algorithm by replacing κ by another positive definite kernel κ'. [18]

We can therefore build new models by simply exchanging one kernel with another.

It makes intuitive sense that the distance to data points is a useful way to compute a prediction for a new data point. However, there is more to kernel functions than just distance measures. A kernel can be interpreted as an inner product of data that has been transformed by a *feature map* $\phi : \mathcal{X} \to \mathcal{F}$,

$$k(x, x') = \langle \phi(x), \phi(x') \rangle.$$

The *feature space* $\mathcal{F}$ is a Hilbert space, a complete vector space with an inner product, and ϕ as a—usually nonlinear—transformation that projects into $\mathcal{F}$. This is interesting because data can be easier to classify in higher dimensional spaces, for example when it becomes linearly separable.

To illustrate how a feature map can make non-linearly separable data separable, consider the following examples:

Example 2.5 (*XOR function*) The full dataset of the XOR function is given by

$$\mathcal{D} = \{((-1, -1)^T, 1), ((-1, 1)^T, -1), ((1, -1)^T, -1), ((1, 1)^T, 1)\}$$

and is clearly not linearly separable. A feature map of the form $\phi((x_1, x_2)^T) = (x_1, x_2, x_1 x_2)^T$ allows for a hyperplane cutting through the 3-dimensional space to separate both classes (see Fig. 2.9).

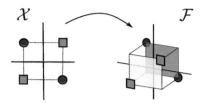

Fig. 2.9 While four data points labelled according to the XOR function (zeros as red circles and ones as blue squares in opposite corners of the unit square) cannot be separated by a linear decision boundary in $\mathcal{X}$, the feature map to $\mathcal{F}$ from Example 2.5 separates the two classes by introducing a third dimension

Example 2.6 (*Concentric circles*) Consider a dataset of two 2-dimensional concentric circles, which is impossible to separate by a linear decision boundary (see Fig. 2.10). A polynomial feature map

$$\phi((x_1, x_2)^T) = (x_1, x_2, 0.5(x_1^2 + x_2^2))^T$$

transforms the data into a linearly separable dataset in a 3-dimensional space.

Example 2.7 (*Infinite dimensional feature spaces*) Feature maps can even map into spaces of infinite dimension. Consider the Gaussian kernel from Table 2.2 with $x, x' \in \mathbb{R}^N$ and use the series expansion of the exponential function to get

$$\begin{aligned} \kappa(x, x') &= e^{-\frac{1}{2}\|x-x'\|^2} \\ &= \sum_{j=0}^{\infty} \frac{(x^T x')^k}{k!} e^{-\frac{1}{2}|x|^2} e^{-\frac{1}{2}|x'|^2} \\ &= \langle \phi(x), \phi(x') \rangle. \end{aligned}$$

The Gaussian kernel effectively implements a feature map, for example leading to feature vectors with entries [19]

$$\phi(x) = \left(\frac{1}{\sqrt{k!^{\frac{1}{k}}}} e^{-\frac{1}{2k}\|x\|^2} \binom{k}{n_1, \ldots, n_N}^{\frac{1}{2}} x_1^{n_1} \ldots x_N^{n_N} \right),$$

with $k = 0, \ldots, \infty$ and $n_1, \ldots, n_N$ such that $\sum_{i=1}^{N} n_i = k$. Of course one never has to calculate this rather ugly expression, and can instead evaluate the much simpler Gaussian kernel function using the original inputs. Note that there are of course other feature maps that lead to a Gaussian kernel [20].

In summary, many models can be expressed in terms of kernel expansions, which makes them data-dependent. This shifts the optimisation problem to a different set of parameters, namely the weights of the kernel functions. By replacing one kernel function κ by another kernel function κ', we effectively change the feature map

2.2 Models

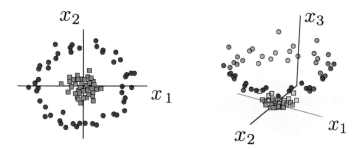

Fig. 2.10 Data points of two classes arranged as concentric circles in the 2-d plane cannot be separated by a linear decision boundary (left), but the feature map from Example 2.6 projects the data onto a square cone in 3 dimensions allows them to be divided by a hyperplane

and can potentially go into very high-dimensional spaces of many different shapes, and hope that our data gets easier to classify in that space. That way simple linear models can get the power of nonlinear classifiers. Of course, designing a good model translates into finding a good feature space, which in turn means to find a good kernel. Excellent introductions to kernel methods are found in [18, 21]. A number of kernel methods will be introduced in Sect. 2.4.

2.3 Training

The goal of training is to select the best model for prediction from a model family. This means we have to define an objective function that quantifies the quality of a model given a set of parameters, and training becomes an optimisation problem over the objective. Machine learning tends to define rather difficult optimisation problems that require a lot of computational resources. This is why optimisation "lies at the heart of machine learning" [22] and often defines the limits of what is possible. In fact, some major breakthroughs in the history of machine learning came with a new way of solving an optimisation problem. For example, the two influential turning points in neural networks research were the introduction of the backpropagation algorithm in the 1980s [23],[4] as well as the use of Boltzmann machines to train deep neural networks in 2006 [24]. However, as outlined in the preceding section, optimisation is only a means to generalisation.

---

[4]The original paper [23], a Technical Report by Rumelhart, Hinton and Williams, has close to 20,000 citation on *Google Scholar* at the time of writing. It is widely known today that the algorithm had been invented by others long before this upsurge of attention.

2.3.1 Cost Functions

As mentioned above, the *cost function* $C(\theta)$ is the function that quantifies the objective according to which we train the parameters of a model. We also said that the cost function consists of different building blocks. The most important blocks are the *loss* $L(\mathcal{D}, \theta)$, which was introduced above to measure how well a model does in predicting the training data, and the *regulariser* $R(\theta)$ which imposes extra constraints on the optimisation. Regularisation has been found to help against overfitting. However, many rather complex and high-dimensional models in use today seem to circumvent this problem in a natural way, and theories of regularisation and overfitting are subject to new questions [25].

Let us have a look at some popular loss functions and regularisers. Possibly the most common choice for the loss function is the *squared loss* which compares the outputs $f(x^m; \theta)$ produced by the model when fed with training inputs x^m with the target outputs y^m,

$$L(\theta) = \frac{1}{2} \sum_{m=1}^{M} (f(x^m; \theta) - y^m)^2. \tag{2.8}$$

The resulting optimisation problem is known in statistics as *least-squares optimisation*. In classification problem f and y are binary values, and the squared loss would be a non-continuous function that quantifies the number of misclassified training points. To make the loss continuous, f is during training often defined to be the continuous output of the model, which only gets binarised when the model is used to make predictions. The squared loss is also called *l2 loss* (not to be mixed up with the L_2 regulariser), and replacing $(f(x^m; \theta) - y^m)^2$ with the absolute value $|f(x^m; \theta) - y^m|$ gives rise to the *l1 loss*.

Another way to quantify the distance between outputs and targets is via the expression $y^m f(x^m; \theta)$, which is positive when the two numbers have the same sign and negative if they are different. One can use this to compute the *hinge loss*,

$$L(\theta) = \sum_{m=1}^{M} \max(0, 1 - y^m f(x^m; \theta)),$$

the *logistic loss*

$$L(\theta) = \sum_{m=1}^{M} \log(1 + e^{-y^m f(x^m; \theta)}),$$

or, if the output of the model is a probability $f(x^m; \theta) = p^m$, the cross entropy loss

$$L(\theta) = \sum_{m=1}^{M} -y^m \log p^m - (1 - y^m) \log(1 - p^m),$$

2.3 Training

Fig. 2.11 Different loss functions and their dependence on the similarity measure $f(x, \theta) - y$ or $yf(x, \theta)$

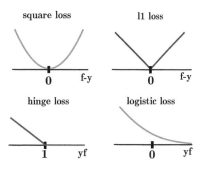

of a model (see Fig. 2.11). For multi-label classification with one-hot encoding, the *cross entropy* can also be used to compare the multi-dimensional target $y = (y_1, \ldots, y_D)^T$, $y \in \{0, 1\}$ with the output probability distribution over predictions $f(x; \theta) = (p_1, \ldots, p_D)^T$,

$$L(\theta) = -\sum_{m=1}^{M} \sum_{d=1}^{D} y_d^m \log p_d^m.$$

Estimating the parameters of a probabilistic model given a dataset is typically done through *maximum likelihood estimation*, a concept with a long-standing tradition in statistics [4]. In fact, the square loss can be shown to be the maximum likelihood solution under the assumption of Gaussian noise [3]. The underlying idea is to find parameters θ so that there is a high probability that the data have been drawn from the distribution $p(x, y; \theta)$. For this, we want to maximise the probability $p(\mathcal{D}|\theta)$ of sampling our data $\mathcal{D}$ given parameter θ. It is standard practice to take the logarithm of the likelihood and maximise the *log-likelihood*, as it does not change the solution of the optimisation task while giving it favourable properties. If the data is independently and identically distributed (which means that samples do not influence each other and are drawn from the same distribution), we have

$$\log p(\mathcal{D}|\theta) = \log \prod_m p(x^m, y^m|\theta) = \sum_m \log p(x^m, y^m|\theta).$$

From Sect. 2.2.3 we know that $p(x^m, y^m|\theta) \propto p(x^m, y^m; \theta)$, and since constant factors also have no influence on the optimisation problem, the maximum log-likelihood estimation problem means to find parameters θ^* which maximise

$$\theta^* = \max_{\theta} \sum_{m=1}^{M} \log p(x^m, y^m; \theta). \tag{2.9}$$

This idea has already been encountered in Example 2.3, as well as in the Bayesian learning framework.

In terms of regularisers, there are two common choices, L_1 and L_2 regularisation (named after the norm they are based on). We will call them in our notation the R_{L_1} and R_{L_2} regulariser. The name comes from the respective norm to use. If we write the parameters as a vector w for the moment and denote by $|\cdot|$ the absolute value, we can define them as

$$R_{L_1}(w) = \lambda ||w||_1^2 = \sum_i |w_i|,$$

and

$$R_{L_2}(w) = \lambda ||w||_2^2 = \sum_i w_i^2,$$

respectively. The L_2 regulariser adds a penalty for the length of the parameter vector and therefore favours parameters with a small absolute value. Adding the L_1 regulariser to the cost function favours sparse parameter vectors instead. The hyperparameter λ regulates how much the regularisation term contributes towards the cost function.

2.3.2 Stochastic Gradient Descent

Once the cost function is constructed, machine learning reduces to the mathematical problem of optimising the cost function, which naturally becomes a computational problem for any realistic problem size. Mathematical optimisation theory has developed an extensive framework to classify and solve optimisation problems [26] which are called *programmes*, and there are important distinctions between types of programmes that roughly define how difficult it is to find a global solution with a computer. For some problems, even local or approximate solutions are hard to compute. The most important distinction is between convex problems for which a number of algorithms and extensive theory exists, and non-convex problems that are a lot harder to treat [27]. Convexity thereby refers to the objective function and possible inequality constraint functions. Roughly speaking, a set is convex if a straight line connecting any two points in that set lies inside the set. A function $f : \mathcal{X} \to \mathbb{R}$ is convex if $\mathcal{X}$ is a convex domain and if a straight line connecting any two points of the function lies 'above' the function (for more details see [26]).

To give an example, least-squares optimisation together with a model function that is linear in the parameters forms a rather simple convex quadratic optimisation problem that has a closed-form solution as we will show in the next section. For general non-convex problems much less is known, and many machine learning problems fall into this category. Popular methods are therefore iterative searches such as *stochastic gradient descent*, which performs a stepwise search for the minimum on batches of the data. Especially in the field of neural networks, gradient-based search methods stand almost without an alternative at the moment.

Gradient descent updates the parameters θ of a cost function $C(\theta)$ successively towards the direction of steepest descent,

2.3 Training

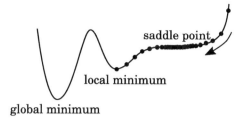

Fig. 2.12 Steps of the training updates in a 1-dimensional parameter space. Gradient descent gets stuck in local minima and convergence becomes very slow at saddle points

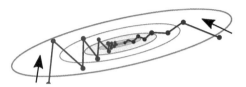

Fig. 2.13 Steps of the training updates in a 2-dimensional parameter space with a thin valley. Lower values of the objective function are shaded darker. While conventional gradient descent converges slower due to strong oscillations (red line starting on the left), the momentum method of considering the direction of past gradients leads to a straighter path (blue line starting on the right). Note that if the learning rate is too high, the valley can be missed altogether and training updates could in fact increase the cost function

$$\theta^{(t+1)} = \theta^{(t)} - \eta \nabla C(\theta^{(t)}), \qquad (2.10)$$

where η is an external parameter called the *learning rate* and t an integer keeping track of the current iteration. The gradient $\nabla C(\theta^{(t)})$ always points towards the ascending direction in the landscape of C, and following its negative means to descend into valleys. As one can imagine, this method can get stuck in local minima if they exist (see Fig. 2.12).[5] Convergence to a minimum can take a prohibitively long time due to saddle points in which gradients are vanishingly small and the steps therefore only gradual. Another problem appears in thin valleys where the search path oscillates heavily.

A solution to increase the convergence is to include a momentum, or to make the change in parameters of step t dependent on the change we made for the previous steps $t - 1$ (see Fig. 2.13). This effectively dampens oscillations and chooses a straighter path. Also the choice of a learning rate provides a lot of possibilities to shape the convergence of the training algorithm. For example, one can choose a step-dependent learning rate $\eta^{(t)}$ that decreases with each iteration, or adapt the learning rate to the current gradient.

---

[5]For large neural networks of thousands of parameters, local minima seem in fact only a minor problem since the search direction is likely to find a dimension in which a local minima in a subset of the dimensions can be escaped.

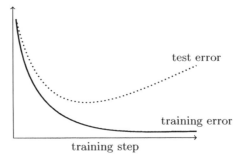

Fig. 2.14 Typical behaviour of the error on the training and test set during training. While the model learns to reduce the error on the training set, the test error usually begins to increase after a certain number of training steps, an indicator that the model begins to overfit. Overfitting means to learn the specific structure of the training set, so that the classifier looses its ability to generalise

Stochastic gradient descent uses only a subset of the training data for each update to calculate the cost function. While the original definition of stochastic gradient descent in fact only considered one randomly sampled training input per iteration, the more common version uses mini-batches of randomly sampled data, and the batch size becomes a hyperparameter for training. (Full-)batch gradient descent decreases the cost function in each iteration (unless the learning rate is large enough to 'jump over minima'). Stochastic gradient descent shows fluctuations in training that get stronger the smaller the batch size. However, the stochastic nature of the gradient direction tends to help in finding minima faster and escaping local minima. An important reason to still consider larger batch sizes is the prospect of parallelising computations.

One last practical remark on overfitting may be useful. An iterative method allows us to recognise overfitting by monitoring the error on the training and test set during each step of the optimisation (see Fig. 2.14). Typically, test and training error start to decrease together. When the training starts to fit the particulars of the training set, thereby losing generalisation ability, the test error begins to rise while the training error continues to decrease. The optimal point to stop the optimisation is just before the renewed increase of the test error, and one should consider regularisation techniques to avoid such behaviour altogether.

2.4 Methods in Machine Learning

Let us briefly summarise the previous sections. A method in machine learning specifies a general ansatz for a *model function or distribution* that can be used for prediction and a *training strategy* of how to use the data to construct a specific model that generalises from the data. In supervised learning, the training objective is to minimise the error between target outputs and predicted outputs for the training set. However, this objective serves the overarching purpose of minimising the generalisation error, which is usually measured on a test or validation set. This section

2.4 Methods in Machine Learning

Table 2.3 Summary of the model functions $f(x, w)$ and distributions $p(x)$ of machine learning methods presented in this section. As further established in the text, x is a model input, s, v, h and o denote different kinds of units or random variables in probabilistic models, while w, W, θ and γ_m are learnable parameters. We denote by φ a nonlinear (activation) function, t a time step, c a specific class and by M_c the number of training samples from that class. $\tilde{\kappa}$ is a scalar kernel and $\tilde{k}$ a vector of kernel functions, and K a kernel Gram matrix

Method	Model function/distribution
Data fitting	
Linear regression	$f(x; w) = w^T x$
Nonlinear regression	$f(x; w) = \varphi(w, x)$
Artificial neural networks	
Perceptron	$f(x; w) = \varphi(w^T x)$
Feed-forward neural network	$f(x; W_1, W_2, \ldots) = \cdots \varphi_2(W_2 \varphi_1(W_1 x)) \cdots$
Recurrent neural network	$f(x^{(t)}; W) = \varphi(W f(x^{(t-1)}; W))$
Boltzmann machine	$p(v; \theta) = \frac{1}{Z} \sum_h e^{-E(v,h;\theta)}$
Graphical models	
Bayesian network	$p(s) = \prod_k p(s_k \mid \pi_k; \theta)$
Hidden Markov model	$p(V, O) = \prod_{t=1}^{T} p(v^{(t)} \mid v^{(t-1)}) \prod_{t=1}^{T} p(o^{(t)} \mid v^{(t)})$
Kernel methods	
Kernel density estimation	$p(x \mid y = c) = \frac{1}{M_c} \sum_{m \mid y^m = c}^{M} \kappa(x - x^m)$
K-nearest neighbour	$p(x \mid y = c) = \frac{\#NN_c}{k}$
Support vector machine	$f(x) = \sum_{m=1}^{M} \gamma_m y^m \kappa(x, x^m) + w_0$
Gaussian process	$p(y \mid x) = \mathcal{N}\left[\tilde{k}^T K^{-1} y; \tilde{\kappa} - \tilde{k}^T K^{-1} \tilde{k}\right]$

introduces examples of supervised machine learning methods that become important in the context of quantum machine learning. They are summarised in Table 2.3 for quick reference. Note that the categorisation of different methods into *data fitting*, *neural networks*, *graphical models* and *kernel methods* is to some extend arbitrary: Boltzmann machines are graphical models as much as neural networks are a type of nonlinear regression, while Gaussian processes and support vector machines can be derived from linear models.

2.4.1 Data Fitting

Most physicists are familiar with statistical methods for data fitting such as *linear* and *nonlinear* regression.[6] These are well-established in statistics and data science

---

[6] In this context the term "regression" is both used for the problem of regression as well as the model.

2.4.1.1 Linear Regression

Needless to say, linear regression tackles the *problem of regression* outlined in Definition 2.1 and is based on a deterministic linear model function (see Definition 2.2),

$$f(x; w) = w^T x + w_0, \tag{2.11}$$

where the vector $w \in \mathbb{R}^N$ contains the parameters and $x \in \mathbb{R}^N$ is the input as usual. The bias w_0 can be included into $w^T x$ by adding it as an extra dimension to w while padding x with an extra value $x_0 = 1$, and will therefore be neglected in the following. Note that the term 'linear' refers to linearity in the model parameters only. A nonlinear feature map on the original input space can turn linear models into powerful predictors that can very well be used to model nonlinear functions. A well known example is the feature map

$$\phi : x \in \mathbb{R} \to (1, x, x^2, \ldots, x^d)^T, \tag{2.12}$$

so that f in Eq. (2.11) becomes

$$f(\phi(x); w) = w_0 + w_1 x + w_2 x^2 + \cdots + w_d x^d . \tag{2.13}$$

We illustrate this example in Fig. 2.15. According to the Weierstrass approximation theorem [28], any real single-valued function that is continuous on a real interval $[a, b]$ can be arbitrarily closely approximated by a polynomial function. Equation (2.13) can therefore model any function for the limit $d \to \infty$. In practice, approximating a function to a small error might involve a large number of parameters w_i, and other methods such as nonlinear regression or more intricate feature maps might be preferable.

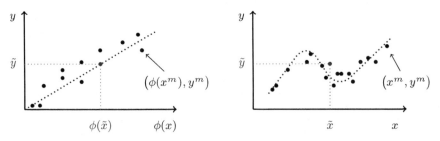

Fig. 2.15 Illustration of linear regression with a feature map. The linear model can fit data well in feature space (left), which in the original space appears as a nonlinear function (right)

2.4 Methods in Machine Learning

Learning in linear regression means to find the parameters w that fit f to the training data in order to predict new data points. A very successful approach to find the optimal[7] parameters is *least squares estimation*. This approach uses the square loss

$$L(w) = \sum_{m=1}^{M} (w^T x^m - y^m)^2. \tag{2.14}$$

It is convenient to express this as a matrix equation,

$$L(w) = (Xw - y)^T (Xw - y),$$

where we introduced the notation

$$y = \begin{pmatrix} y^1 \\ \vdots \\ y^M \end{pmatrix}, \quad X = \begin{pmatrix} x_1^1 & \cdots & x_N^1 \\ \vdots & \ddots & \vdots \\ x_1^M & \cdots & x_N^M \end{pmatrix}, \quad w = \begin{pmatrix} w_1 \\ \vdots \\ w_N \end{pmatrix}.$$

We call X *data* or *design* matrix. If $X^T X$ is of full rank, the estimated parameter vector can be calculated by the closed-form equation

$$w = (X^T X)^{-1} X^T y. \tag{2.15}$$

To show this, write $(Xw - y)^2 = (Xw - y)^T (Xw - y) = y^T y - 2w^T X^T y + w^T X^T X w$ and calculate the derivative $\partial_w L(w)$, which results in $-2X^T y + 2X^T X w$. At the minimum, this expression is zero. The solution to the least squares optimisation problem for a linear regression model is therefore

$$w = X^+ y \tag{2.16}$$

with

$$X^+ = (X^T X)^{-1} X^T. \tag{2.17}$$

The matrix X^+ is also called a *pseudoinverse*, a generalisation of the inverse for non-square or square singular matrices. In computational terms, training a linear regression model reduces to the inversion of a $M \times K$ dimensional data matrix, where usually the dimension of the feature space K is larger than the dimension of the input vectors. The fastest classical algorithms take time $\mathcal{O}(K^\delta)$, where for the best current algorithms the constant δ is bounded by $2 \leq \delta \leq 3$.

An alternative way to solve this problem is to write the pseudoinverse as a singular value decomposition. A singular value decomposition is the generalisation of an eigendecomposition $A = SDS^T$, but for general (i.e. singular) matrices [29]. Any real matrix can be written as $A = U \Sigma V^T$, where the orthogonal real matrices U, V

---

[7]Least squares can be shown to produce an unbiased estimator with minimum variance [4].

carry the *singular vectors* $u_r, v_r, \ r = 1, \ldots, R$ as columns, where R is the rank of A. Σ is a diagonal matrix of appropriate dimension containing the R nonzero singular values σ_r. The inverse of A is calculated by inverting the singular values on the diagonal and taking the transpose of U, V. Using the singular value decomposition to decompose X in Eq. (2.17) yields

$$\begin{aligned} X^+ &= (X^T X)^{-1} X^T \\ &= (V\Sigma U^T \ U\Sigma V^T)^{-1} V\Sigma U^T \\ &= V\Sigma^{-2} V^T V\Sigma U^T \\ &= V\Sigma^{-1} U^T, \end{aligned}$$

where we used $U^T U = \mathbb{1}$ which is true for orthogonal matrices. With this formulation of the pseudoinverse, the solution (2.16) can be expressed in terms of the singular vectors and values,

$$w = \sum_{r=1}^{R} \frac{1}{\sigma_r} u_r^T y \ v_r. \tag{2.18}$$

The computational problem is now to find the singular value decomposition of X.

2.4.1.2 Nonlinear Regression

While in linear regression the model function has a linear dependency on the parameters, this condition gets relaxed in nonlinear regression. One way of deriving nonlinear regression from the linear version is by replacing the nonlinear feature map ϕ (see for example Eq. (2.13)) by a nonlinear function in the inputs *and* parameters,

$$f(x) = \varphi(w, x). \tag{2.19}$$

Currently, the most successful nonlinear regression models in machine learning are neural networks.

2.4.2 Artificial Neural Networks

From a mathematical perspective, neural networks can be seen as a nonlinear regression model with a specific choice for the model function φ in Eq. (2.19) [9], namely where the input to the nonlinear function is given by a linear model $\varphi(w, x) = \varphi(w^T x)$, an expression which then gets nested several times. Historically these models were derived from biological neural networks [30, 31] and they have a beautiful graphical representation reminiscent of neurons that are connected by synapses. The nonlinear function originally corresponded to the 'integrate-and-fire' principle found in biological neurons, which prescribes that a neuron fires once the incoming signal surpasses a certain threshold value [32]. Neural network research

2.4 Methods in Machine Learning

was abandoned and revived a number of times during its history. Important milestones were when Hopfield showed in 1982 that a certain type of network recovers properties of a memory [31], the rediscovery of the backpropagation algorithm in the late 80s [33], as well as recent developments in 'deep' neural network structures [24].

2.4.2.1 Perceptrons

Perceptrons are the basic building block of artificial neural networks. Their model function is given by

$$f(x; w) = \varphi(w^T x), \tag{2.20}$$

where the inputs and outputs are either real or binary numbers. Sometimes the mathematical structure makes it convenient to choose $\{-1, 1\}$ rather than $\{0, 1\}$. The nonlinear function φ is called an *activation function* and in the original proposal of a perceptron it referred to the sign or step function

$$\mathrm{sgn}(a) = \begin{cases} 1, & \text{if } a \geq 0, \\ -1, & \text{else.} \end{cases}$$

The perceptron model can be trained by iterating through the data and updating the weights according to

$$w_i^{(t+1)} = w_i^{(t)} + \eta(y^m - \varphi(x^T w^{(t)}))x_i^m,$$

where η is the learning rate. The computational properties of a perceptron have been studied from as early as the 1960s [32, 34], and show that the learning rule always converges to the optimal weights. However, after the initial excitement it was found that this is only true for linearly separable datasets,[8] excluding a simple XOR gate from its scope (see Example 2.5). Only when perceptrons were combined to build more complex structures did their power become apparent, and the perceptron model is the core unit of artificial neural networks.

Figure 2.16 shows a perceptron in the typical graphical representation of neural networks, in which inputs and outputs are understood as units with certain values that are updated by the units that feed into them. The connections between units are associated with a weight. The activation function for the output node is not always drawn.

2.4.2.2 Feed-Forward Neural Networks

Feed forward neural networks have a model function of the form

---

[8]Remember that a dataset is linearly separable if it can be divided by a hyperplane in input space.

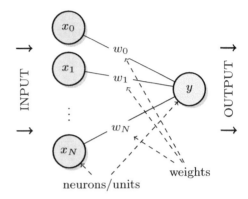

Fig. 2.16 Illustration of the perceptron model. The input features are each displayed as nodes of a graph, called units or neurons. The input units are connected to an output unit by weighed edges. Together with an activation function φ for the output unit, this graph defines a model function $y = \varphi(w^T x)$

$$f(x, W_1, W_2, \ldots) = \cdots \varphi_2 \left(W_2 \, \varphi_1(W_1 x) \right) \cdots . \qquad (2.21)$$

Here $\varphi_1, \varphi_2, \ldots$ are nonlinear activation functions between appropriate spaces. The parameters are summarised as weight matrices $W_i, i = 1, 2, \ldots$. The dots indicate that we could extend the model by an arbitrary number of such nested activations, as long as the outermost activation function has to map onto the output space $\mathcal{Y}$.

This model function is a concatenation of 'linear models' and activation functions, and as the dots suggest, the concatenation can be repeated many times. An interesting perspective on the success of neural networks is that they combine the flexibility of nonlinear dynamics with the data processing power of linear algebra. An important existence theorem by Hornik, Stincombe and White from 1989 [35] proves that only one concatenation of the form

$$f(x, W_1, W_2) = W_2 \, \varphi_1(W_1 x)$$

suffices to make the model a universal function approximator, meaning that up to finite precision it can be used to express any function on a compact domain (similar to the polynomial from Eq. (2.13)). This might however require weight matrices of very large dimensions.

In terms of their graphical representation, feed-forward neural networks connect multiple perceptrons in layers so that the units of each layer are connected to the units of the following layer (see Fig. 2.17). The first layer is made up of the input units $x_1, \ldots, x_N$, the following L layers contain the hidden units $h_1^l, \ldots, h_J^l$ (where $l = 1, \ldots, L$). The last layer contains the output unit(s). Each neuron is updated by an activation function depending on all neurons feeding into it, and the update protocol prescribes that each layer is updated after the previous one. This way, information is 'fed forward', and an input fed into the first layer is mapped onto an output that can be read out from the last layer. The model for the single-hidden-layer feed-forward neural network reads

$$f(x; W_1, W_2) = \varphi_2(\, W_2 \, \varphi_1(\, W_1 x \,) \,). \qquad (2.22)$$

2.4 Methods in Machine Learning

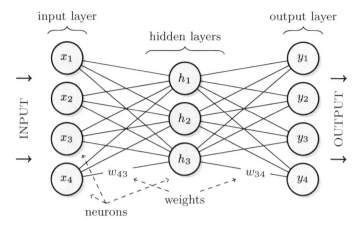

Fig. 2.17 A feed-forward neural network (here with only one hidden layer) feeds information forward through several layers

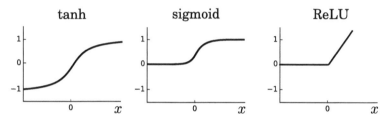

Fig. 2.18 Different options for activation functions: a hyperbolic tangent, a sigmoid function and rectified linear units (ReLU)

A feed-forward neural network can be combined with a variety of possible activation functions. Common examples for these functions are a hyperbolic tangent, a sigmoid $\varphi(a) = \frac{1}{1+e^{-a}}$, as well as 'rectified linear units' [36] that are zero for negative inputs and linear functions for positive inputs (see Fig. 2.18). But also radial basis functions have been used, and other activation functions might prove interesting in future.

The square loss for a single-hidden-layer neural network for some labelled training data $\mathcal{D} = \{(x^m, y^m)\}_{m=1}^{M}$ and the weight matrices W_1, W_2 reads

$$C(W_1, W_2, \mathcal{D}) = \sum_{m=1}^{M} (\varphi_2(W_2^T \, \varphi_1(W_1^T x^m)) - y^m)^2. \tag{2.23}$$

For nonlinear activation functions this is in general a non-convex, nonlinear (and hence difficult) optimisation problem. By far the most common training algorithm for feed-forward neural networks is therefore gradient descent with

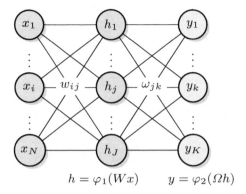

Fig. 2.19 Neural network model used to explain the backpropagation algorithm. The green nodes denote the input layer encoding the input features $x_1, \ldots, x_N$, followed by the blue hidden layer with hidden units $h_1, \ldots, h_J$ and the yellow output layer of units $y_1, \ldots, y_K$. The activation function of the hidden and output units are φ_1, φ_2, respectively. The weights are summarised in weight matrices W and Ω

$$w_{ij}^{(t+1)} = w_{ij}^{(t)} - \eta \frac{\partial C(W_1, W_2)}{\partial w_{ij}}. \tag{2.24}$$

The updated weight $w_{ij}^{(t+1)}$ is the current weight $w_{ij}^{(t)}$ minus a step down towards the direction of the steepest gradient with a step size or learning rate η. Computing gradients for neural networks is called *backpropagation* [33] and although a notational nightmare, we shall now walk through it. We will have a look at the weight update for a neural network with a single hidden layer only, but the generalisation to more layers is straightforward.

Let us first assemble all relevant expressions (see Fig. 2.19). Assume a cost function with squared loss and no regulariser (where $m = 1, \ldots, M$ can refer to the entire dataset or the current training data batch),

$$C(W_1, W_2) = \frac{1}{2} \sum_{m=1}^{M} \left(f(x^m; W, \Omega) - y^m \right)^2,$$

with the neural network model function

$$f(x; W, \Omega) = \varphi_2 \left(\Omega \varphi_1(W^T x) \right),$$

where $W \in \mathbb{R}^{J \times N}$ and $\Omega \in \mathbb{R}^{K \times J}$, and N, J, K are the dimensions of the input, hidden and output layer. For notational simplicity, we have renamed $W_1 = W$ and $W_2 = \Omega$, and their entries are respectively denoted by w_{ij} and ω_{jk}, while the kth row vector is given by a single index, W_k and Ω_k. We denote the vector summarising the values of the input layer as x, and use h for the hidden layer and y for the output layer. The linear 'net' inputs fed into an activation function will be denoted as $\text{net}_k^y = \Omega_k h$

2.4 Methods in Machine Learning

Fig. 2.20 Updating a hidden-to-output layer weight ω_{ab} only requires gradients of the output unit y_b it leads to. For notation, see Fig. 2.19

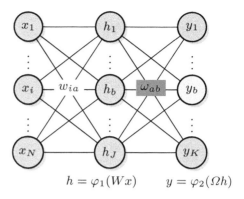

$$h = \varphi_1(Wx) \qquad y = \varphi_2(\Omega h)$$

for the output neurons and $\text{net}_j^h = W_j x$ for the hidden layer. The goal is to compute $\frac{\partial C}{\partial \omega_{jk}}, \frac{\partial C}{\partial w_{ij}}$ for all $i = 1 \ldots N, j = 1 \ldots J, k = 1 \ldots K$.

First calculate the derivatives of the hidden-to-output layer weights for e general weight matrix element ω_{ab} (see Fig. 2.20),

$$\frac{\partial C}{\partial \omega_{ab}} = \underbrace{\frac{\partial C}{\partial y_b} \frac{\partial y_b}{\partial \text{net}_b^y}}_{\delta_{y_b}} \frac{\partial \text{net}_b^y}{\partial \omega_{ab}}.$$

The first two terms are summarised by δ_{y_b}. The derivatives in the above expression are

$$\frac{\partial C}{\partial y_b} = -\sum_{m=1}^{M}\left(f_b - y_b^m\right),$$

$$\frac{\partial y_b}{\partial \text{net}_b^y} = (\varphi_2)'_b,$$

$$\frac{\partial \text{net}_b^y}{\partial \omega_{ab}} = h_a.$$

Taking the derivative of the b'th component of the second activation function, $(\varphi_2)'_b = \frac{d\varphi(z)_b}{dz}$, requires the activation function to be differentiable. It is no surprise that the sigmoid, tanh and 'rectified linear units' (ReLU) activation functions are popular, since their derivatives are rather simple. For example, the derivative of the sigmoid function is given by $\varphi'(z) = \varphi(z)(1 - \varphi(z))$. Putting it all together we get

$$\frac{\partial C}{\partial \omega_{ab}} = -\sum_{m=1}^{M}\left(f_b - y_b^m\right)(\varphi_2)'_b h_a = \delta_{y_b} h_a.$$

Fig. 2.21 Updating an input-to-hidden layer weight requires gradients of all output units and the hidden unit it leads to. For notation, see Fig. 2.19

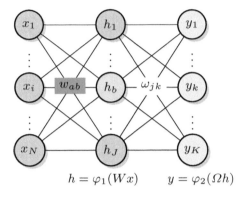

$$h = \varphi_1(Wx) \qquad y = \varphi_2(\Omega h)$$

Now we know how to update the parameters in the second weight matrix Ω which connects the hidden to the output layer. In order to calculate the update of the input-to-hidden layer with weights w_{ab} we have to consider not only h_b but also the derivatives of all neurons that depend on h_b (see Fig. 2.21), which in this case are all neurons of the output layer, and that gets a little more messy.

$$\frac{\partial C}{\partial w_{ab}} = \sum_{k=1}^{K} \underbrace{\frac{\partial C}{\partial y_k} \frac{\partial y_k}{\partial \text{net}_k^y}}_{\delta_{y^k}} \underbrace{\frac{\partial \text{net}_k^y}{\partial h_b}}_{\omega_{bk}} \underbrace{\frac{\partial h_b}{\partial \text{net}_b^h}}_{(\varphi_1)'_b} \underbrace{\frac{\partial \text{net}_b^h}{\partial w_{ab}}}_{x_a}$$

$$= \underbrace{\left(\sum_{k=1}^{K} \delta_{y^k} \omega_{bk} \right) (\varphi_1)'_b}_{\delta_{h_b}} x_a$$

$$= \delta_{h_b} x_a .$$

The δ's are also called the "errors" of a neuron, and one can now see that in order to compute the error of a hidden neuron, one requires the errors of all following neurons in the direction of the forward pass. In other words, the error has to be 'backpropagated' from right to left in the network, which is the reverse direction of the classification.

The computational steps needed for one iteration in backpropagation grow polynomial with the number of connections $NJ + JK$ [8] which can be costly for larger architectures, especially in 'deep' networks of many layers. Also the convergence time for the gradient algorithm can be very long, and training state-of-the art networks can take weeks on a supercomputer. Numerous tricks like layer-wise pre-training can be used to facilitate training.

2.4.2.3 Recurrent Neural Networks

While feed-forward neural networks are organised in layers, the graphical representation of recurrent neural networks is an all-to-all connected graph of units which are collectively updated in discrete time steps (Fig. 2.22). Information is therefore not fed-forward through layers, but 'in time'. The input can be understood as the state that some units are set to at time $t = 0$, while the output can be read from one or more designated units at time T. The units that are neither fed with input, nor used to read out the output, are called 'hidden units' and have a mere computational function.

Let $s^{(t)} = (s_1^{(t)}, \ldots, s_G^{(t)})^T$ describe the state of the G (hidden or visible) units of a recurrent neural network at time t. The edge between s_i and s_j is associated to a weight w_{ij}. The state of the network after each update is given by

$$s^{(t+1)} = \varphi(W^T s^{(t)}),$$

where the $\mathbb{R}^{G \times G}$ matrix W contains the weights and φ is a nonlinear activation function.

A recurrent neural network can be 'unfolded' to a feed-forward structure with T layers by interpreting every time step as a separate layer. It can then be trained by *backpropagation through time* which works exactly like the standard backpropagation algorithm. This method has some unwanted properties, such as exploding or vanishing gradients during training [37]. Proposals to improve on training range from artificially 'clipping' the exploding gradients [37], to introducing time-delays between units [38], to the use of unitary weight matrices [39]. Recurrent neural networks gained relatively little interest from the machine learning community for a long time. However, and possibly due to the resemblance of the time-unfolded structure to deep neural networks, this has been changing in the last few years [40]. A prominent example for an application of a recurrent neural network is sequence-to-sequence modeling for the translation of text.

One relatively simple class of recurrent neural networks for a task called 'pattern matching' or 'associative memory' are *Hopfield neural networks* [31].[9] Hopfield networks have binary units, symmetric all-to-all connections with $w_{ii} = 0$ for $i = 1, \ldots, G$, as well as a threshold activation function. One can easily show that for a given training point $s_1^m, \ldots, s_G^m$, choosing the weights w_{ij} proportional to $\sum_m s_i^m s_j^m$ leads to the s^m being stable states or 'attractors' of the network [8]. This means that an update of these states acts as the identity, $\varphi(W^T s^m) = s^m$. Moreover, the dynamics of consecutive updates of the units decreases an Ising-type energy function,

$$E(s; W) = -s^T W s = -\frac{1}{2} \sum_{i,j=1}^{G} w_{ij} s_i s_j,$$

---

[9]Although they are not first and foremost used for supervised learning, Hopfield networks appear frequently in the quantum machine learning literature and are therefore mentioned here.

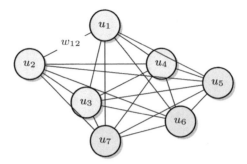

Fig. 2.22 A recurrent neural network is represented by an all-to-all connected graph. The units $u_1, \ldots, u_4$ are used to represent inputs, and at a later time, the outputs of the model (green and yellow nodes). The hidden units u_5, u_6, u_7 (blue) are 'not accessible' and can be understood as pure computational units, adding to the complexity of the dynamics of a recurrent neural net. The network here follows the design of a Hopfield model which does not have self-connections of nodes and where connections have symmetric weights

until one reaches one of these stable states. In other words, the updates drive the system state from the initial configuration to the closest memorised training pattern [41].

An important characteristic of Hopfield networks is their storage capacity indicating how many randomly selected patterns can be stably stored in the model of N neurons. Without going much into detail [8], one can say that the ratio of storable patterns to the size of the network is around 0.15. A network of $N = 100$ neurons can consequently only store around 15 states. This is not much, considering that it can represent 2^{100} patterns.

2.4.2.4 Boltzmann Machines

Boltzmann machines are probabilistic recurrent neural networks. As probabilistic models, they do not exhibit a feed-forward or time dynamics, but define a probability distribution over the states of the binary units $s = (s_1, \ldots, s_G)$. Again, the units can be divided into *visible units* $v = (v_1, \ldots, v_{N+K})$ and *hidden units* $h = (h_1, \ldots, h_J)$. The visible units can be further divided into input and output units if we deal with a supervised pattern recognition task, so that the first N visible units encode the inputs $x_1, \ldots, x_N$, and the last K visible units $v_{N+1}, \ldots, v_{N+K}$ encode the output y. In more general, Boltzmann machines do not necessarily require the distinction between inputs and outputs and are primarily used for unsupervised learning.

Boltzmann machines can be understood as Hopfield neural networks [42] for which every neuron carries a probability of being in state -1 or 1. Given a state $s_1, .., s_i, .., s_G$ of a Hopfield network, where all neurons but s_i are in a given state, the probability of neuron s_i to be in state 1 is given by

2.4 Methods in Machine Learning

$$p(s_i = 1) = \frac{1}{1 + e^{-\sum_{i'} \gamma_{ii'} s_i s_{i'}}}. \tag{2.25}$$

The term $\sum_{i'} \gamma_{ii'} s_i s_{i'}$ with the inter-neuron weights $\gamma_{ii'}$ is the energy associated with the given state and $s_{i'} = 1$. Overall, a Boltzmann machine assigns the following parametrised probability distribution to possible values (or 'states') $s = v, h$:

$$p(v, h; \theta) = \frac{1}{Z} e^{-E(v,h)} \tag{2.26}$$

where the *partition function* Z sums over all (v, h)

$$Z = \sum_{v,h} e^{-E(v,h)}, \tag{2.27}$$

The energy E is an Ising-type energy function that depends on a set of weights θ, and distinguishing between visible and hidden units makes the energy read

$$E(v, h) = -\sum_{i,j} w_{ij} \, v_i h_j - \sum_{j,j'} u_{jj'} \, h_j h_{j'} - \sum_{i,i'} z_{ii'} \, v_i v_{i'}$$
$$- \sum_i a_i \, v_i - \sum_j b_j h_j,$$

where the weights γ above are now divided into weights w_{ij} between visible and hidden units, weights $u_{jj'}$ between hidden units and weights $z_{ii'}$ between visible units (summarised by θ). We also added the 'local fields' a_i and b_j which give each separate unit a energy contribution. The joint probability distribution is thus given by a Boltzmann distribution. This is nothing other than a spin-glass model in statistical physics, where physical spins interact with different interaction strengths.

Since only the visible state $v = (v_1, \ldots, v_{N+K})$ is of interest, one can marginalise the joint probability distribution over the hidden units by summing over all their configurations,

$$p(v; \theta) = \frac{1}{Z} \sum_h e^{-E(v,h)}, \tag{2.28}$$

which is the probability distribution $p(v)$ of a generative probabilistic machine learning model.[10] Training a Boltzmann machine means to determine connection and local field strengths with which the model distribution is likely to generate the distribution of training data points.

It turns out that general Boltzmann machines are hard to train and the model only became popular when an efficient training algorithm for *restricted Boltzmann machines* was found [43] (see Fig. 2.23). In a restricted Boltzmann machine (RBM),

---

[10] Note that this expression can be written in terms of the *free energy* $F(s)$ known from physics, and training a Boltzmann machine minimises the free energy of the model.

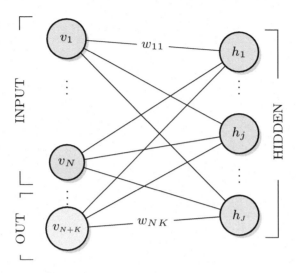

Fig. 2.23 Graphical representation of a restricted Boltzmann machine with visible units $v_1, \ldots, v_{N+K}$ and hidden units $h_1, \ldots, h_J$, in which the visible units include input and output units. A full Boltzmann machine would also allow for connections in between hidden units as well as in between visible units

visible units are only connected to hidden units and vice versa, so that $u_{ii'} = z_{jj'} = 0$ and

$$E_{\text{RBM}}(v, h) = -\sum_{ij} w_{ij}\, v_i h_j - \sum_i a_i v_i - \sum_j b_j h_j.$$

The objective function is chosen according to maximum log-likelihood estimation,

$$C(\theta) = \sum_{m=1}^{M} \log\, p(v^m; \theta),$$

where s^m represents the mth training data point. Inserting the formula for the probabilities from Eq. (2.28), the gradient for the interaction weights w_{ij} becomes

$$\begin{aligned}
\frac{\partial C(W)}{\partial w_{ij}} &= \frac{\partial}{\partial w_{ij}} \sum_{m=1}^{M} \log\, p(v^m) \\
&= \sum_m \left(\frac{\partial}{\partial w_{ij}} \log \sum_h e^{-E(v^m, h)} \right) - \frac{\partial}{\partial w_{ij}} \log Z \\
&= \sum_m \frac{1}{\sum_h e^{-E(v^m, h)}} \frac{\partial}{\partial w_{ij}} \sum_h e^{-E(v^m, h)} - \frac{1}{Z} \frac{\partial}{\partial w_{ij}} \sum_{v, h} e^{-E(v, h)}
\end{aligned}$$

2.4 Methods in Machine Learning

$$= \sum_m \sum_h \frac{e^{-E(v^m,h)}}{Z_\mathcal{D}} v_i^m h_j - \sum_{v,h} \frac{e^{-E(v,h)}}{Z} v_i h_j, \qquad (2.29)$$

where $Z_\mathcal{D} := \sum_{h,m} e^{-E(v^m,h)}$ is the partition function over the data set only, while Z includes a sum over all possible patterns v. Similar expressions can be derived for the local fields. This result is not surprising: The first term is the expectation value of the correlation between hidden and visible units over the data set distribution, while the second term is the expectation value over the model distribution, and the goal of learning is to find a model where the two are as close as possible. The result of Eq. (2.29) is usually abbreviated as

$$\frac{\partial C(W)}{\partial w_{ij}} = \langle v_i h_j \rangle_\text{data} - \langle v_i h_j \rangle_\text{model}. \qquad (2.30)$$

Calculating $\langle v_i h_j \rangle_\text{data}$ is relatively straightforward, since it is an average over all data samples [44]. But even getting samples of $\langle v_i h_j \rangle_\text{model}$ is intractable due to the partition function Z that involves computing a number of states which grows exponentially with the number of units. One approach would be to approximate it with Gibbs sampling. This is a *Markov Chain Monte Carlo* method [45] in which, starting with an initial state, the values of the random variables $v_1, \ldots, v_{N+K}, h_1, \ldots, h_J$ are iteratively updated by drawing samples from the probability distribution (2.25). After a while the process 'thermalises' and values of (s, h) (with a sufficient number of updates between them to avoid correlations) can be interpreted as samples for the Boltzmann distribution $\langle s_i h_j \rangle_\text{model}$. However, thermalisation can be very slow and there is no method that indicates without fail whether an equilibrium is reached [46]. Also mean-field approximations known from statistical physics perform in most cases rather poorly [6]. This was why Boltzmann machines were replaced by neural networks with backpropagation training algorithms in the 1980s [47], until in 2002 *contrastive divergence* was proposed [43] as a rather rough but effective approximation method. A number of quantum machine learning algorithms refer to this training method, which is why it shall be sketched briefly.

Contrastive divergence is surprisingly simple. The idea is to use a Markov Chain Monte Carlo sampling method, but stop the chain prematurely after only a few steps. The Markov chain successively samples the state of the visible units as well as the state of the hidden units. The sampling is designed in a way so that after many steps in the chain, samples are approximately drawn according the the Boltzmann distribution.

In the first step of the Markov chain one sets the visible units $v^{(0)}$ to a randomly picked training sample v^m. With the values of the visible units fixed, one samples the hidden units $h_j^{(0)}$ one by one with the probability of Eq. (2.25). Since hidden units are only connected to visible units in the restricted Boltzmann machine, this probability is fully defined by the states of the visible units. Now fix the hidden units to the sampled value and sample the visible units $v^{(1)}$ from Eq. (2.25) to start the first step in the Markov chain. Again, the restriction in the connectivity makes the sampling

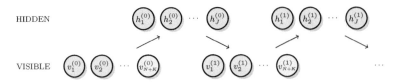

Fig. 2.24 In contrastive divergence, the visible units of the Boltzmann machine are initialised with a randomly drawn training sample v^m in step $t = 0$. The hidden units $h_1, \ldots, h_J$ are sampled from a certain probability that depends on the state of the visible units. This completes one step in the Markov chain, after which the visible units are again resampled with the fixed values for the hidden units. This procedure is repeated for T steps and the final pair $v^{(T)}, h^{(T)}$ is used to approximate a sample from the model distribution

only depend on the fixed hidden units. To finish the first step, resample the hidden units fixing the visible state. This procedure is repeated T times after which one ends up with a overall state $v^{(T)}, h^{(T)}$ of the Boltzmann machine (see Fig. 2.24). This state is used as an approximate sample from the model distribution. The weight update in Eq. (2.30) is therefore replaced by $v^{(0)}h^{(0)} - v^{(T)}h^{(T)}$. Against all intuition, only one single step of this sampling and resampling procedure ($T = 1$) is sufficient because the weights are updated in many iterations which overall induces an average of sorts[11] Although the contrastive divergence procedure actually does not lead to an update of the parameters according to the gradient of an existing objective function [48], it works well enough in many applications and became important for the training of deep (i.e., multi-layer) neural networks.

2.4.3 Graphical Models

Graphical models are probabilistic models that use graphical representations to display and simplify probability distributions [7]. Again, let $s = \{s_1, \ldots, s_G\}$ denote a set of visible and hidden binaey random variables, where the hidden variables represent the input x and output y of a supervised machine learning task.

2.4.3.1 Bayesian Networks

A *Bayesian network* or *belief network* is a probabilistic model with conditional independence assumptions that simplify the model distribution $p(s)$. In probability theory, a general joint probability distribution over s can be expressed by the chain rule

---

[11] The idea for this approach originated from the attempt to approximate an altogether different objective function, the difference between two Kullback-Leibler (KL) divergences [46]. The Kullback-Leibler divergence measures the similarity of two distributions can be interpreted as a free energy [47].

2.4 Methods in Machine Learning

$$p(s_1, \ldots, s_G) = p(s_1)p(s_2|s_1)p(s_3|s_1, s_2) \ldots p(s_G|s_1, \ldots, s_{G-1}), \quad (2.31)$$

which follows directly from the definition of a conditional probability

$$p(s_G|s_{G-1}, \ldots, s_1) = \frac{p(s_G, \ldots, s_1)}{p(s_{G-1}, \ldots, s_1)}.$$

Two random variables a, b are *conditionally independent* given another random variable z if $p(a, b|z) = p(a|z)p(b|z)$. Assuming externally given conditional independences between the variables s_i reduces the conditional probabilities in Eq. (2.31) to $p(s_i|\pi_i)$, where π_i is the set of variables that s_i conditionally depends on. This reduces the original probability distribution to

$$p(s_1, \ldots, s_G) = \prod_{i=1}^{G} p(s_i|\pi_i). \quad (2.32)$$

For example, the factor $p(s_3|s_1, s_2)$ in Eq. (2.31) reduces to $p(s_3|s_2)$ if s_3 is conditionally independent of s_1.

To use the model for inference, the conditional probabilities $p(s_i|\pi_i)$ have to be derived from the data with methods discussed before. If they are parametrised by parameters θ_i, learning means to find the optimal parameters given the data, for example with maximum (log)-likelihood estimation,

$$\max_{\theta} \sum_{i,m} \log p(s_i^m|\pi_i, \theta_i). \quad (2.33)$$

Here, $s^m = (s_1^m, \ldots, s_G^m)$ is the m'th training point. To use Bayesian networks for prediction in the supervised learning case, one conditions $p(s)$ on the given input x to get the class-conditional probability distribution $p(y|x_1 \ldots x_N)$ over the corresponding output variable x.

The important graphical representation of Bayesian nets as a directed acyclic graph makes these independence relations a lot clearer (see Fig. 2.25). Each random variable corresponds to a node in the graph. The *parents* of a node are all nodes with a directed connection to it. The *non-descendants* of a node s_i are all nodes that cannot be reached by following the connections starting from s_i. The connectivity of a graph representing a Bayesian net follows the *Markov condition*: Any node is conditionally independent of its non-descendants given its parents. The parents of a node s_i therefore correspond to the variables π_i. The conditional probabilities $p(s_i|\pi_i)$ are "attached" to each node of the graph, for example as *local conditional probability tables*.

Note that the graph architecture can be understood as a hyperparameter similar to the choice of the number and size of layers in neural networks. Not only the local probabilities can be learnt, but also the structure of the graph. Structure learning is very difficult, and even with an infinitely large dataset one can only learn directed connec-

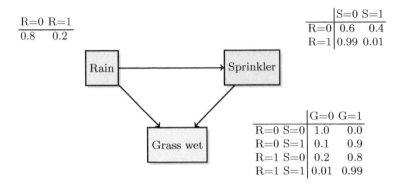

Fig. 2.25 An example of a Bayesian network with local conditional probability tables indicating the conditional probabilities of a child node given the state of its parents. The variable 'Rain' is conditionally independent from the other two variables, while the variable 'Sprinkler' is only conditionally independent from the variable 'Grass wet'. If one knows the value of 'Rain' (the input) and one wants to derive the probability of 'Grass wet' (the output), the 'Sprinkler' becomes a hidden variable over which one has to marginalise. Each variable comes with a *local probability table* listing the probability of the state of this variable given the state of the parents

tions up to a property called *Markov equivalence* [49]. This stems from the fact that different directed graphs encode the same conditional (in)dependence statements. Many algorithms define a scoring metric that measures the quality of a structure given the data and find better graphs by brute force search [50].

Also inference in Bayesian nets is generally a hard problem. In typical applications of Bayesian nets for inference, one observes values for some of the variables while the hidden variables remain unknown. In the example in Fig. 2.25 one might want to know the probability for the grass to be wet (output) given that it rained (input). Mathematically speaking, this means that the remainder of the s_i (the sprinkler) are hidden units h over which one has to marginalise, so that

$$p(y|x) = \sum_h p(y|x; h_1 \ldots h_J).$$

For binary units, the sum over h grows exponentially with J and inference becomes a NP-hard problem [51]. Some efficient inference algorithms for restricted classes of Bayesian nets are known, such as the *message passing algorithm* which is in $\mathcal{O}(J)$ (see [52] and references therein). Other solutions are approximate inference methods such as Monte Carlo sampling, in most cases with unknown accuracy and running time [6].

2.4.3.2 Hidden Markov Models

Hidden Markov models are graphical models whose properties are a bit different from the preceding methods, as they describe sequences of 'chained events'. Consider the problem of having a speech recording (i.e., a sequence of frequencies), and the task

2.4 Methods in Machine Learning

is to find the most likely sequence of words that correspond to the audio signal. The sequence of words can be modeled as a sequence of values for a random variable 'word', while the audio signal is a sequence of frequencies that can be represented by another random variable 'signal'. In that sense, the inputs and outputs of the hidden Markov model are sequences of values of random variables.

Let $\{h^{(t)}\}$ be a collection of random variables indexed by time t, and each random variable can take values in $\{h_1, \ldots, h_J\}$. Let $p(h_i|h_j)$ with $i, j = 1, \ldots, J$ be the time-independent transition probability that indicates how likely it is that the state of h changes from value h_j to h_i. A (first order) *Markov process* is a stochastic process where the state of the system changes according to the transition probabilities. The Markov property refers to the fact that the current state only depends on the previous one, but not on the history of the sequence. A possible sequence of a Markov process from time 0 to T shall be denoted as $H(T) = h^{(0)}, \ldots, h^{(T)}$.

We add a layer of complexity. In hidden Markov models the states h of the system are unknown at any time (the corresponding units are hidden, and hence the name). The only known values are the 'observations' at time t modeled by a second random variable $o^{(t)}$ with possible values $\{o_1, \ldots, o_S\}$ [53]. What is also known are the probabilities $p(o_k|h_j)$ of an observation o_s being made given that the system is in state h_j. Sequences of observations up to time T are denoted by $O(T) = o^{(1)}, \ldots, o^{(T)}$. Hidden Markov models are therefore 'doubly embedded' stochastic processes. An example for a trajectory of the process is illustrated in Fig. 2.26 on the right, while the left sketches a graph for the two different kinds of transition probabilities.

The motivation behind this model are machine learning tasks in which we have data which is a signature or hint of the actual information that we are interested in. In the speech recognition example, the states h may be the words uttered while the observation is the signal in the recording of a word. Given a recording of a speech as data, we are actually interested in the word sequence. The word sequence itself is modeled by a Markov process using transition probabilities that define how likely it is to have one word following another. The hidden Markov model can then be employed to find the sequence of words that is the most likely given the recording of a speech. Hidden Markov models also play an important role in many other applications such as text prediction, DNA analysis and online handwriting recognition [3].

The machine learning task for hidden Markov models is is to find the most likely state sequence $\tilde{H}(T)$ given an observation $\tilde{O}(T)$, which is a typical supervised pattern recognition problem (called *state estimation* in this context [6]). The probabilistic model distribution of a hidden Markov model is given by

$$p(H(T), O(T)) = \prod_{t=1}^{T} p(h^{(t)}|h^{(t-1)}) \prod_{t=1}^{T} p(o^{(t)}|h^{(t)}),$$

where $p(h^{(0)}|h^{(-1)}) = p(h^{(0)})$ is an initial value. In words, to find the probability of a sequence of states $H(T)$ and a sequence of observations $O(T)$ to occur together, one has to calculate the product of transitions between the states in the sequence, $p(h^{(0)})p(h^{(1)}|h^{(0)}) \ldots p(h^{(T)}|h^{(T-1)})$ multiplied by the product of probabilities of the

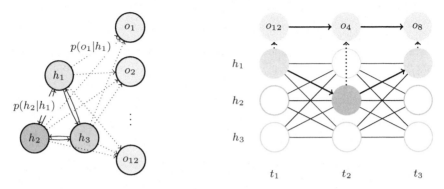

Fig. 2.26 Illustration of a Hidden Markov model with three possible states h_1, h_2, h_3 and a set of possible observations $o_1, \ldots, o_{12}$. The transition probabilities $p(h_i|h_j)$, $i, j = 1, 2, 3$ between the states as well as the probabilities $p(o_s|h_i)$ of an observation made given a state (with $i = 1, 2, 3$ and $s = 1, \ldots, 12$) define the model and are illustrated by the graph on the left. A possible trajectory of the doubly stochastic Markov process unfolded in three time steps is sketched on the right: While the state jumps from state h_1 to h_2 and back to h_1, the observer 'receives' observations o_{12}, o_4 and o_8

observations made given the state, $p(o^{(0)}|h^{(0)})p(o^{(1)}|h^{(1)}) \ldots p(o^{(T)}|h^{(T)})$. Learning in this context means to infer the transition probabilities $\{p(h_i|h_j), p(o_k|h_j)\}$ from a training data set.

2.4.4 Kernel Methods

The following are some important methods for supervised learning that make use of the concept of kernels, which we introduced in Sect. 2.2.4. The model functions and distributions in kernel methods typically depend on the data. Training in this case does not absorb the information from the data into parameters, but finds parameters that define the importance of a data point for a decision.

2.4.4.1 Kernel Density Estimation

Kernel density estimation [6] for pattern classification derives a class-conditional model distribution from data based on a simple idea. Given a similarity measure or kernel on the input space, define the class-conditional distribution for class $y = c$ as

$$p(x|y=c) = \sum_{m|y^m=c}^{M} \frac{1}{M_c} \kappa(x - x^m). \tag{2.34}$$

2.4 Methods in Machine Learning

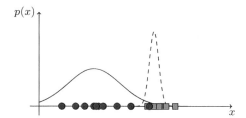

Fig. 2.27 A kernel density estimator defines a probability distribution over data points that assigns high probabilities to regions with lots of data. Here we have two different classes of data (red circles represent Class 0 while blue squares represent Class 1). The full line shows the learned probability distribution $p(x|y=0)$ over the inputs for Class 0, while the dashed line shows $p(x|y=1)$. The kernel function κ defines the 'smoothness' of the overall distribution

This is the sum of a distance measure κ between all M_c 'class c' training inputs (see Fig. 2.27). The total sum is larger if the training inputs from this class are close to the new input. This classifier is therefore based on the notion of 'similar inputs have similar outputs'. Also called a *Parzen window estimator* [6], the distribution is a smoothed version of a histogram over the data. We have used a variation of this method in our Titanic example in the introduction and it is a simple example of how to define a probabilistic model via kernels. Remember that the class-conditional probability is related to the desired distribution for prediction, $p(y|x)$, via Bayes formula in the Bayesian learning framework (see Sect. 2.2.3).

2.4.4.2 k-Nearest Neighbour

The k-nearest neighbour method can be understood as a kernel density estimator with a uniform kernel which is a non-zero constant in the radius that encircles the k nearest neighbours and zero else [3]. Given the dataset, one selects the k closest training inputs relative to the new input $\tilde{x}$ and according to a predefined distance metric on the input space (examples are given in Table 2.4). The predicted class label $\tilde{y}$ can be chosen according to the majority class amongst the neighbours when we consider a classification task (see Fig. 2.28), or as the average of their target outputs for regression tasks. Variations to this simple algorithm include weighting

Table 2.4 Examples of distance measures between data points x and x' for the nearest neighbour classifier

Distance measure	Data type	Formula
Euclidean distance	$x, x' \in \mathbb{R}^N$	$\sqrt{\sum_{i=1}^{N}(x_i - x'_i)^2}$
Squared distance	$x, x' \in \mathbb{R}^N$	$\sum_{i=1}^{N}(x_i - x'_i)^2$
Cosine distance	$x, x' \in \mathbb{R}^N$	$\frac{x^T x'}{|x||x'|}$
Hamming distance	$x, x' \in \{0, 1\}^{\otimes N}$	Number of differing bits

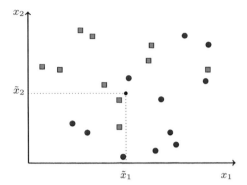

Fig. 2.28 Illustration of k-nearest neighbour using the Euclidean distance measure. The symbols show the 2-dimensional inputs that have each a class attribute 'circle' or 'rectangle'. A new input $\tilde{x} = (\tilde{x}_1, \tilde{x}_2)$ is classified according to its $k = 6$ nearest neighbours (i.e., taking the class label of the majority of its neighbours, in this case 'rectangle')

the neighbours according to their distance [54], or replacing the training inputs of each class by their centroids.

2.4.4.3 Support Vector machines

Support vector machines were very popular throughout the 1990s, when they took over from neural networks as the method of choice. They can be derived from linear models for which one tries to minimise the distance between the separating hyperplane and the training vectors (see Fig. 2.29). In other words, the model function is very simple, but the optimisation problem is more involved than for example in the perceptron model, where we just want to find a hyperplane that classifies all data correct, irrespective of the margin. The power of support vector machines only

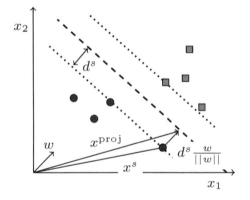

Fig. 2.29 A support vector machine is derived from the problem of finding the discriminant hyperplane with the largest margin to the input vectors. The figure shows the geometric construction of the expression for the distance d^s of the margin

2.4 Methods in Machine Learning

becomes clear when the inner product kernels appearing in the objective function are replaced by other kernels that effectively implement feature maps [55].

The following outlines the somewhat technical derivation of the model following the description of [3] and [10]. We consider a binary classification problem with input domain $\mathcal{X} = \mathbb{R}^N$ and output domain $\mathcal{Y} = \{-1, 1\}$ and assume for now that the data is linearly separable. Start with a linear model,

$$f(x; w) = w^T x + w_0. \tag{2.35}$$

If the data is linearly separable, there is a set of parameters w which defines a hyperplane that separates the training data so that points of the two different classes are on either side of the hyperplane. Mathematically this can be expressed by the restriction $f(x^m; w) > 0$ for training inputs with $y^m = 1$, and $f(x^m; w) < 0$ for data points of class $y^m = -1$. In summary, one can write

$$f(x^m; w) y^m > 0, \tag{2.36}$$

for $m = 1, .., M$. The goal is to find the decision boundary that maximises the *margin*, which is the distance between the closest training points (the *support vectors*) x^s and the separating hyperplane. To formulate the margin via the model and weight vector, decompose the input x^s into its orthogonal projection onto the hyperplane x^{proj}, plus a vector of length d^s that is orthogonal to the hyperplane (see Fig. 2.29). This orthogonal direction is given by the unit vector $\frac{w}{||w||}$, so that

$$x^s = x^{\text{proj}} + d^s \frac{w}{||w||}.$$

Calculating the output of this point for our model, $f(x^s) = w^T x^{\text{proj}} + d^s \frac{w^2}{||w||}$, and noting that w and x^{proj} are orthogonal by construction leads to the distance

$$d^s = \frac{f(x^s)}{||w||}$$

Here, $|| \cdot ||$ is the Euclidean norm. The support vector machine defines the decision boundary such that the margin is maximised while ensuring condition (2.36) is fulfilled.

Since the decision boundary is the same for any length of w, there are infinitely many solutions. We fix the length with the condition $d^s ||w|| = 1$, so that that the optimisation problem becomes equivalent to minimising $||w||$ under the constraint $f(x^m; w) y^m \geq 1$. Alternatively—and more common—one minimises $\frac{1}{2}||w||^2$ under the same constraint. Using Lagrangian multipliers $\gamma_1, \ldots, \gamma_M$, this constrained quadratic optimisation problem can be turned into an unconstrained version in which we need to minimise the Lagrange function

$$\mathcal{L}(w, w_0, \{\gamma\}) = \frac{1}{2}w^T w - \sum_{m=1}^{M} \gamma_m \left(y^m (w^T x^m + w_0) - 1 \right).$$

The last step is to formulate the dual problem which does not depend on the weights. Setting the derivative of $\mathcal{L}$ with respect to w to zero leads to the relation

$$w = \sum_{m=1}^{M} \gamma_m y^m x^m, \tag{2.37}$$

while setting the derivative of L with respect to w_0 to zero leads to $\sum_m \gamma^m y^m = 0$. Resubstituting these results in the objective function yields the Lagrangian function

$$\mathcal{L}_d(\{\gamma\}) = \sum_{m=1}^{M} \gamma_m - \frac{1}{2} \sum_{m,m'=1}^{M} \gamma_m \gamma_{m'} y^m y^{m'} \langle x^m, x^{m'} \rangle, \tag{2.38}$$

that has to be minimised with respect to the Lagrangian multipliers γ_m and subject to the constraints $\gamma_m \geq 1\ \forall m$ and $\sum_m \gamma_m y^m = 0$. We used the $\langle \cdot, \cdot \rangle$ notation for the inner product here to emphasise that this expression can be interpreted as a kernel. Note that if the data is not separable, one can introduce slack variables ξ^m that change the right side of the inequality (2.36) to $1 - \xi^m$, and minimise the slack variables along with the other parameters in the Lagrange function.

Once the Lagrangian multipliers are found, the weights are determined by the relation (2.37) and the new output is found as

$$\tilde{y} = \sum_{m=1}^{M} \gamma_m y^m \langle x^m, \tilde{x} \rangle + w_0. \tag{2.39}$$

Note that Eq. (2.38) defines the optimisation problem connected to a support vector machine. The scalar product $\langle x^m, x^{m'} \rangle$ is a linear kernel, and the kernel trick (Sect. 2.2.4) can be applied to let the support vector machine find a decision hyperplane in a feature space. By this trick, support vector machines can find nonlinear decision boundaries in input space. Solving this optimisation problem by convex optimisation methods discussed in the context of linear models takes time in roughly $\mathcal{O}(M^3)$, but in practice specialised linear solvers are used.

2.4.4.4 Gaussian Processes

Gaussian processes are a kernel method with again a rather different idea. Assume we have a supervised regression task with $\mathcal{X} = \mathbb{R}^N, \mathcal{Y} = \mathbb{R}$. The idea behind the method of Gaussian processes is to assign a probability to every possible model function $f(x)$, favouring those we consider more likely, such as smooth functions.

2.4 Methods in Machine Learning

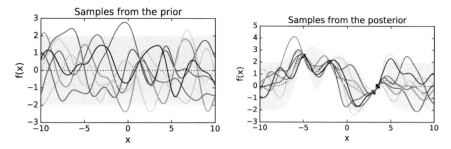

Fig. 2.30 Prior (left) functions drawn from a Gaussian process. The covariance determines the smoothness of the sample functions. The training data points reduce the space of possible functions and introduce certainty where the data values are. Samples from the posterior (right) are therefore reduced to model functions that agree with the data points. *The plots were made with the python infpy package written by John Reid. Accessed through http://pythonhosted.org/infpy/*

Out of the resulting 'probability distribution over functions' one selects only those that agree with the data points in $\mathcal{D}$ to get a 'posteriori over functions' in the Bayesian learning framework. The mean of this posteriori can be understood as the best guess for $f(\tilde{x})$ at a certain point $\tilde{x}$ given the data, and the variance is the uncertainty of this guess (see Fig. 2.30).

The details of this approach are based on rather technical details of which we try to give a short overview. A Gaussian process (GP) is formally a "collection of random variables, any finite number of which have a joint Gaussian distribution" [21]. In this context the random variables are the outputs of the model function $f(x)$. Given a covariance function $\kappa(x, x')$ and a number of inputs $x^1, \ldots, x^M$, the definition of a Gaussian process states that the outputs $f(x^1), \ldots, f(x^M)$ are random variables sampled from the joint normal distribution $\mathcal{N}$,

$$\begin{pmatrix} f(x^1) \\ \vdots \\ f(x^M) \end{pmatrix} \sim \mathcal{N}\left[\begin{pmatrix} m(x^1) \\ \vdots \\ m(x^M) \end{pmatrix}, \begin{pmatrix} \kappa(x^1, x^1) & \cdots & \kappa(x^1, x^M) \\ \vdots & \ddots & \vdots \\ \kappa(x^M, x^1) & \cdots & \kappa(x^M, x^M) \end{pmatrix}\right]. \quad (2.40)$$

As the notation suggests, the covariance function used in the covariance matrix is nothing other than a kernel and responsible for the type of distribution over functions defined by a Gaussian process. The Gaussian process is fully defined by the mean and kernel, and can be written as

$$f(x) \sim GP(m(x), \kappa(x, x')).$$

The mean is usually, and in the following, set to zero.

Model selection for a Gaussian process refers to finding suitable kernels. Common choices are the dot product have been shown in Table 2.2. The covariance matrix made up of the kernel functions is positive semi-definite and therefore a kernel Gram matrix.

Equation (2.40) shows how to draw samples of model outputs $f(x)$ for a set of inputs, and if we increase the number of inputs we can get a good idea of what the function sampled from the Gaussian process looks like. We can use the expression in Eq. (2.40) for prediction. Instead of some general inputs, we consider the joint distribution for the training data outputs as well as the new output $f(\tilde{x})$,

$$\begin{pmatrix} f(x^1) \\ \vdots \\ f(x^M) \\ f(\tilde{x}) \end{pmatrix} \sim \mathcal{N}\left[\begin{pmatrix} 0 \\ \vdots \\ 0 \end{pmatrix}, \begin{pmatrix} \kappa(x^1,x^1) & \cdots & \kappa(x^1,x^M) & \kappa(x^1,\tilde{x}) \\ \vdots & \ddots & \vdots & \vdots \\ \kappa(x^M,x^1) & \cdots & \kappa(x^M,x^M) & \kappa(x^M,\tilde{x}) \\ \kappa(\tilde{x},x^1) & \cdots & \kappa(\tilde{x},x^M) & \kappa(\tilde{x},\tilde{x}) \end{pmatrix}\right]. \quad (2.41)$$

To get the desired probability distribution $p(\tilde{y}|\tilde{x})$ over the predicted output $f(\tilde{x}) = \tilde{y}$ we have to marginalise the distribution in Eq. (2.41) for the fixed values $f(x^1) = y^1, \ldots, f(x^M) = y^M$. This can be thought of as 'cutting' through the joint probability distribution in order to reduce the multivariate Gaussian distribution to a univariate (one-dimensional) distribution. The technical algebraic details of obtaining a marginal Gaussian distribution can be found in most statistics textbooks and we only write down the result here[12]:

$$p(y|x) = \mathcal{N}\left[y \mid \underbrace{\tilde{\kappa}^T K^{-1} y}_{\text{mean } \mu}, \underbrace{\tilde{k} - \tilde{\kappa}^T K^{-1} \tilde{\kappa}}_{\text{covariance } \delta}\right], \quad (2.42)$$

with the notation

$$\tilde{\kappa} = \begin{pmatrix} \kappa(x^1,\tilde{x}) \\ \vdots \\ \kappa(x^M,\tilde{x}) \end{pmatrix}, \quad (2.43)$$

$$K = \begin{pmatrix} \kappa(x^1,x^1) & \cdots & \kappa(x^1,x^M) \\ \vdots & \ddots & \vdots \\ \kappa(x^M,x^1) & \cdots & \kappa(x^M,x^M) \end{pmatrix}, \quad (2.44)$$

$$\tilde{k} = \kappa(\tilde{x},\tilde{x}), \quad (2.45)$$

$$y = \begin{pmatrix} y^1 \\ \vdots \\ y^M \end{pmatrix}. \quad (2.46)$$

Besides model selection, which involves choosing the kernel and its hyperparameters, Gaussian processes do not have a distinct learning phase. With Eq. (2.42), prediction is mainly a computational problem of efficiently inverting the

---

[12] Note that we only consider noise free data here, however, including Gaussian noise simply adds a variance parameter to the matrix K. For more details, see [21].

2.4 Methods in Machine Learning

$M \times M$-dimensional kernel matrix K. Gaussian processes do take a lot of computational resources when large datasets have to be processed [6], and many proposals for approximations have been put forward [21].

References

1. Russell, S.J., Norvig, P., Canny, J.F., Malik, J.M., Edwards, D.D.: Artificial Intelligence: A Modern Approach, vol. 3. Prentice Hall, Englewood Cliffs (2010)
2. Lee, H., Ekanadham, C., Ng, A.Y.: Sparse deep belief net model for visual area V2. In: Advances in Neural Information Processing Systems, pp. 873–880 (2008)
3. Bishop, C.M.: Pattern Recognition and Machine Learning, vol. 1. Springer (2006)
4. Hastie, T., Friedman, J., Tibshirani, R.: The Elements of Statistical Learning, vol. 1. Springer, Berlin (2001)
5. Duda, R.O., Hart, P.E., Stork, D.G.: Pattern Classification. Wiley, New York (2012)
6. Murphy, K.P.: Machine Learning. A Probabilistic Perspective. MIT Press (2012)
7. Koller, D., Friedman, N.: Probabilistic Graphical Models: Principles and Techniques. MIT Press (2009)
8. Hertz, J.A., Krogh, A.S., Palmer, R.G.: Introduction to the Theory of Neural Computation, vol. 1. Westview Press, Redwood City (California) (1991)
9. Bishop, C.M.: Neural Networks for Pattern Recognition, vol. 1. Clarendon Press, Oxford (1995)
10. Alpaydin, E.: Introduction to Machine Learning. MIT Press, Cambridge (2004)
11. Silver, D., et al.: Mastering the game of go without human knowledge. Nature **550**(7676), 354 (2017)
12. Domingos, P.: A few useful things to know about machine learning. Commun. ACM **55**(10), 78–87 (2012)
13. Ng, A.Y., Jordan, A.: On discriminative vs. generative classifiers: a comparison of logistic regression and naive Bayes. Adv. Neural Inf. Process. Syst. **14**, 841–846 (2002)
14. Griffiths, T., Yuille, A.: A primer on probabilistic inference. In: Chater, N., Oaksford, M. (eds.) The Probabilistic Mind: Prospects for Bayesian Cognitive Science, pp. 33–57. Oxford University Press (2008)
15. Ghahramani, Z.: Probabilistic machine learning and artificial intelligence. Nature **521**(7553), 452–459 (2015)
16. Smola, A.J., Schölkopf, B., Müller, K.-R.: The connection between regularization operators and support vector kernels. Neural Netw. **11**(4), 637–649 (1998)
17. Schölkopf, B., Smola, A.J.: Learning with Kernels: Support Vector Machines, Regularization, Optimization, and Beyond. MIT Press (2002)
18. Schölkopf, B., Herbrich, R., Smola, A.: A generalized representer theorem. In: Computational Learning Theory, pp. 416–426. Springer (2001)
19. Shashua, A.: Introduction to machine learning: Class notes 67577. arXiv:0904.3664 (2009)
20. Steinwart, I., Hush, D., Scovel, C.: An explicit description of the reproducing kernel Hilbert spaces of Gaussian RBF kernels. IEEE Trans. Inf. Theory **52**(10), 4635–4643 (2006)
21. Rasmussen, C.E.: Gaussian Processes for Machine Learning. MIT Press (2006)
22. Bennett, K.P., Parrado-Hernández, E.: The interplay of optimization and machine learning research. J. Mach. Learn. Res. **7**(Jul):1265–1281 (2006)
23. Rumelhart, D.E., Hinton, G.E., Williams, R.J.: Learning internal representations by error propagation. Technical report, DTIC Document (1985)
24. Hinton, G., Osindero, S., Teh, Y.-W.: A fast learning algorithm for deep belief nets. Neural Comput. **18**(7), 1527–1554 (2006)
25. Zhang, C., Bengio, S., Hardt, M., Recht, B., Vinyals, O.: Understanding deep learning requires rethinking generalization. In: Proceedings of the International Conference on Learning Representations (2017)

26. Boyd, S., Vandenberghe, L.: Convex Optimization. Cambridge University Press (2004)
27. Vavasis, S.A.: Nonlinear Optimization: Complexity Issues. Oxford University Press (1991)
28. Weierstrass, K.: Über die analytische Darstellbarkeit sogenannter willkürlicher Functionen einer reellen Veränderlichen. Sitzungsberichte der Königlich Preußischen Akademie der Wissenschaften zu Berlin **2**, 633–639 (1885)
29. Trefethen, L.N., Bau III, D.: Numerical Linear Algebra, vol. 50. Siam (1997)
30. McCulloch, W.S., Pitts, W.: A logical calculus of the ideas immanent in nervous activity. Bull. Math. Biol. **5**(4), 115–133 (1943)
31. Hopfield, J.J.: Neural networks and physical systems with emergent collective computational abilities. Proc. Natl. Acad. Sci. **79**(8), 2554–2558 (1982)
32. Minsky, M., Papert, S.: Perceptrons: An Introduction to Computational Geometry. MIT Press, Cambridge (1969)
33. Rumelhart, D.E., Hinton, G.E., Williams, R.J.: Learning representations by back-propagating errors. Nature **323**(9), 533–536 (1986)
34. Novikoff, A.B.J.: On convergence proofs on perceptrons. In: Proceedings of the Symposium on the Mathematical Theory of Automata, vol. 12, pp. 615–622 (1962)
35. Hornik, K., Stinchcombe, M., White, H.: Multilayer feedforward networks are universal approximators. Neural Netw. **2**(5), 359–366 (1989)
36. Nair, V., Hinton, G.E.: Rectified linear units improve restricted Boltzmann machines. In: Proceedings of the 27th International Conference on Machine Learning (ICML-10), pp. 807–814 (2010)
37. Pascanu, R., Mikolov, T., Bengio, Y.: On the difficulty of training recurrent neural networks. ICML **3**(28), 1310–1318 (2013)
38. El Hihi, S., Bengio, Y.: Hierarchical recurrent neural networks for long-term dependencies. In: NIPS'95 Proceedings of the 8th International Conference on Neural Information Processing Systems, vol. 400, pp. 493–499. MIT Press, Cambridge, MA, USA (1995)
39. Arjovsky, M., Shah, A., Bengio, Y.: Unitary evolution recurrent neural networks. J. Mach. Learn. Res. **48** (2016)
40. Bengio, Y., Boulanger-Lewandowski, N., Pascanu, R.: Advances in optimizing recurrent networks. In: 2013 IEEE International Conference on Acoustics, Speech and Signal Processing, pp. 8624–8628. IEEE (2013)
41. Duda, R.O., Hart, P.E., Stork, D.G.: Pattern Classification, 2 ed. Wiley (2000)
42. Rojas, R.: Neural Nets: A Systematic Introduction. Springer, New York (1996)
43. Hinton, G.: Training products of experts by minimizing contrastive divergence. Neural Comput. **14**(8), 1771–1800 (2002)
44. Hinton, G.: A practical guide to training restricted Boltzmann machines. UTML TR 2010-003, Version 1 (2010)
45. Gilks, W.R., Richardson, S., Spiegelhalter, D.J.: Markov Chain Monte Carlo in Practice. Chapman & Hall, London (1996)
46. Carreira-Perpinan, M.A., Hinton, G.: On contrastive divergence learning. In: Cowell, R., Ghahramani, Z. (eds.) AISTATS 2005: Proceedings of the Tenth International Workshop on Artificial Intelligence and Statistics, vol. 10, pp. 33–40. The Society for Artificial Intelligence and Statistics (2005)
47. Bengio, Y.: Learning deep architectures for AI. Found. Trends Mach. Learn. **2**(1), 1–127 (2009)
48. Sutskever, I., Tieleman, T.: On the convergence properties of contrastive divergence. In: International Conference on Artificial Intelligence and Statistics, pp. 789–795 (2010)
49. Pearl, J.: Causality. Cambridge University Press (2009)
50. Heckerman, D., Geiger, D., Chickering, D.M.: Learning Bayesian networks: the combination of knowledge and statistical data. Mach. Learn. **20**(3):197–243 (1995)
51. Dagum, P., Luby, M.: Approximating probabilistic inference in Bayesian belief networks is NP-hard. Artif. Intell. **60**(1), 141–153 (1993)
52. Ben-Gal, I.: Bayesian networks. In: Encyclopedia of Statistics in Quality and Reliability (2007)
53. Rabiner, L.R.: A tutorial on hidden Markov models and selected applications in speech recognition. Proc. IEEE **77**(2), 257–286 (1989)

References

54. Dudani, S.A.: The distance-weighted k-nearest-neighbor rule. IEEE Trans. Syst. Man Cybern. **4**, 325–327 (1976)
55. Boser, B.E., Guyon, I.M., Vapnik, V.N.: A training algorithm for optimal margin classifiers. In: COLT'92 Proceedings of the Fifth Annual Workshop on Computational Learning Theory, pp. 144–152. ACM (1992)

Chapter 3
Quantum Information

While the last chapter introduced readers without a firm background in machine learning into the foundations, this chapter targets readers who are not closely familiar with quantum information. Quantum theory is notorious for being complicated, puzzling and difficult (or even impossible) to understand. Although this impression is debatable, giving a short introduction into quantum theory is indeed a challenge for two reasons: On the one hand its backbone, the mathematical apparatus with its objects and equations requires some solid mathematical foundations in linear algebra, and is usually formulated in the Dirac notation [1] that takes a little practice to get familiar with. On the other hand the physical reality this apparatus seeks to describe is difficult to grasp from the point of our daily intuition about physical phenomena, and the interpretation of the results this mathematical apparatus yields, unchallenged and reproduced by thousands of experiments up to today, are still controversially debated.

There are many different ways to introduce quantum theory.[1] Scott Aaronson, in his online lecture notes[2] remarks sarcastically that in order to learn about quantum theory,

> [...] you start with classical mechanics and electrodynamics, solving lots of grueling differential equations at every step. Then you learn about the "blackbody paradox" and various strange experimental results, and the great crisis these things posed for physics. Next you learn a complicated patchwork of ideas that physicists invented between 1900 and 1926 to try to make the crisis go away. Then, if you're lucky, after years of study you finally get around to the central conceptual point: that nature is described not by probabilities (which are always non-negative), but by numbers called amplitudes that can be positive, negative, or even complex.

---

[1]Common didactic approaches are the historical account of discovering quantum theory, the empirical account of experiments and their explanations, the Hamiltonian path from formal classical to quantum mechanics, the optical approach of the wave-particle-dualism, and the axiomatic postulation of its mathematical structure [2].

[2]The lecture notes are available at http://www.scottaaronson.com/democritus/, and led to the book "Quantum Computing since Democritus" [3].

From the perspective of quantum computing this comment certainly contains some truth, since a central concept is the *qubit*, which is a rather simple quantum system. This is why we will neglect many details of quantum theory and focus on finite, discrete quantum systems.

In the following we will start with an intuition for what operators, states and unitary evolutions are, and proceed to a more rigorous introduction to quantum theory based on the Dirac notation. We then introduce quantum computing together with the concepts of gates, qubits and quantum algorithms. As a preparation for later chapters we then discuss different ways to encode information into quantum systems and present some core quantum routines used in later chapter.

3.1 Introduction to Quantum Theory

3.1.1 What Is Quantum Theory?

Quantum theory (used synonymously with the term *quantum mechanics*) is "first and foremost a calculus for computing the probabilities of outcomes of measurements made on physical systems" [4] called quantum systems. A quantum system is typically a collection of microscopic physical objects for which classical Newtonian mechanics does not explain experimental observations. Examples are a hydrogen atom, light of very low intensity, or a small number of electrons in a magnetic field.

Quantum theory incorporates classical mechanics as a limit case [5] and is often celebrated as the most accurate physical theory ever developed.[3] The reason why despite this 'superiority' over classical mechanics the latter is still commonly used in science and education is that quantum mechanics is impractical to apply to large systems, as the calculations very soon become too complex to execute. At the same time, quantum effects in macroscopic systems are usually small enough to be neglected without a visible error and can for many purposes be replaced by easier, higher aggregated classical models. However, and similar to machine learning in which a lot of problems are uncomputable in their exact formulation, a range of solvable models as well as powerful approximation techniques have been developed which lead to a range of technological applications in the fields of medicine, chemistry, biology and engineering.

Figure 3.1 shows a brief historical overview of quantum theory. What Aaronson describes as a "complicated patchwork of ideas that physicists invented between 1900 and 1926" forms the beginnings of quantum theory as it is still taught today, and which was then used to rethink the entire body of physics knowledge. The initial step is commonly attributed to the year 1900 when Max Planck introduced the

---

[3] It shall be remarked that quantum theory has no notion of the concept of gravity, an open problem troubling physicists who dream of the so called 'Grand Unified Theory' of quantum mechanics and general relativity, and a hint towards the fact that quantum mechanics still has to be developed further.

3.1 Introduction to Quantum Theory

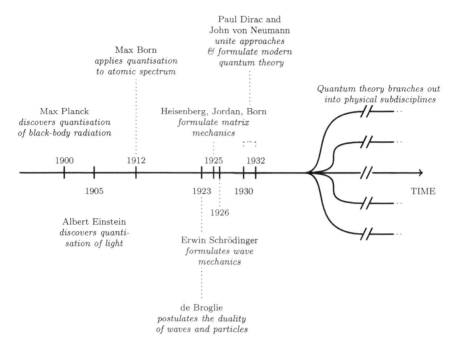

Fig. 3.1 Timeline of the early years of quantum theory

idea that energy (in this case so called *black-body radiation*) can only take discrete values as if existing as 'quanta' or small energetic portions. With this assumption he was able to resolve a heated debate regarding the spectrum of black-body radiation [6]. Almost in parallel, Albert Einstein made a similar discovery derived from the statistical mechanics of gases and derived the concept of a photon—a portion or energy quantum of light [7, 8].

In the following years, these early ideas of a 'theory of energy quanta' were applied to atomic spectroscopy (most notably by Niels Bohr), and to light (by Louis de Broglie) but still based on rather ad-hoc methods [6]. Werner Heisenberg followed by Jordan, Born and, independently, Paul Dirac formulated the first mathematically consistent version of quantum theory referred to as matrix mechanics in 1925, with which Wolfgang Pauli was able to derive the experimental results of measuring the spectrum of a hydrogen atom. Heisenberg postulated his uncertainty principle shortly after, stating that certain properties of a quantum system cannot be measured accurately at the same time. In 1926, following a slightly different and less abstract route, Erwin Schrödinger proposed his famous equation of motion for the 'wave function' describing the state of a quantum system. These two approaches were shown to be equivalent in the 1930s, and a more general version was finally proposed by Dirac and Jordan. In the following years, quantum theory branched out into many sub-disciplines such as quantum many-body systems, quantum thermodynamics, quantum optics, and last but not least, quantum information processing.

3.1.2 A First Taste

Quantum theory can be understood as a generalisation of classical probability theory to make statements about the probabilities of the outcomes of measurements on quantum systems [9]. The rigorous formulation of this generalisation requires some serious mathematics (see for example [10]) and is beyond the scope of this book. However, instead of directly diving into Hilbert spaces and Hermitian operators, this section starts with a simple classical stochastic system and shows how to construct the description for a very similar quantum system.

3.1.2.1 States and Observables

Consider a set of N mutually exclusive events $\mathcal{S} = \{s_1, \ldots, s_N\}$. In physics, events usually refer to the outcomes of measurements ('observations') on physical properties of a system, but one may also think of other examples such as 'it is raining' and 'it is not raining'. One can associate each event with a random variable X that can take the values $\{x_1, \ldots, x_N\}$, and define a probability distribution over these values quantifying our knowledge on how likely an event is to occur. The probabilities related to the N events, $\{p_1, \ldots, p_N\}$, are real numbers, and they fulfill $p_i \geq 0$ and $\sum_{i=1}^{N} p_i = 1$. The expectation value of the random variable is defined as

$$\langle X \rangle = \sum_{i=1}^{N} p_i x_i, \tag{3.1}$$

and is the weighed average of all values the random variable can take.

Example 3.1 (*Particle in a rectangular box*)
Consider the model of a particle in a 2-dimensional rectangular box which is divided into four sections as illustrated in Fig. 3.2. The event space is given by the four events of the particle being found in the first, second, third or fourth section, $\mathcal{S}_{\text{particle}} = \{s1, s2, s3, s4\}$. The random variable X for the measurement result can take values $\{1, 2, 3, 4\}$, and each measurement outcome has a probability $p_i, i = 1, \ldots, 4$ to occur. For example, if $p_1 = 0.2$, $p_2 = 0.2$, $p_3 = 0.2$, $p_4 = 0.4$ the expectation value is given by $\langle X \rangle = 2.8$ which reveals that the particle has a higher probability to be in the right half than in the left.

From here one needs two steps in order to arrive at the description of a quantum system with N different possible configurations or measurement outcomes: First,

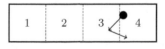

Fig. 3.2 Illustration of the particle in a box from Example 3.1

3.1 Introduction to Quantum Theory

summarise the probabilities and the outcomes by matrices and vectors, and second, replace real positive probabilities with complex amplitudes.[4]

For the first step, consider a vector of the square roots of probabilities $p_1, \ldots, p_N$,

$$q = \begin{pmatrix} \sqrt{p_1} \\ \vdots \\ \sqrt{p_N} \end{pmatrix} = \sqrt{p_1} \begin{pmatrix} 1 \\ \vdots \\ 0 \end{pmatrix} + \cdots + \sqrt{p_N} \begin{pmatrix} 0 \\ \vdots \\ 1 \end{pmatrix}, \tag{3.2}$$

as well as a diagonal matrix X corresponding to the random variable with the same name,

$$X = \begin{pmatrix} x_1 & \cdots & 0 \\ \vdots & \ddots & \vdots \\ 0 & \cdots & x_N \end{pmatrix}.$$

The expectation value of Eq. (3.1) can now be written as

$$\langle X \rangle = q^T X q = \sum_{i=1}^{N} p_i x_i. \tag{3.3}$$

So far we just made the expression of the expectation value more complicated. In Eq. (3.3), the matrix X contributes the values of the random variable, and the vector q contains the square roots of the probabilities. For simplicity, both objects were expressed in the standard basis of $\mathbb{R}^N$. One can see that the basis vector $(1, 0, \ldots, 0)^T$ forms a subspace of the $\mathbb{R}^N$ that is associated with the first event. Moreover, the outer product of this basis vector with itself

$$\begin{pmatrix} 1 & \cdots & 0 \end{pmatrix} \begin{pmatrix} 1 \\ \vdots \\ 0 \end{pmatrix} = \begin{pmatrix} 1 & \cdots & 0 \\ \vdots & \ddots & \vdots \\ 0 & \cdots & 0 \end{pmatrix} \tag{3.4}$$

is a *projector* onto this subspace. Multiplied to a vector, it 'picks out' the first element. This is a motivation why in quantum probability theory, the measurement events are represented by projectors onto subspaces which we will see in the next section. The sum of all projectors to the subspaces of $x_1, \ldots, x_N$ is the identity.

Note that one does not have to use the standard basis, but that any other basis $\{v_i\}$ for which the eigenvalue equations $X v_i = x_i v_i$ hold would have done the job, as can be confirmed by decomposing $q = q_1 v_1 + \cdots + q_N v_N$ and employing the eigenvalue equation to calculate the expectation value.

---

[4]In this sense, Aaronson is right in saying that the "central conceptual point" of quantum theory is "that nature is described not by probabilities [...], but by numbers called amplitudes that can be positive, negative, or even complex".

It turns out that Eq. (3.3) is already very close to how quantum mechanics describes expectation values, but some mathematical properties of the vectors and matrix are slightly different. In order to turn Eq. (3.3) into the formula of computing expectation values of measurements on quantum systems, we have to replace q with a complex *amplitude vector* $\alpha = (\alpha_1, \ldots, \alpha_N)^T \in \mathbb{C}^N$, and X by a complex, self-adjoint matrix $O \in \mathbb{C}^{N \times N}$. Another term for self-adjoint is *Hermitian*. Hermitian operators have the property that they are equal to their complex-conjugate transpose, $O = (O^*)^T$, where in quantum mechanics, the symbol for the complex-conjugate transpose is the dagger, $(O^*)^T = O^\dagger$.

Let $\{v_1, \ldots, v_N\}$ be a basis of orthonormal eigenvectors of O which spans the $\mathbb{C}^N$, fulfilling eigenvalue equations $O v_i = o_i v_i$ and normalisations $v_i^\dagger v_j = \delta_{ij}$ for all $i, j = 1, \ldots, N$. Any amplitude vector α can be written as a linear combination of this basis,[5]

$$\alpha = \alpha_1 v_1 + \cdots + \alpha_N v_N.$$

The expectation value of the random variable O corresponding to the matrix O now becomes

$$\langle O \rangle = \alpha^\dagger O \alpha = \sum_{i=1}^{N} |\alpha_i|^2 v_i^\dagger O v_i = \sum_{i=1}^{N} |\alpha_i|^2 o_i. \quad (3.5)$$

One can see that the $|\alpha_i|^2$ take the role of the probabilities in the classical example, and we therefore demand that $\sum_i |\alpha_i|^2 = 1$ to ensure that the probabilities sum up to one. If the basis $\{v_1, \ldots, v_N\}$ is the standard basis $\{(1, 0, \ldots, 0)^T, \ldots, (0, 0, \ldots, 1)^T\}$, the α_i will just be the entries of α, and the normalisation condition therefore means that the amplitude vector has unit length.

The eigenvalues o_i of O correspond to the values of the random variable or the outcomes of measurements. For example, when measuring the energy configurations of an atom, o_i would simply be an energy value in a given unit. However, without further constraints the eigenvalues of a complex matrix can be complex, which does not make any sense when looking at physically feasible variables (i.e., what is a complex energy?). This is the reason to choose O to be a Hermitian operator, since the eigenvalues of Hermitian operators are always real and therefore physical.

Since O (just like X in the classical example) contains information on the values of the random variable, which in a physical setup correspond to possible observations in a measurement, it is called an *observable*. The vector α tells us something about the probability distribution of the measurement outcome and is a representation of a so-called *quantum state*, since probabilistic information is everything we know about the configuration a quantum system is in. Together, observables and states allow us to calculate probabilities and expectation values of measurement outcomes.

---

[5]The *spectral theorem* of linear algebra guarantees that the eigenvectors of a Hermitian operator O form a basis of the Hilbert space, so that every amplitude vector can be decomposed into eigenvectors of O.

3.1.2.2 Unitary Evolutions

Physical theories define equations for the evolution of the state of a system over time. Let us once more start with a classical stochastic description and show how to arrive at quantum evolutions.

A map of a discrete probability distribution to another discrete probability distribution (describing the change of our knowledge over time) has to preserve the property that the probabilities of all elementary events sum up to one. Consider the case that we know the probability of transition from one state to the other in one discrete time step. Then such an update can be modelled by applying a matrix containing these transition probabilities, to a probability vector $p = (p_1, \ldots, p_N)^T$,

$$\begin{pmatrix} m_{11} & \cdots & m_{1N} \\ \vdots & \ddots & \vdots \\ m_{N1} & \cdots & m_{NN} \end{pmatrix} \begin{pmatrix} p_1 \\ \vdots \\ p_N \end{pmatrix} = \begin{pmatrix} p'_1 \\ \vdots \\ p'_N \end{pmatrix}, \quad \sum_{i=1}^{N} p_i = \sum_{i=1}^{N} p'_i = 1. \quad (3.6)$$

The transition matrix $M = (m)_{ij}$ has to ensure that the probability vector remains normalised, which implies that M is a so-called *stochastic matrix* whose columns sum up to one. The stochastic process resulting from the transition matrix is also called a *Markov process*.

Example 3.2 (*Stochastic description of the weather*) We observe the weather on consecutive days and know the probability that it is raining tomorrow if it was raining today is 60% while the probability for the weather to change is 40%. Likewise, if it was sunny, it rains tomorrow with 40% and it stays sunny with 60%. Let $p_{\text{tod}} = (p_1, p_2)^T$ describe the probability p_1 that it is raining, or the probability p_2 that it is sunny today. We can now calculate the probability of the weather tomorrow by applying the stochastic or transition matrix to the probabilistic state,

$$\begin{pmatrix} 0.6 & 0.4 \\ 0.4 & 0.6 \end{pmatrix} p_{\text{tod}} = p_{\text{tom}}$$

If we know that it was sunny today, $p_{\text{tod}} = (0, 1)^T$, the probability of sun will be 0.6 and of rain will be 0.4.

In quantum theory, we work with amplitude vectors with the property that the entries' sum of absolute squares has to sum up to one. Consequently, an evolution has to be described by a *unitary* matrix (the complex equivalent of an orthogonal transformation which is known to maintain lengths),

$$\begin{pmatrix} u_{11} & \cdots & u_{1N} \\ \vdots & \ddots & \vdots \\ u_{N1} & \cdots & u_{NN} \end{pmatrix} \begin{pmatrix} \alpha_1 \\ \vdots \\ \alpha_N \end{pmatrix} = \begin{pmatrix} \alpha'_1 \\ \vdots \\ \alpha'_N \end{pmatrix}, \quad \sum_{i=1}^{N} |\alpha_i|^2 = \sum_{i=1}^{N} |\alpha'_i|^2 = 1. \quad (3.7)$$

A unitary matrix has the property that its inverse is its complex conjugated, $U^{-1} = U^\dagger$.

Besides unitary evolutions, one can perform *measurements* on a quantum system, which change the state. Although there are many subtleties, in their basic nature quantum measurements are equivalent to classical statistics. If we had a probabilistic description $p_{\text{tom}} = (0.6, 0.4)^T$ of the weather tomorrow as in Example 3.2, and we wait one day to observe that the weather is sunny, our knowledge would be best described by the updated probability state vector $p'_{\text{tom}} = (1, 0)^T$. This discrete transition of $(0.6, 0.4)^T \rightarrow (1, 0)^T$ is in quantum theory sometimes referred to as the 'collapse of the wavefunction' (where the term 'wavefunction' refers to the quantum state), but is in essence similar to a classical update of our state of knowledge: If the weather is replaced by a quantum spin described by an amplitude vector $\alpha = (\sqrt{0.6}, \sqrt{0.4})^T$ (where the first entry corresponds to the event 'spin up' and the second to 'spin down', and we observe 'spin up', the amplitude vector after the measurement would be given by $\alpha' = (1, 0)^T$.

3.1.2.3 Composite Systems and Subsystems

In classical probability theory, sample spaces are combined using a tensor product. Take the weather example: Let $p = (p_1, p_2)^T$ again describe the probability of rain or sun, while $q = (q_1, q_2)^T$ denotes the probability of the wind blowing strong or not. A combined description is a 4-dimensional vector $c = (c_1, c_2, c_3, c_4)^T$ containing the probability for the events 'wind & sun', 'no wind & sun', 'wind & rain', 'no wind & rain'. These four events form the joint event space. If the events are independent of each other, the probabilities of each elementary event multiply and $c = (q_1 p_1, q_1 p_2, q_2 p_1, q_2 p_2)^T$, which can be called a *product state*. If the probabilities do not factorise, or in other words c cannot be expressed by a tensor product of p and q, the two variables are non-separable. Creating a joint sample space in the quantum case is done in exactly the same manner, but using amplitude vectors instead of probability vectors.

Let us look at the opposite direction of how to *marginalise* over parts of the system. Marginalising over variables basically means to give up information about them, a change in our state of knowledge towards more uncertainty. Given a multivariate probability distribution, marginalising means to consider 'cuts' or 'subspaces' of the distribution. Again, the marginalisation process in quantum mechanics is structurally the same in the classical setting, and has to be modelled 'on top' of the quantum probability. For this one uses a slightly different description of quantum states than amplitude vectors.

3.1.2.4 Density Matrices and Mixed States

Consider an amplitude vector $\alpha = (\alpha_1, \alpha_2)^T$. We have perfect knowledge that α is the state of the system, the state is *pure*. A stochastic description of this system means

3.1 Introduction to Quantum Theory

that we are uncertain if our system is in state $\alpha = (\alpha_1, \alpha_2)^T$ or $\beta = (\beta_1, \beta_2)^T$, and the overall state is called a statistical *mixture* of two pure states, or a *mixed state*.

We can incorporate statistical mixtures of quantum states into the formalism of quantum mechanics by switching from amplitude vectors to so called *density matrices*. A density matrix ρ that corresponds to a pure quantum state α is given by the outer product $\rho = \alpha \alpha^T$. To statisticians, this is known as the covariance matrix of α, which for our 2-d example $\alpha = (\alpha_1, \alpha_2)^T$ reads

$$\rho = \alpha \alpha^\dagger = \begin{pmatrix} |\alpha_1|^2 & \alpha_1 \alpha_2^* \\ \alpha_2 \alpha_1^* & |\alpha_2|^2 \end{pmatrix}.$$

If we do not know the quantum state of the system, but we know that the system is in state α with probability p_1 and in state β with probability p_2, the density matrix to describe such a state reads

$$\rho = p_1 \alpha \alpha^\dagger + p_2 \beta \beta^\dagger = \begin{pmatrix} p_1 |\alpha_1|^2 + p_2 |\beta_1|^2 & p_1 \alpha_2 \alpha_1^* + p_2 \beta_2 \beta_1^* \\ p_1 \alpha_1 \alpha_2^* + p_2 \beta_1 \beta_2^* & p_1 |\alpha_2|^2 + p_2 |\beta_2|^2 \end{pmatrix}.$$

In other words, the density matrix formulation allows us to combine 'quantum statistics' expressed by a quantum state, with 'classical statistics' or a probabilistic state that describes our lack of knowledge. This is possible because ρ's diagonal contains probabilities and not amplitudes.

The density matrix notation lets us elegantly exclude a subsystem from the description, and the corresponding mathematical operation is to calculate the *partial trace* over the subsystem. We will later see that this is closely related to the idea of marginalising over distributions. The partial trace operation in matrix notation is the sum over the diagonal elements in the corresponding space. Assume a state describing two joint 2-dimensional system in density matrix notation is given by

$$\rho_{AB} = \begin{pmatrix} \rho_{11} & \rho_{12} & \rho_{13} & \rho_{14} \\ \rho_{21} & \rho_{22} & \rho_{23} & \rho_{24} \\ \rho_{31} & \rho_{32} & \rho_{33} & \rho_{34} \\ \rho_{41} & \rho_{42} & \rho_{43} & \rho_{44} \end{pmatrix},$$

and the partial trace over the first subsystem is

$$\text{tr}_A \{\rho_{AB}\} = \begin{pmatrix} \rho_{11} + \rho_{33} & \rho_{12} + \rho_{34} \\ \rho_{21} + \rho_{43} & \rho_{22} + \rho_{44} \end{pmatrix},$$

while the state of the second subsystem is given by

$$\text{tr}_B \{\rho_{AB}\} = \begin{pmatrix} \rho_{11} + \rho_{22} & \rho_{13} + \rho_{24} \\ \rho_{31} + \rho_{42} & \rho_{33} + \rho_{44} \end{pmatrix}.$$

3.1.3 The Postulates of Quantum Mechanics

Let us switch gears after this more intuitive introduction of the formalism of quantum theory, and formulate the framework of quantum mechanics more rigorously for finite-dimensional systems with the help of a number of postulates. These postulates pick up various concepts introduced in the previous section, to which we will refer throughout, but generalise and formalise them with help of the Dirac notation. This section is meant to serve as a reference for the remainder of the book, and may be not immediately accessible for readers without a background in quantum physics. If this is the case, we recommend to continue with Sect. 3.1.1 and return here when the need for more depth arises.

The Dirac notation may be confusing to newcomers to quantum mechanics. Most topics of quantum computing can actually be expressed in terms of matrices and vectors, and we will frequently switch between the two notations. However, some quantum systems have operators with a continuous eigenbasis, where discrete and finite vectors are not very useful. Dirac notation is an abstraction that incorporates such cases, and is the standard formalism in quantum computing. Table 3.1 compares the matrix and Dirac formalisms to illustrate their equivalence from a mathematical perspective.

3.1.3.1 State Space

A quantum mechanical state lives in a Hilbert space $\mathcal{H}$, which is a complex separable vector space. As mentioned, for the purposes of this book it suffices to consider

Table 3.1 Comparison of the two different formulations of quantum theory, Heisenberg's matrix notation and Dirac's abstract formalism for discrete finite systems

Vector-matrix formalism	Dirac formalism
Quantum state	
Vector $\alpha = (\alpha_1, \ldots, \alpha_N)^T \in \mathbb{C}^N$	State vector in Hilbert space $|\psi\rangle \in \mathcal{H}$
Observables	
Hermitian matrix $O \in \mathbb{C}^{N \times N}$	Hermitian operator O
Basis	
$\{v_i\}$ basis of $\mathbb{C}^N$	$\{|v_i\rangle\}$ basis of $\mathcal{H}$
Eigenvalues	
$Ov_i = o_i v_i$ o_i eigenvalue to v_i	$O|v_i\rangle = o_i|v_i\rangle$ o_i eigenvalue to $|v_i\rangle$
Evolution	
$U\alpha = \alpha'$ U unitary matrix	$U|\psi\rangle = |\psi'\rangle$ U unitary operator

3.1 Introduction to Quantum Theory

discrete and finite-dimensional Hilbert spaces. These are isomorphic to the $\mathbb{C}^n$, for $n = 2, 3, \ldots$ and a quantum state therefore has a representation as a complex vector.

Vectors in Hilbert space are usually denoted by $|\psi\rangle$, where $|\cdot\rangle$ is a 'ket'. The complex vector representation of a ket from a discrete and finite-dimensional Hilbert space is exactly the amplitude vector $\alpha = (\alpha_1, \ldots, \alpha_N)^T$ that we introduced in the previous section. The norm of a vector is defined in terms of the inner product on $\mathcal{H}$, which in Dirac notation is denoted by $\langle \cdot | \cdot \rangle$. The 'left side' of the inner product, $\langle \cdot |$ is called a 'bra'. The corresponding object in vector notation is a complex conjugate row vector, $\alpha^\dagger = (\alpha_1, \ldots, \alpha_N)$. The inner product is linear in the 'ket' $|\cdot\rangle$ argument and anti-linear in the 'bra' $\langle \cdot |$. Thus, for two vectors $|\psi_1\rangle$ and $|\psi_2\rangle$ in $\mathcal{H}$ we have

$$\langle \psi_2 | \psi_1 \rangle^* = \langle \psi_1 | \psi_2 \rangle.$$

The norm of a vector $|\psi\rangle \in \mathcal{H}$ is then

$$||\psi|| = \sqrt{\langle \psi | \psi \rangle}.$$

For every vector in $\mathcal{H}$ there is a complete orthonormal basis $\{|e_i\rangle\}$, $i \in \mathbb{N}$, with $\langle e_i | e_j \rangle = \delta_{ij}$, such that

$$|\psi\rangle = \sum_i |e_i\rangle\langle e_i|\psi\rangle = \sum_i \langle e_i|\psi\rangle |e_i\rangle. \tag{3.8}$$

A particular choice for such a complete orthonormal basis for the vector representation of quantum states is the standard basis used in Eq. (3.2). Note that Eq. (3.8) describes a *basis change*, since it expresses $|\psi\rangle$ in terms of the basis $\{|e_i\rangle\}$. Furthermore, $|e_i\rangle\langle e_i|$ denotes a projector on a one-dimensional sub-space. We have already seen a projector for the standard basis in Eq. (3.4). Equation (3.8) expresses the property of *separability* of Hilbert spaces. The same equation implies that the *completeness relation*

$$\mathbb{1} = \sum_i |e_i\rangle\langle e_i| \tag{3.9}$$

holds (i.e., to make the first equality sign work).

In Sect. 3.1.2.4 we saw that the state of a quantum mechanical system is sometimes not completely known, in which case we have to express it as a density matrix. Let us revisit this concept in Dirac notation. Imagine a system is in one of the states $|\psi_k\rangle$, $k = 1, \ldots, K$, with a certain probability p_k. The density operator ρ describing this *mixed* quantum mechanical state is defined as

$$\rho = \sum_{k=1}^{K} p_k |\psi_k\rangle\langle \psi_i|, \tag{3.10}$$

where all $p_k \geq 0$ and $\sum_k p_k = 1$.

3.1.3.2 Observables

As we saw before, observables of a quantum mechanical system are realised as self-adjoint operators O in $\mathcal{H}$ that act on the ket $|\psi\rangle$ characterizing the system. Such self-adjoint operators have a diagonal representation

$$O = \sum_i o_i |e_i\rangle\langle e_i|,$$

in terms of the set of eigenvalues $\{o_i\}$ and the corresponding eigenvectors $|e_i\rangle$.[6] This so-called spectral representation is unique and allows to calculate analytic functions f of the observable according to the formula

$$f(O) = \sum_i f(o_i) |e_i\rangle\langle e_i|.$$

In particular, if the observable is a Hermitian operator H, then for a real scalar ϵ

$$U = \exp(i\epsilon H)$$

is a unitary operator.

In general, expectation values of a quantum mechanical observable O can be calculated as

$$\langle O \rangle = \text{tr}\{\rho O\},$$

where 'tr' computes the trace of the operator. The trace operation in Dirac notation 'sandwiches' the expression inside the curly brackets by a sum over a full basis,

$$\text{tr}\{A\} = \sum_i \langle e_i | A | e_i \rangle.$$

For a system in a pure state $\rho = |\psi\rangle\langle\psi|$, the expression for the expectation value of an observable reduces to

$$\langle O \rangle = \sum_i \langle e_i|\psi\rangle\langle\psi|O|e_i\rangle = \sum_i \langle\psi|e_i\rangle\langle e_i|O|\psi\rangle = \langle\psi|O|\psi\rangle,$$

where we used the completeness relation (3.9). We already encountered this formula in matrix notation in Eq. (3.5).

---

[6] From this form of an operator one can see that the density matrix is also an operator on Hilbert space, although it describes a quantum state.

3.1.3.3 Time Evolution

The time evolution of a quantum mechanical system is described by the Schrödinger equation

$$i\hbar \frac{d}{dt}|\psi\rangle = H|\psi\rangle,$$

where H denotes the Hamiltonian of the system and $\hbar$ is Planck's constant. In the next chapter we will introduce several Hamiltonians relevant to quantum information processing. For time-independent Hamiltonians the solutions of the Schrödinger equation for an initial condition $|\psi(t=0)\rangle = |\psi_0\rangle$ can be written as

$$|\psi(t)\rangle = U(t)|\psi_0\rangle, \tag{3.11}$$

where

$$U(t) = \exp(-i\frac{t}{\hbar}H),$$

is the unitary time-evolution operator. The unitary operator in Eq. (3.11) corresponds to the unitary matrix we encountered in Eq. (3.7).

The time evolution of a quantum system in a mixed state described by the density operator ρ follows immediately from its definition (3.10),

$$\begin{aligned} i\hbar \frac{d}{dt}\rho(t) &= \frac{d}{dt} \sum_i p_i |\psi_i\rangle\langle\psi_i| \\ &= [H, \rho], \end{aligned} \tag{3.12}$$

where we have used the Schrödinger equation and its Hermitian conjugate to derive the second above equation, which is called the *von Neumann equation*. The formal solution of the von Neumann equation for initial condition $\rho(t_0)$ follows along similar lines from Eq. (3.11)

$$\rho(t) = U(t, t_0)\rho(t_0)U^\dagger(t, t_0).$$

In other words, to evolve a density operator, we have to apply the unitary from the left, and its complex conjugate from the right. Note that in theoretical physics literature, as well as in this book, the constant $\hbar$ is usually set to 1 in order to simplify the equations.

3.1.3.4 Quantum Measurement

For a quantum mechanical system in a pure state $|\psi\rangle$ we consider an observable O, with a discrete spectrum, i.e.,

$$O = \sum_i o_i P_i,$$

where P_i denotes the projector on the eigenspace spanned by the i-th eigenvector. The o_i are the measurement results of a measurement associated with projectors P_i. (In Sect. 3.1.2.1 these were the events $x_1, \ldots, x_N$.) Such a measurement is called a *projective measurement*. According to the *Born rule*, the probability to obtain the measurement result o_n is given by

$$p(o_n) = \text{tr}\{P_n |\psi\rangle\langle\psi|\} = ||P_n|\psi\rangle||^2.$$

The state of the system after the measurement is given by

$$|\psi\rangle \longrightarrow \frac{P_n|\psi\rangle}{\sqrt{\langle\psi|P_n|\psi\rangle}}.$$

In general, if the system is described by a density matrix ρ, the probability of measuring o_n will be

$$p(o_n) = \text{tr}\{P_n \rho\}$$

and the state of the system after the measurement will be[7]

$$\frac{P_n \rho P_n}{\text{tr}\{P_n \rho\}}.$$

Consider now a measurement defined by a collection of operators $\{A_m\}$ that do not necessarily have to be projectors. This is a more general case. If, again, the state of the system is described by the pure state $|\psi\rangle$, the probability that the result o_m related to operator A_m occurs is given by

$$p(a_m) = \langle\psi|A_m^\dagger A_m|\psi\rangle$$

and the state after the measurement is

$$\frac{A_m|\psi\rangle}{\sqrt{\langle\psi|A_m^\dagger A_m|\psi\rangle}}.$$

The measurement operators satisfy the completeness relation $\sum_m A_m^\dagger A_m = \mathbb{1}$. If we define

$$E_m = A_m^\dagger A_m,$$

it follows that $\sum_m E_m = \mathbb{1}$ and $p(o_m) = \langle\psi|E_m|\psi\rangle$. The operators E_m are called Positive Operator-Valued Measure (POVM) elements and their complete set is known as a Positive Operator-Valued Measurement. This is the most common framework to describe quantum measurements.

---

[7] This formalism is also known as the Lüders postulate.

3.1.3.5 Composite Systems

We saw in Sect. 3.1.2.3 that in vector notation, a composite state of two joint quantum systems is a tensor product of two quantum states, which of course carries over to Dirac notation. Let us assume that our quantum system of interest Σ is composed of two sub-systems, say Σ_1 and Σ_2. The sub-system Σ_i, $i = 1, 2$, is in a Hilbert space $\mathcal{H}_i$ with $\dim(\mathcal{H}_i) = N_i$, and is spanned by an orthonormal basis set $\{|e_j^i\rangle\}$, $j \in \mathbb{N}$. Then, the Hilbert space $\mathcal{H}$ of the total system Σ is given by the tensor product of the Hilbert spaces of the two sub-systems, i. e., $\mathcal{H}_1$ and $\mathcal{H}_2$,

$$\mathcal{H} = \mathcal{H}_1 \otimes \mathcal{H}_2.$$

The states in this joint Hilbert space can be written as

$$\sum_{j=1}^{N_1} \sum_{k=1}^{N_2} c_{jk} |e_j^1\rangle \otimes |e_k^2\rangle$$

and have $\dim(\mathcal{H}) = N_1 N_2$. The coefficients c_{jk} are complex scalars and satisfy $\sum_{jk} |c_{jk}|^2 = 1$.

Composite quantum systems have the famous quantum feature of *entanglement*. If a state $|\psi\rangle$ can be expressed as

$$|\psi\rangle = |\psi_1\rangle \otimes |\psi_2\rangle,$$

with $|\psi_1\rangle \in \mathcal{H}_1$ and $|\psi_2\rangle \in \mathcal{H}_2$, it is called *separable*. If the above representation is not possible, the state is called *entangled*. Separability and entanglement work similarly for density matrices. If the two subsystems Σ_1 and Σ_2 are uncorrelated and described by density matrices ρ_1 and ρ_2, respectively, then the total density matrix ρ of the composite system Σ is given by

$$\rho = \rho_1 \otimes \rho_2.$$

Operators O acting on the total Hilbert space $\mathcal{H}$ have the structure

$$O = O_1 \otimes O_2,$$

where the operators O_1 and O_2 act on Σ_1 and Σ_2, respectively. In general, the expectation value of such an operator can be calculated as

$$\langle O \rangle = \text{tr}\{O\rho\} = \text{tr}_1\{O_1 \rho_1\} \cdot \text{tr}_2\{O_2 \rho_2\}, \tag{3.13}$$

where tr_i, $i = 1, 2$ denotes the partial trace, which only sums over basis states of the subsystem.

Sometimes, one is interested in observables O that operate only on one of the two subsystems, i.e.

$$O = O_1 \otimes \mathbb{1}_2.$$

In this case, the above expression for the expectation value (3.13) simplifies to

$$\langle O \rangle = \mathrm{tr}_1\{O_1 \rho_1\}.$$

3.1.3.6 Open Quantum Systems

The above discussion is at the core of the theory of *open quantum systems* [11], where one identifies Σ_1 with the *open* system of interest (sometimes also referred to as the reduced system) and Σ_2 with its environment. The aim of the theory of open quantum systems is to find an effective dynamical description for the open system, through an appropriate elimination of the degrees of freedom of the environment. The starting point is, of course, the Hamiltonian evolution of the total system. The density matrix ρ of the total system Σ evolves according to the von Neumann equation (3.12)

$$\frac{d}{dt}\rho = -i[H, \rho].$$

Introducing the reduced density matrix ρ_1 of the open system Σ_1 as the trace over the degrees of freedom of system Σ_2, i.e.,

$$\rho_1 = \mathrm{tr}_2\{\rho\},$$

the dynamics of the open system can be obtained from (3.12)

$$\frac{d}{dt}\rho_1 = -i\,\mathrm{tr}_2\{[H, \rho]\}.$$

Of course, the above equation is as difficult to evaluate as the dynamical equation of motion for the total system and appropriate approximations are needed to make it useful. In the case of weak-coupling between the sub-systems Σ_1 and Σ_2, a situation that is typical in quantum optical applications, one can often assume (according to the Born approximation) that the density matrix of the total system factorizes at all times and that the density matrix of the environment is unchanged

$$\rho(t) \approx \rho_1(t) \otimes \rho_2.$$

If one further assumes that changes in the environment occur on time scales that are not resolved, then the dynamics of the open system is described by the so-called Markovian Quantum Master equation in Gorini-Kossakowski-Sudarshan-Lindblad form [11], which reads

3.1 Introduction to Quantum Theory

$$\frac{d}{dt}\rho_1 = -i[H, \rho_1] + \sum_k \gamma_k \left(L_k \rho_1 L_k^\dagger - \frac{1}{2} L_k^\dagger L_k \rho_1 - \frac{1}{2} \rho_1 L_k^\dagger L_k \right). \quad (3.14)$$

The operators L_k introduced above are usually referred to as Lindblad operators and describe the transitions in the open system induced by the interaction with the environment. The constants γ_k are the relaxation rates for the different decay modes k of the open system.

3.2 Introduction to Quantum Computing

We will now turn to quantum computing, an application of quantum theory, where the abstract concepts introduced in the previous section are filled with more meaning.

3.2.1 What Is Quantum Computing?

As quantum theory motivated physicists to rethink all aspects of their discipline from a new perspective, it was only a matter of time before the question arose what quantum mechanics means to information processing. In a way, with the debates of non-locality and Einstein's "spooky action at a distance" this has already been part of the early days of quantum mechanics. However, it took until the late 1980s for quantum information processing research to form an independent sub-discipline. Quantum computing is a subfield of quantum information processing, whose central questions are: *What is quantum information? Can we build a new type of computer based on quantum systems? How can we formulate algorithms on such machines? What does quantum theory mean for the limits of what is computable? What distinguishes quantum computers from classical ones?*

In the media (and possibly much fueled by researchers themselves hunting for grants), quantum computing is still portrayed as the cure for all, mirrored in the following remark from a machine learner's perspective:

> Much like artificial intelligence in its early days, the reputation of quantum computing has been tarnished by grand promises and few concrete results. Talk of quantum computers is often closely flanked by promises of polynomial time solutions to NP-Hard problems and other such implausible appeals to blind optimism. [12, p. 3]

It is true that after more than 20 years of establishing an independent research discipline, and not-surprisingly, there is still no final answer to many of the questions posed. Most prominently, we still do not know exactly how a quantum Turing machine compares to a classical Turing machine. Nevertheless, there is a lot more we know. For example, we know that there are quantum algorithms that grow slower in runtime with the size of the input than known classical algorithms solving the same problem [13–15]. Relative to a black-box function or oracle, quantum algorithms are even

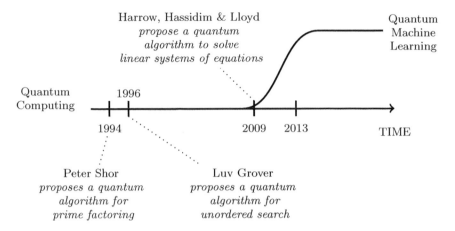

Fig. 3.3 Timeline of quantum computing and quantum machine learning

proven to be faster than any *possible* classical algorithm. Another important result is that every classical algorithm can be implemented on a quantum computer with only polynomial overhead, so in theory a quantum computer is at least as good as a classical computer [16] from the perspective of asymptotic complexity.

A rich landscape of quantum computational models and routines has been formulated [17], and a database of algorithms is frequently updated by Stephen Jordan at the American National Institute of Standards and Technology.[8] Two influential quantum algorithms were Shor's factorisation algorithm [15] and Grover's search algorithm [13]. For one branch of the quantum machine learning literature, Harrow, Hassidim and Lloyd's quantum algorithm for linear systems of equations in 2009 set another milestone. From roughly 2013 onwards, quantum machine learning became its own subdiscipline of quantum computing (see Fig. 3.3).

Although there are a variety of computational models that formalise the idea of quantum computation, most of them are based on the concept of a *qubit*, which is a quantum system associated with two measurable events, and in some sense similar to a random bit or biased coin toss as we saw in the introduction. Many popular notions of quantum information claim that the power of quantum computers stems from the fact that qubits can be in a linear combination of 0 and 1, so it can take 'all states in between'. But also a classical random bit (i.e. a classical coin toss) has this property to a certain extent. This is why sampling algorithms are often the most suitable competitors to quantum routines. One major difference however is that coefficients of the linear combination—the amplitudes—can be complex numbers in the quantum case. In Chap. 1 we saw that a certain evolution can make amplitudes even cancel each other out, or *interfere* with each other. This fact, together with other subtleties can lead to measurement statistics that are *non-classical*, which means that they cannot be reproduced by classical systems.

---

[8]http://math.nist.gov/quantum/zoo/.

3.2 Introduction to Quantum Computing

A quantum computer can be understood as a physical implementation of n qubits (or other basic quantum systems) with a precise control on the evolution of the state. A quantum algorithm is a targeted manipulation of the quantum system with a subsequent measurement to retrieve information from the system. In this sense a quantum computer can be understood as a special sampling device: We choose certain experimental configurations—such as the strength of a laser beam, or a magnetic field—and read out a distribution over possible measurement outcomes.

An important theorem in quantum information states that any quantum evolution (think again of manipulating the physical system in the lab) can be approximated by a sequence of only a handful of elementary 'manipulations', called *quantum gates*, which only act on one or two qubits at a time [18]. Based on this insight, quantum algorithms are widely formulated as *quantum circuits* of these elementary gates. A universal quantum computer consequently only has to know how to perform a small set of operations on qubits, just like classical computers are build on a limited number of logic gates. Runtime considerations usually count the number of quantum gates it takes to implement the entire quantum algorithm. *Efficient* quantum algorithms are based on evolutions whose decomposition into a circuit of gates grows at most polynomially with the input size of the problem. We will speak more about runtime and speedups in Chap. 4.

As much as a classical bit is an abstract model of a binary system that can have many different physical realisations in principle, a qubit can be used as a model of many different physical quantum systems. Some current candidates for implementations are superconducting qubits [19], photonic setups [20], ion traps [21] or topological properties of quasi-particles [22]. Each of them has advantages and disadvantages, and it is not unlikely that future architectures use a mixture of these implementations.

3.2.2 Bits and Qubits

A qubit is realised as a quantum mechanical two-level system and as such can be measured in two states, called the basis states. Traditionally, they are denoted by the Dirac vectors $|0\rangle$ and $|1\rangle$.

For our purposes it therefore suffices to describe the transition from a classical to a quantum mechanical description of information processing—from bits to qubits—by means of a very simple quantization procedure, namely

$$0 \longrightarrow |0\rangle,$$
$$1 \longrightarrow |1\rangle.$$

The vectors $|0\rangle$, $|1\rangle$ form a orthonormal basis of a 2-dimensional Hilbert space which is also called the *computational basis*. As a consequence of the superposition principle of quantum mechanics the most general state of a qubit is

$$|\psi\rangle = \alpha_1|0\rangle + \alpha_2|1\rangle, \tag{3.15}$$

where $\alpha_1, \alpha_2 \in \mathbb{C}$ and because of the normalisation of the state vector $|\alpha_1|^2 + |\alpha_2|^2 = 1$. The Hermitian conjugate to the above ket-state, the bra-state, is then

$$\langle\psi| = \alpha_1^*\langle 0| + \alpha_2^*\langle 1|,$$

where * denotes complex conjugation.

As discussed in the previous sections, such a Dirac vector has a vector representation, since N-dimensional, discrete Hilbert spaces are isomorphic to the space of complex vectors $\mathbb{C}^N$. In vector notation, a general qubit is expressed as

$$|\psi\rangle = \begin{pmatrix} \alpha_1 \\ \alpha_2 \end{pmatrix}.$$

The Hermitian conjugate of this amplitude column vector is the transposed and conjugated row vector

$$(\alpha_1^*, \alpha_2^*) \in \mathbb{C}^2.$$

Furthermore, we can represent the two states $|0\rangle$ and $|1\rangle$ as the standard basis vectors of the $\mathbb{C}^2$,

$$|0\rangle = \begin{pmatrix} 1 \\ 0 \end{pmatrix} \in \mathbb{C}^2,$$

$$|1\rangle = \begin{pmatrix} 0 \\ 1 \end{pmatrix} \in \mathbb{C}^2.$$

Vector notation can be very useful to understand the effect of quantum gates. However, as common in quantum computing, we will predominantly use Dirac notation.

It is sometimes useful to have a geometric representation of a qubit. A generic qubit in the pure state (3.15) can be parametrised as

$$|\psi\rangle = \exp(i\gamma)\left(\cos\frac{\theta}{2}|0\rangle + \exp(i\phi)\sin\frac{\theta}{2}|1\rangle\right),$$

where θ, ϕ and γ are real numbers with $0 \leq \theta \leq \pi$ and $0 \leq \phi \leq 2\pi$. The global phase factor $\exp(i\gamma)$ has no observable effect and will be omitted in the following. The angles θ and ϕ have the obvious interpretation as spherical coordinates, so that the Hilbert space vector $|\psi\rangle$ can be visualised as the $\mathbb{R}^3$ vector $(\sin\theta\cos\phi, \sin\theta\sin\phi, \cos\phi)$ pointing from the origin to the surface of a ball, the so-called Bloch sphere. The Bloch sphere is illustrated in Fig. (3.4).[9]

The Dirac notation allows also for a compact description of the inner product of two vectors in Hilbert space that was introduced in Sect. 3.1.3.1. Consider for

---

[9] Adapted from https://tex.stackexchange.com/questions/345420/how-to-draw-a-bloch-sphere.

3.2 Introduction to Quantum Computing

Fig. 3.4 The Bloch sphere representation of a qubit

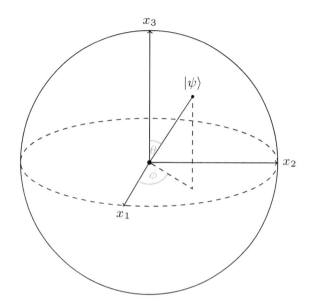

example, two vectors in $\mathbb{C}^2$, $|\psi_1\rangle = \alpha_1|0\rangle + \alpha_2|1\rangle$ and $|\psi_2\rangle = \beta_1|0\rangle + \beta_2|1\rangle$, with $\alpha_i, \beta_i \in \mathbb{C}$ for $i = 1, 2$, $|\alpha_1|^2 + |\alpha_2|^2 = 1$ and $|\beta_1|^2 + |\beta_2|^2 = 1$. Since $|0\rangle, |1\rangle$ are orthonormal, we have that

$$\langle 0|0\rangle = \langle 1|1\rangle = 1, \quad \langle 0|1\rangle = \langle 1|0\rangle = 0.$$

The inner product of $|\psi_1\rangle$ and $|\psi_2\rangle$ is therefore given by

$$\langle\psi_1|\psi_2\rangle = \alpha_1^*\beta_1 + \alpha_2^*\beta_2.$$

Of course, this is equivalent the scalar or vector product of the two corresponding amplitude vectors. Similarly, the outer product of two states can be compactly written as

$$|\psi_1\rangle\langle\psi_2| = \begin{pmatrix} \alpha_1\beta_1^* & \alpha_1\beta_2^* \\ \alpha_2\beta_1^* & \alpha_2\beta_2^* \end{pmatrix},$$

which is the outer product of the amplitude vectors.

According to Sect. 3.1.3.5, n unentangled qubits are described by a tensor product of single qubits,

$$|\psi\rangle = |q_1\rangle \otimes \cdots \otimes |q_n\rangle.$$

If the qubits are entangled, state $|\psi\rangle$ is no longer separable, and in the computational basis reads

$$|\psi\rangle = \alpha_1|0\ldots00\rangle + \alpha_2|0\ldots01\rangle + \cdots + \alpha_{2^n}|1\ldots11\rangle,$$

with $\alpha_i \in \mathbb{C}$, and $\sum_{i=1}^{2^n} |\alpha_i|^2 = 1$. Here we introduce the common shorthand which writes the tensor product $|a\rangle \otimes |b\rangle$ as $|ab\rangle$. The basis $\{|0\ldots00\rangle, \ldots, |1\ldots11\rangle\}$ is the computational basis for n qubits. Note that for some algorithms the qubits are divided into certain *registers*, which have different functions in the computation.

We will make heavy use of an elegant notation that summarises a Dirac vector in computational basis as

$$|\psi\rangle = \sum_{i=1}^{2^n} \alpha_i|i\rangle. \tag{3.16}$$

The expression $|i\rangle$ for indices 1 to 2^n refers to the ith computational basis state from the basis $\{|0\ldots0\rangle, \ldots, |1\ldots1\rangle\}$. The sequence of zeros and ones in the basis state at the same time corresponds to the binary representation of integer $i - 1$ (see Table 3.2). As a rule from now on, an index i between the Dirac bracket for a system of $n > 1$ qubits refers to the ith computational basis state. The only exception is a single qubit state, whose basis is traditionally referred to as $|0\rangle, |1\rangle$. The reason why we introduce a differing convention for multi-qubit states is that we will use $|i\rangle$ often as an index register that refers to the ith data input, and it is rather unnatural (and becomes inconvenient) to number data points starting from 0.

Using this notation, a pure density matrix is given by

$$\rho_{\text{pure}} = |\psi\rangle\langle\psi| = \sum_{i,j=1}^{N} \alpha_i^* \alpha_j |i\rangle\langle j|. \tag{3.17}$$

Table 3.2 The index i of an amplitude from the 2^3-dimensional quantum state vector refers to the computational basis state $|i\rangle$. The binary string of the basis state has an integer representation of $i-1$

Amplitude	Binary/integer	Basis state	Shorthand
α_1	000 ↔ 0	$|000\rangle$	$|1\rangle$
α_2	001 ↔ 1	$|001\rangle$	$|2\rangle$
α_3	010 ↔ 2	$|010\rangle$	$|3\rangle$
α_4	011 ↔ 3	$|011\rangle$	$|4\rangle$
α_5	100 ↔ 4	$|100\rangle$	$|5\rangle$
α_6	101 ↔ 5	$|101\rangle$	$|6\rangle$
α_7	110 ↔ 6	$|110\rangle$	$|7\rangle$
α_8	111 ↔ 7	$|111\rangle$	$|8\rangle$

3.2 Introduction to Quantum Computing

For a general mixed state the coefficients do not factorise and we get

$$\rho_{\text{mixed}} = \sum_{i,j=1}^{N} \alpha_{ij} |i\rangle\langle j|, \quad \alpha_{ij} \in \mathbb{C}. \tag{3.18}$$

It is useful to consider the measurement of a qubit in the computational basis to illustrate the more abstract content of Sect. 3.1.3.4. Let us assume that the state being measured is the generic (normalised) qubit state $|\psi\rangle = \alpha_1 |0\rangle + \alpha_2 |1\rangle$. In this case the projectors on the two possible eigenspaces are $P_0 = |0\rangle\langle 0|$ and $P_1 = |1\rangle\langle 1|$. The probability of obtaining the measurement outcome 0 is then

$$p(0) = \text{tr}\{P_0 |\psi\rangle\langle\psi|\} = \langle\psi|P_0|\psi\rangle = |\alpha_1|^2.$$

Similarly, one finds that $p(1) = |\alpha_2|^2$. After the measurement, say of outcome 0, the qubit is in the state

$$|\psi\rangle \to \frac{P_0 |\psi\rangle}{\sqrt{\langle\psi|P_0|\psi\rangle}} = |0\rangle.$$

A computational basis measurement can be understood as drawing a sample of a binary string of length n—where n is the number of qubits—from a distribution defined by the quantum state.

3.2.3 Quantum Gates

Postulates 3.1.3.3 and 3.1.3.4 highlighted the evolution of quantum states, as well as their measurement. In analogy, there are two basic operations on qubits that are central to quantum computing,

- quantum logic gates,
- computational basis measurements.

Quantum logic gates are realised by unitary transformations introduced in Eqs. (3.7) and (3.11). As we have seen, after a projective measurement in the computational basis the state of the qubit $|\psi\rangle = \alpha_1 |0\rangle + \alpha_2 |1\rangle$, will be either $|0\rangle$ with probability $|\alpha_1|^2$ or $|1\rangle$ with probability $|\alpha_2|^2$.

Single qubit gates are formally described by 2×2 unitary transformations. It is illustrative to write these transformations as matrices. As an example we consider the so-called X gate, which is the quantum equivalent of the classical NOT gate, as it acts as

$$|0\rangle \mapsto |1\rangle, \tag{3.19}$$
$$|1\rangle \mapsto |0\rangle. \tag{3.20}$$

It is represented by the matrix

$$X = \begin{pmatrix} 0 & 1 \\ 1 & 0 \end{pmatrix}.$$

In vector-matrix notation it is easy to check that applying the X gate to the state $|0\rangle$ results in the state $|1\rangle$

$$X|0\rangle = \begin{pmatrix} 0 & 1 \\ 1 & 0 \end{pmatrix} \begin{pmatrix} 1 \\ 0 \end{pmatrix} = \begin{pmatrix} 0 \\ 1 \end{pmatrix} = |1\rangle.$$

In other words, applied to a generic single qubit state, the X gate swaps the amplitudes of the $|0\rangle$ and $|1\rangle$ components. Obviously, X is unitary,

$$XX^\dagger = XX^{-1} = \mathbb{1}.$$

Some useful single qubit gates are summarised in the Table 3.3. The first three gates X, Y, Z, are equivalent to the *Pauli matrices*

$$\sigma_x = \begin{pmatrix} 0 & 1 \\ 1 & 0 \end{pmatrix}, \sigma_y = \begin{pmatrix} 0 & -i \\ i & 0 \end{pmatrix}, \sigma_z = \begin{pmatrix} 1 & 0 \\ 0 & -1 \end{pmatrix}, \quad (3.21)$$

which will often appear in later chapters.

One gate that we will make frequent use of is the Hadamard gate H (which is not to be confused with the Hamiltonian of the same symbol). The Hadamard was

Table 3.3 Some useful single qubit logic gates and their representations

Gate	Circuit representation	Matrix representation	Dirac representation
X	—[X]—	$\begin{pmatrix} 0 & 1 \\ 1 & 0 \end{pmatrix}$	$\|1\rangle\langle 0\| + \|0\rangle\langle 1\|$
Y	—[Y]—	$\begin{pmatrix} 0 & -i \\ i & 0 \end{pmatrix}$	$i\|1\rangle\langle 0\| - i\|0\rangle\langle 1\|$
Z	—[Z]—	$\begin{pmatrix} 1 & 0 \\ 0 & -1 \end{pmatrix}$	$\|1\rangle\langle 0\| - \|0\rangle\langle 1\|$
H	—[H]—	$\frac{1}{\sqrt{2}}\begin{pmatrix} 1 & 1 \\ 1 & -1 \end{pmatrix}$	$\frac{1}{\sqrt{2}}(\|0\rangle + \|1\rangle)\langle 0\| + \frac{1}{\sqrt{2}}(\|0\rangle - \|1\rangle)\langle 1\|$
S	—[S]—	$\frac{1}{\sqrt{2}}\begin{pmatrix} 1 & 0 \\ 0 & i \end{pmatrix}$	$\frac{1}{\sqrt{2}}\|0\rangle\langle 0\| + \frac{1}{\sqrt{2}}i\|1\rangle\langle 1\|$
R	—[R]—	$\frac{1}{\sqrt{2}}\begin{pmatrix} 1 & 0 \\ 0 & \exp(-i\pi/4) \end{pmatrix}$	$\frac{1}{\sqrt{2}}\|0\rangle\langle 0\| + \frac{1}{\sqrt{2}}\exp^{-i\pi/4}\|1\rangle\langle 1\|$

already introduced in Chap. 1. Its effect on the basis states $|0\rangle$ and $|1\rangle$ is given as

$$|0\rangle \mapsto \frac{1}{\sqrt{2}}(|0\rangle + |1\rangle), \tag{3.22}$$

$$|1\rangle \mapsto \frac{1}{\sqrt{2}}(|0\rangle - |1\rangle). \tag{3.23}$$

As made clear from the above expression, the role of H is to create superpositions of qubits.

Of course, it is important to operate on more qubits at the same time as well. The paradigmatic 2-qubit gate is the so-called CNOT gate, which is an example of a controlled gate. The state of a qubit is changed, based on the value of another, control, qubit. In the case of the CNOT gate, the NOT operation (or X operation) is performed, when the first qubit is in state $|1\rangle$; otherwise the second qubit is unchanged

$$|00\rangle \mapsto |00\rangle, |01\rangle \mapsto |01\rangle, |10\rangle \mapsto |11\rangle, |11\rangle \mapsto |10\rangle. \tag{3.24}$$

Accordingly, the matrix representation of the CNOT gate is given by

$$CNOT = \begin{pmatrix} 1 & 0 & 0 & 0 \\ 0 & 1 & 0 & 0 \\ 0 & 0 & 0 & 1 \\ 0 & 0 & 1 & 0 \end{pmatrix}.$$

The CNOT gate (3.24) is a special case of a more general controlled U gate

$$|00\rangle \mapsto |00\rangle, |01\rangle \mapsto |01\rangle, |10\rangle \mapsto |1\rangle U|0\rangle, |11\rangle \mapsto |1\rangle U|1\rangle, \tag{3.25}$$

where U is an arbitrary single qubit unitary gate. For the CNOT, we obviously have $U = X$. Any multiple qubit gate may be composed by a sequence of single qubit gates and CNOT gates [18]. In Table 3.4 we summarize some useful multi-qubit gates.

Tables 3.3 and 3.4 reveal also the circuit notation of the respective gates. Circuit notation allows the graphical visualisation of a quantum routine, and we will introduce it with our first little quantum algorithm, the entangling circuit shown in Fig. (3.5).

Example 3.3 (*Entangling circuit*) We will compute the action of the quantum circuit from Fig. (3.5) on the initial state of the two qubits, namely $|0\rangle_2|0\rangle_1$. For clarity, we have labelled the upper qubit by the subscript 2 and the lower qubit by the subscript 1. We read the circuit from left to right and write

$$CNOT((H_2 \otimes \mathbb{1}_1)(|0\rangle_2 \otimes |0\rangle_1)).$$

Table 3.4 Some useful and common multi-qubit gates: The 2-qubit CNOT and SWAP gate, a well as the 3-qubit Toffoli gate

Gate	Circuit representation	Matrix representation
CNOT	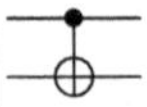	$\begin{pmatrix} 1 & 0 & 0 & 0 \\ 0 & 1 & 0 & 0 \\ 0 & 0 & 0 & 1 \\ 0 & 0 & 1 & 0 \end{pmatrix}$
SWAP		$\begin{pmatrix} 1 & 0 & 0 & 0 \\ 0 & 0 & 1 & 0 \\ 0 & 1 & 0 & 0 \\ 0 & 0 & 0 & 1 \end{pmatrix}$
T	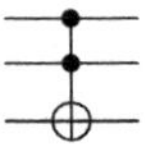	$\begin{pmatrix} 1 & 0 & 0 & 0 & 0 & 0 & 0 & 0 \\ 0 & 1 & 0 & 0 & 0 & 0 & 0 & 0 \\ 0 & 0 & 1 & 0 & 0 & 0 & 0 & 0 \\ 0 & 0 & 0 & 1 & 0 & 0 & 0 & 0 \\ 0 & 0 & 0 & 0 & 1 & 0 & 0 & 0 \\ 0 & 0 & 0 & 0 & 0 & 1 & 0 & 0 \\ 0 & 0 & 0 & 0 & 0 & 0 & 0 & 1 \\ 0 & 0 & 0 & 0 & 0 & 0 & 1 & 0 \end{pmatrix}$

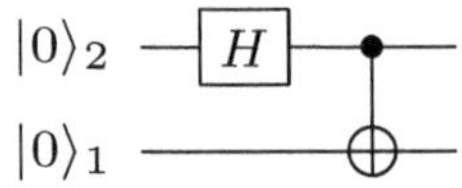

Fig. 3.5 A quantum circuit that entangles two qubits displayed in graphical notation. A line denotes a so called 'quantum wire' that indicates the time evolution of a qubit. The initial state of the qubit is written to the left of the quantum wire, and sometimes the final state is written towards the right. A gate sits on the wires that correspond to the qubits it acts on

Recalling the effect of the Hadamard gate H (3.22), the above expression becomes a fully entangled state

$$CNOT(\frac{1}{\sqrt{2}}|0\rangle_2 \otimes |0\rangle_1 + \frac{1}{\sqrt{2}}|1\rangle_2 \otimes |0\rangle_1)$$
$$= \frac{1}{\sqrt{2}}(|0\rangle_2 \otimes |0\rangle_1 + |1\rangle_2 \otimes |1\rangle_1).$$

This state is also known as the Bell state (Fig. 3.5).

3.2.4 Quantum Parallelism and Function Evaluation

As a first larger example of a quantum algorithm we want to construct a quantum logic circuit that evaluates a function $f(x)$ (see also [17]). This simple algorithm will already exhibit one of the salient features of quantum algorithms: *quantum parallelism*. Roughly speaking, we will see how a quantum computer is able to evaluate many different values of the function $f(x)$ at the same time, or in superposition.

To be specific we consider a very simple function $f(x)$, that has a single bit as input and a single bit as output, i.e. a function with a one bit domain and range,

$$f(x) : \{0, 1\} \rightarrow \{0, 1\}.$$

Examples of such a function are the identity function

$$f(x) = 0, \quad \text{if} \quad x = 0 \quad \text{and} \quad f(x) = 1, \quad \text{if} \quad x = 1,$$

the constant functions

$$f(x) = 0 \quad \text{or} \quad f(x) = 1,$$

and the bit flip function

$$f(x) = 1, \quad \text{if} \quad x = 0 \quad \text{and} \quad f(x) = 0, \quad \text{if} \quad x = 1.$$

The idea is to construct a unitary transformation U_f such that

$$(x, y) \xrightarrow{U_f} (x, y \oplus f(x)). \tag{3.26}$$

In the above equation the symbol $\oplus$ denotes mod 2 addition, i.e. $0 \oplus 0 = 1 \oplus 1 = 0$ and $0 \oplus 1 = 1 \oplus 0 = 1$. Note, that the more straightforward approach $x \rightarrow f(x)$ is not suitable for quantum computation, as it is not unitary in general. The unitary transformation U_f is represented as a circuit diagram in Fig. 3.6.

For the initial value $y = 0$, Eq. (3.26) reduces to

$$(x, 0) \xrightarrow{U_f} (x, f(x)).$$

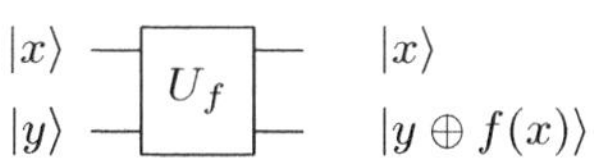

Fig. 3.6 The schematic representation of the gate U_f

It is easy to verify that U_f is unitary, we just check that $U_f^2 = \mathbb{1}$

$$(x, [y \oplus f(x)]) \xrightarrow{U_f} (x, [y \oplus f(x)] \oplus f(x)) = (x, y),$$

because $f(x) \oplus f(x) = 0$, for any x. Expressing U_f in operator notation, this reads

$$U_f(|x\rangle \otimes |y\rangle) = |x\rangle \otimes |y \oplus f(x)\rangle,$$

and we obtain the useful expression $U_f(|x\rangle \otimes |0\rangle) = |x\rangle \otimes |f(x)\rangle$.

We are now in the position to demonstrate quantum parallelism. We consider the quantum circuit of 2 qubits sketched in Fig. 3.7. The circuit acts in the following way:

i. We apply the Hadamard gate to the first qubit in state $|0\rangle$ to obtain

$$H|0\rangle = \frac{1}{\sqrt{2}}(|0\rangle + |1\rangle).$$

ii. If the second qubit is in state $|0\rangle$, the state $|\psi\rangle$ after the application of the unitary U_f is

$$|\psi\rangle = U_f(H|0\rangle \otimes |0\rangle) = U_f \frac{1}{\sqrt{2}}(|0\rangle \otimes |0\rangle + |1\rangle \otimes |0\rangle)$$
$$= \frac{1}{\sqrt{2}}(|0\rangle \otimes |f(0)\rangle + |1\rangle \otimes |f(1)\rangle).$$

It is now evident that the state $|\psi\rangle$ contains simultaneously the information on $f(0)$ and $f(1)$. The circuit has produced a superposition in one single step state that contains $f(0)$ and $f(1)$.

It is important to realize that the above procedure does not (yet) give us an advantage over a classical computation. Although $|\psi\rangle$ is a superposition of $f(0)$ and $f(1)$, in order to access the information we need to perform a measurement. If we measure $\sum_{x=0,1} |x\rangle|f(x)\rangle$ with a computational basis measurement, we obtain only one value of x and $f(x)$. In fact, we are only able to get a value of our function at a random argument. The real power of quantum algorithms will be made evident in the next example where we follow [17].

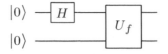

Fig. 3.7 A simple circuit to demonstrate quantum parallelism. The two-qubit gate is the unitary U_f described in the text

3.3 An Example: The Deutsch-Josza Algorithm

3.3.1 The Deutsch Algorithm

The Deutsch algorithm exploits what we have learned so far to obtain information about a global property of a function $f(x)$. A function of a single bit can be either *constant* ($f(0) = f(1)$), or *balanced* ($f(0) \neq f(1)$). These properties are *global*, because in order to establish them we need to calculate both $f(0)$ and $f(1)$ and compare the results. As we will see a quantum computer can do better.

The quantum circuit of the Deutsch algorithm is depicted in Fig. 3.8. Essentially, the Deutsch algorithms computes

$$|\psi_3\rangle = (H \otimes \mathbb{1}) U_f (H \otimes H) |01\rangle, \tag{3.27}$$

for an initial state $|\psi_0\rangle = |01\rangle$. Let us discuss the calculation of (3.27) step by step.

(i) In a first step we calculate $|\psi_1\rangle$

$$|\psi_1\rangle = (H \otimes H)|01\rangle = \frac{1}{\sqrt{2}}(|0\rangle + |1\rangle) \frac{1}{\sqrt{2}}(|0\rangle - |1\rangle)$$
$$= \frac{1}{2}(|00\rangle - |01\rangle + |10\rangle - |11\rangle).$$

(ii) Next we apply U_f to $|\psi_1\rangle$. It is convenient to write $|\psi_1\rangle$ as

$$|\psi_1\rangle = \frac{1}{2}\left(\sum_{x=0}^{1} |x\rangle\right) \otimes (|0\rangle - |1\rangle).$$

In order to evaluate the action of U_f on $|\psi_1\rangle$ we consider separately the case $f(x) = 0$ and $f(x) = 1$.
For the case $f(x) = 0$ we find

$$U_f(|x\rangle \otimes (|0\rangle - |1\rangle)) = |x, 0 \oplus f(x)\rangle - |x, 1 \oplus f(x)\rangle = |x, 0 \oplus 0\rangle - |x, 1 \oplus 0\rangle$$
$$= |x\rangle(|0\rangle - |1\rangle).$$

Similarly, for $f(x) = 1$ we find

Fig. 3.8 The quantum circuit for the implementation of the Deutsch algorithm

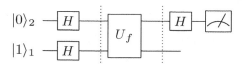

$$U_f(|x\rangle \otimes (|0\rangle - |1\rangle)) = |x, 0 \oplus 1\rangle - |x, 1 \oplus 1\rangle$$
$$= -|x\rangle(|0\rangle - |1\rangle).$$

The two above cases can elegantly be summarised as

$$|\psi_2\rangle = U_f|\psi_1\rangle = \frac{1}{2}\left(\sum_{x=0}^{1}(-1)^{f(x)}|x\rangle\right) \otimes (|0\rangle - |1\rangle) = |\phi\rangle \otimes \frac{1}{\sqrt{2}}(|0\rangle - |1\rangle).$$

In other words, the net result after the application of U_f on the input register is

$$|\phi\rangle = \frac{1}{\sqrt{2}}\left((-1)^{f(0)}|0\rangle + (-1)^{f(1)}|1\rangle\right).$$

(iii) In the last step, just before the measurement, we apply a Hadamard gate to the input qubit

$$|\psi_3\rangle = (H \otimes \mathbb{1})\left(|\phi\rangle \otimes \frac{1}{\sqrt{2}}(|0\rangle - |1\rangle)\right).$$

It is easy to calculate

$$H|\phi\rangle = \frac{1}{2}\left((-1)^{f(0)} + (-1)^{f(1)}\right)|0\rangle + \frac{1}{2}\left((-1)^{f(0)} - (-1)^{f(1)}\right)|1\rangle.$$

If the measurement of the qubit gives the state $|0\rangle$, then we know that $f(0) = f(1)$, as the coefficient in front of $|1\rangle$ vanishes, and, hence the function $f(x)$ is constant. If the measurement gives the state $|1\rangle$, then $f(0) \neq f(1)$, and the function is balanced.

Quantum parallelism has allowed the calculation of the global properties of a function without having to evaluate explicitly the values of the function.

3.3.2 The Deutsch-Josza Algorithm

The Deutsch algorithm can be generalised to functions with multiple input values. Before explaining the general idea behind the so-called Deutsch-Josza algorithm it is useful to briefly discuss how to generalize the input register from one qubit to n qubits. Let us start by considering the example of the case of $n = 3$ qubits. In the notation $|x\rangle$, x is one of the 8 numbers

$$000, 001, 010, 011, 100, 101, 110, 111.$$

In order to exemplify the power of this compact notation we consider the state

3.3 An Example: The Deutsch-Josza Algorithm

$$|\psi\rangle = H^{\otimes 3}|000\rangle = (H|0\rangle) \otimes (H|0\rangle) \otimes (H|0\rangle)$$
$$= \left(\frac{1}{\sqrt{2}}\right)^3 (|000\rangle + |001\rangle + |010\rangle + |011\rangle + |100\rangle + |101\rangle + |110\rangle + |111\rangle).$$

With the convention of Eq. (3.16) this state can now be compactly written as

$$|\psi\rangle = H^{\otimes 3}|000\rangle = \frac{1}{\sqrt{2^3}} \sum_{i=1}^{8} |i\rangle.$$

In general we have,

$$H^{\otimes n}|0^{\otimes n}\rangle = \frac{1}{\sqrt{2^n}} \sum_{i=1}^{2^n} |i\rangle.$$

The obvious generalisation of the operator U_f can be defined as

$$U_f |x \otimes z\rangle = |x \otimes [z \oplus f(x)]\rangle.$$

It is important to remark that in this more general case the symbol $\oplus$ denotes the operation mod 2 addition without 'carry over'. As an example, one might consider $1101 \oplus 0111 = 1010$.

Eventually, we are in the position to define the Deutsch-Josza problem [14]. We are given a black box quantum computer (oracle) that implements the function $f : \{0, 1\}^n \longrightarrow \{0, 1\}$. We are promised that the function is either constant (all outputs 0 or all outputs 1) or balanced (returns 1 for half the inputs and 0 for the other half). The task is to determine, whether the function is balanced or constant using the oracle. At worse a classical computer will have to evaluate the function $2^n/2 + 1$ times. The quantum algorithm suggested by Deutsch and Josza requires just one evaluation of the function f [14]. The corresponding quantum circuit can be seen in Fig. 3.9.

In the following we will analyse the Deutsch-Josza algorithm step by step.

i. The input state is $|\psi_0\rangle = |0\rangle^{\otimes n} \otimes |1\rangle$.
ii. After the application of the Hadamard gates on the input state $|\psi_0\rangle$ the state becomes

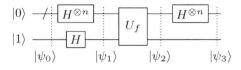

Fig. 3.9 The quantum circuit for the implementation of the Deutsch-Josza algorithm. The backslash symbol indicates that there are n quantum wires summarised as one

$$|\psi_1\rangle = \frac{1}{\sqrt{2^n}} \sum_{i=1}^{2^n} |i\rangle \left[\frac{1}{\sqrt{2}}[|0\rangle - |1\rangle] \right],$$

where we used the compact index notation introduced in Eq. (3.16).

iii. After the evaluation of the function the state evolves to

$$|\psi_2\rangle = \frac{1}{\sqrt{2^n}} \sum_{i=1}^{2^n} (-1)^{f(x)} |i\rangle \frac{1}{\sqrt{2}}[|0\rangle - |1\rangle]$$

iv. Now, we need to apply another Hadamard transformation to all qubits except the last. We start by considering the effect of the Hadamard transformation. For this we change from summing over an index to explicitly writing $i - 1 = x_1, \ldots, x_n$ or $z_1, \ldots, z_n$, where x_k, z_k are binary variables for all $k = 1, \ldots, n$. It is straightforward to check that

$$H^{\otimes n} |x_1, \ldots, x_n\rangle = \frac{1}{\sqrt{2^n}} \sum_{z_1, \ldots, z_n} (-1)^{x_1 z_1 + \cdots + x_n z_n} |z_1, \ldots, z_n\rangle$$

$$= \frac{1}{\sqrt{2^n}} \sum_z (-1)^{x \cdot z} |z\rangle,$$

where we have introduced the compact notation of bitwise inner product mod 2, i.e. $x \cdot z = x_1 z_1 + \ldots x_n z_n$. The third step of the Deutsch-Josza algorithm thus leads to the state

$$|\psi_3\rangle = \frac{1}{2^n} \sum_z \sum_x (-1)^{x \cdot z + f(x)} |z\rangle \frac{1}{\sqrt{2}}[|0\rangle - |1\rangle].$$

v. Lastly, we have to evaluate the probability of measuring the state $|0\rangle^{\otimes n}$

$$p(0 \ldots 0) = \left| \frac{1}{2^n} \sum_x (-1)^{f(x)} \right|^2 .$$

Of course, if $f(x) =$ constant, constructive interference leads to $p(0 \ldots 0) = 1$. Similarly, if the function f is balanced, destructive interference leads to $p(0 \ldots 0) = 0$.

3.3.3 Quantum Annealing and Other Computational Models

Although the circuit model of qubits and gates is by far the most common formalism, there are some other computational models that we want to mention briefly. These

3.3 An Example: The Deutsch-Josza Algorithm

have so far been shown to be equivalent up to a polynomial overhead, which means that efficient translations from one to the other exist.

Prominent in the quantum machine learning literature is a technique called *quantum annealing*, which can be understood as a heuristic to *adiabatic quantum computing*. Adiabatic quantum computing [23] is in a sense the analog version of quantum computing [24] in which the solution of a computational problem is encoded in the ground state (i.e. lowest energy state) of a Hamiltonian which defines the dynamics of a system of n qubits. Starting with a quantum system in the ground state of another Hamiltonian which is relatively simple to realise in a given experimental setup, and slowly adjusting the system so that it is governed by the desired Hamiltonian ensures that the system is afterwards found in the ground state. The Hamiltonian can for example be adjusted by changing the interaction strengths between the qubits.

It turns out that for many problems, to keep the system in the ground state during the adjustment (the 'annealing schedule') requires a very slow evolution from one to the other Hamiltonian, and often a time exponential in the problem size, which shows once more that nature seems to set some universal bounds for computation. Quantum annealing may be seen as a heuristic or 'shortcut' to the adiabatic algorithm whose dynamics work much like simulated annealing in computer science. The main difference between classical and quantum annealing is that thermal fluctuations are replaced by quantum fluctuations which enables the system to *tunnel* through high and thin energy barriers. The probability of quantum tunnelling decreases exponentially with the barrier width, but is independent of its height. That makes quantum annealing especially fit for problems with a sharply ragged objective function (see Fig. 3.10).

The great interest in quantum annealing by the quantum machine learning community has been driven by the fact that an annealing device was the first commercially

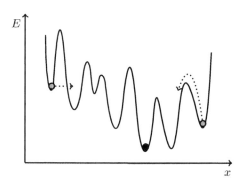

Fig. 3.10 Illustration of quantum annealing in an energy landscape over (here continuous) states x. The ground state is the configuration of lowest energy (black dot). Quantum tunnelling allows the system state to go through high and thin energy barriers (gray dot on the left), while in classical annealing techniques stochastic fluctuations have to be large enough to allow for jumps over peaks (gray dot on the right)

available hardware for quantum computing. Demonstrations of machine learning with the D-Wave quantum annealer[10] date back to as early as 2009 [25]. Current quantum annealers are limited to solving a special kind of problem, so called *quadratic unconstrained binary optimisation problems* which we will introduce in Chap. 7. Measuring the performance of quantum annealing compared to classical annealing schemes is a non-trivial problem, and although advantages of the quantum schemes have been demonstrated in the literature mentioned above, general statements about speed-ups and 'real quantum behaviour' are still controversial [26, 27].

Another famous quantum computational model is *one-way* or *measurement-based quantum computing*. The idea [28] is to prepare a highly entangled state called a *cluster state* and to perform a series of single-qubit measurements which conditionally depend on the output of former single-qubit measurements. This computation is of course not unitary. The result can be either the state of the unmeasured qubits, or the outcome of a final measurement [29]. Many important quantum algorithms have been implemented using one-way computation [30–33]. However, there are not many investigations into quantum machine learning based on this model, and it is an open question whether it offers a particularly suitable framework to approach learning tasks.

While quantum annealing and one-way quantum computing still refer to qubits as the basic computational unit, *continuous-variable quantum computing* [34, 35] shows promising features for quantum machine learning [36, 37]. Continuous variable systems encode information not in a discrete quantum system such as a qubit, but in a quantum state with a continuous basis. The Hilbert space of such a system is infinite-dimensional, which can potentially be leveraged to build machine learning models [38].

3.4 Strategies of Information Encoding

There are different ways to encode information into a n-qubit system described by a state (3.16), and we will introduce some relevant strategies in this section. While for data mining and machine learning the question of information encoding becomes central, this is not true for many other topics in quantum computing, which is presumably why quantum computing so far has no terminology for such strategies. We will therefore refer to them as *basis encoding, amplitude encoding, qsample encoding* and *dynamic encoding*. The encoding methods presented here will be explored in more detail in Chap. 5, where we look at them in relation to quantum machine learning algorithms.

An illustration of the different encoding methods can be found in Fig. 3.11 and a summary of the notation used here can be found in Table 3.5. It is interesting to note that many quantum algorithms - such as the matrix inversion routine introduced below—can be understood as strategies of transforming information from one kind of encoding to the other.

---

[10]http://www.dwavesys.com/.

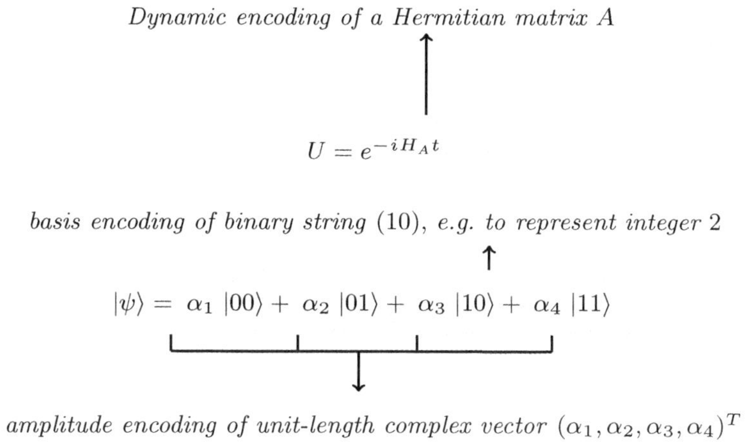

3.4 Strategies of Information Encoding

Dynamic encoding of a Hermitian matrix A
↑
$$U = e^{-iH_A t}$$

basis encoding of binary string (10), e.g. to represent integer 2
↑
$$|\psi\rangle = \alpha_1 |00\rangle + \alpha_2 |01\rangle + \alpha_3 |10\rangle + \alpha_4 |11\rangle$$
↓
amplitude encoding of unit-length complex vector $(\alpha_1, \alpha_2, \alpha_3, \alpha_4)^T$

Fig. 3.11 Illustration of the different encoding strategies for a quantum state of a 2-qubit system

Table 3.5 A summary of the different types of information encoding presented in the text, as well as their possible variations

Classical data	Properties	Quantum state
Basis encoding		
$(b_1, \ldots, b_d)$, $b_i \in \{0,1\}$	b encodes $x \in \mathbb{R}^N$ in binary	$|x\rangle = |b_1, \ldots, b_d\rangle$
Amplitude encoding		
$x \in \mathbb{R}^{2^n}$	$\sum_{i=1}^{2^n} |x_i|^2 = 1$	$|\psi_x\rangle = \sum_{i=1}^{2^n} x_i |i\rangle$
$A \in \mathbb{R}^{2^n \times 2^m}$	$\sum_{i=1}^{2^n} \sum_{j=1}^{2^m} |a_{ij}|^2 = 1$	$|\psi_A\rangle = \sum_{i=1}^{2^n} \sum_{j=1}^{2^m} a_{ij} |i\rangle |j\rangle$
$A \in \mathbb{R}^{2^n \times 2^n}$	$\sum_{i=1}^{2^n} a_{ii} = 1$, $a_{ij} = a_{ji}^*$, A pos.	$\rho_A = \sum_{ij} a_{ij} |i\rangle\langle j|$
Qsample encoding		
$p(x)$, $x \in \{0,1\}^{\otimes n}$	$\sum_x p(x) = 1$	$\sum_x \sqrt{p(x)} |x\rangle$
Dynamic encoding		
$A \in \mathbb{R}^{2^n \times 2^n}$	A unitary	U_A with $U_A = A$
$A \in \mathbb{R}^{2^n \times 2^n}$	A Hermitian	H_A with $H_A = A$
$A \in \mathbb{R}^{2^n \times 2^n}$	–	$H_{\tilde{A}}$ with $\tilde{A} = \begin{pmatrix} 0 & A \\ A^\dagger & 0 \end{pmatrix}$

3.4.1 Basis Encoding

Basis encoding associates a computational basis state of a n-qubit system (such as $|3\rangle = |0011\rangle$) with a classical n-bit-string (0011). In a way, this is the most straight forward way of computation, since each bit gets literally replaced by a qubit, and a 'computation' acts in parallel on all bit sequences in a superposition. We have used basis encoding in the algorithms investigated so far in this chapter.

The value of the amplitudes α_i of each basis state does not carry any other information than to 'mark' the result of the computation with a high enough probability of being measured. For example, if the basis state $|0011\rangle$ with amplitude α_{0011} has a probability $|\alpha_{0011}|^2 > 0.5$ of being measured, repeated execution of the algorithm and measurement of the final state in the computational basis will reveal it as the most likely measurement result, and hence the overall result of the algorithm. For the basis encoding method, the goal of a quantum algorithm is therefore to increase the probability or absolute square of the amplitude that corresponds to the basis state encoding the solution.

Like in classical computers, basis encoding uses a binary representation of numbers. A quantum state $|x\rangle$ with $x \in \mathbb{R}$ will therefore refer to a binary representation of x with the number n of bits that the qubit register encoding $|x\rangle$ provides. There are different ways to represent a real number in binary form, for example by fixed or floating point representations. In the following it is always assumed that such a strategy is given, and we will elaborate more on this point where the need arises.

Example 3.4 (*Basis encoding*) Let us choose a binary fraction representation, where each number in the interval [0, 1) is represented by a τ-bit string according to

$$x = \sum_{k=1}^{\tau} b_k \frac{1}{2^k}. \tag{3.28}$$

To encode a vector $x = (0.1, -0.6, 1.0)$ in basis encoding, we have to first translate it into a binary sequence, where we choose a binary fraction representation with precision $\tau = 4$ and the first bit encoding the sign,

$$0.1 \to 0\ 0001\ldots$$
$$-0.6 \to 1\ 1001\ldots$$
$$1.0 \to 0\ 1111\ldots$$

The vector therefore corresponds to a binary sequence $b = 00001\ 11001\ 01111$ and can be represented by the quantum state $|00001\ 11001\ 01111\rangle$.

3.4.2 Amplitude Encoding

Amplitude encoding is much less common in quantum computing, as it associates classical information such as a real vector with quantum amplitudes, and there are different options to do so. A normalised classical vector $x \in \mathbb{C}^{2^n}$, $\sum_k |x_k|^2 = 1$ can be represented by the amplitudes of a quantum state $|\psi\rangle \in \mathcal{H}$ via

3.4 Strategies of Information Encoding

$$x = \begin{pmatrix} x_1 \\ \vdots \\ x_{2^n} \end{pmatrix} \leftrightarrow |\psi_x\rangle = \sum_{j=1}^{2^n} x_j |j\rangle.$$

In the same fashion a classical matrix $A \in \mathbb{C}^{2^n \times 2^m}$ with entries a_{ij} that fulfil $\sum_{ij} |a_{ij}|^2 = 1$, can be encoded as

$$|\psi_A\rangle = \sum_{i=1}^{2^m} \sum_{j=1}^{2^n} a_{ij} |i\rangle |j\rangle$$

by enlarging the Hilbert space accordingly. This turns out to be rather useful for quantum algorithms, since the 'index registers' $|i\rangle, |j\rangle$ refer to the ith row and the jth column respectively. By fixing either of the register we can therefore address a row or column of the matrix. For Hermitian positive trace-1 matrices $A \in \mathbb{C}^{2^n \times 2^n}$, another option arises: One can associate its entries with the entries of a density matrix ρ_A, so that $a_{ij} \leftrightarrow \rho_{ij}$.

Encoding information into the probabilistic description of a quantum system necessarily poses severe limitations on which operations can be executed. This becomes particularly important when we want to perform a nonlinear map on the amplitudes, which is impossible to implement in a unitary fashion. This has been extensively debated under the keyword of nonlinear quantum theories [39, 40] and it has been demonstrated that assumptions of nonlinear operators would immediately negate fundamental principles of nature that are believed to be true [41, 42].

Another obvious restriction of this method is that only normalised classical vectors can be processed. Effectively this means that quantum states represent the data in one less dimension or with one less degree of freedom: A classical two dimensional vector (x_1, x_2) can only be associated with an amplitude vector (α_1, α_2) of a qubit which fulfils $|\alpha_1|^2 + |\alpha_2|^2 = 1$. This means that it lies on a unit circle, a one-dimensional shape in two dimensional space. Three-dimensional vectors encoded in three amplitudes of a 2-qubit quantum system (where the last of the four amplitudes is redundant and set to zero) will reduce the 3-dimensional space to the surface of a sphere, and so on. A remedy can be to increase the space of the classical vector by one dimension $x_{N+1} = 1$ and normalise the resulting vector. The N-dimensional space will then be embedded in a $N + 1$-dimensional space in which the data is normalised without loss of information (Fig. 3.12).

Example 3.5 (*Amplitude encoding*) To encode the same vector from Example 3.4, $x = (0.1, -0.6, 1.0)$, in amplitude encoding, we have to first normalise it to unit length (rounding to three digits here) and pad it with zeros to a dimension of integer logarithm,

$$x' = (0.073, -0.438, 0.730, 0.000).$$

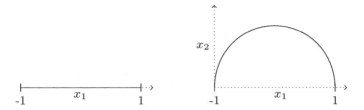

Fig. 3.12 Data points in the one-dimensional interval $[-1, 1]$ (left) can be projected onto normalised vectors by adding a constant value in a second dimension x_2 and renormalising

Now it can be represented by a quantum state of 2 qubits:

$$0.073|00\rangle - 0.438|01\rangle + 0.730|10\rangle + 0|11\rangle.$$

This state would at the same time encode the matrix

$$A = \begin{pmatrix} 0.073 & -0.438 \\ 0.730 & 0.000 \end{pmatrix},$$

if we interpret the first qubit as an index for the row and the second qubit as an index for the column.

3.4.3 Qsample Encoding

Consider an n-qubit quantum state described by an amplitude vector $\alpha = (\alpha_1, \ldots, \alpha_{2^n})^T$. Measuring the n qubits in the computational basis is equivalent to sampling a bit string of length n from a discrete probability distribution $p_1 = |\alpha_1|^2, \ldots, p_{2^n} = |\alpha_{2^n}|^2$. By 'equivalent' we mean that both the measurement and the classical sampling routine draw samples with the same probability. This means that we can use the amplitude vector to represent a classical discrete probability distribution. Given a classical discrete probability distribution over binary strings $p_1, \ldots, p_{2^n}$, the quantum state

$$|\psi\rangle = \sum_{i=1}^{2^n} \sqrt{p_i}|i\rangle$$

is sometimes referred to as a *qsample* [43, 44], a term that we will apply here as well.

3.4 Strategies of Information Encoding

3.4.4 Dynamic Encoding

For some applications it can be useful to encode matrices into the dynamic evolution, for example into unitary operators. While unitary operators obviously restrict the class of matrices that they can represent, a useful option from [45] is to associate a Hamiltonian H with a square matrix A. In case A is not Hermitian, one can sometimes use the trick of instead encoding

$$\tilde{A} = \begin{pmatrix} 0 & A \\ A^\dagger & 0 \end{pmatrix}, \tag{3.29}$$

and only considering part of the output. This way, the eigenvalues of A can be processed in a quantum routine, for example to multiply A or A^{-1} with an amplitude encoded vector. Table 3.5 summarises the different strategies of information encoding mentioned here.

Example 3.6 (*Hamiltonian encoding*) The matrix

$$A = \begin{pmatrix} 0.073 & -0.438 \\ 0.730 & 0.000 \end{pmatrix}$$

from Example 3.5 can also define the dynamics of a quantum system. Since A is not Hermitian we define the Hamiltonian

$$H = \begin{pmatrix} 0 & 0 & 0.073 & -0.438 \\ 0 & 0 & 0.730 & 0.000 \\ 0.073 & 0.730 & 0 & 0 \\ -0.438 & 0.000 & 0 & 0 \end{pmatrix}$$

The effect of applying the Hamiltonian via e^{-iHt} to an amplitude vector α can be calculated from the corresponding eigenvalue equations. The eigenvectors/values of H are

$$v^1 = \begin{pmatrix} -0.108 & -0.699 & 0.704 & -0.065 \end{pmatrix}, \quad \lambda_1 = -0.736$$
$$v^2 = \begin{pmatrix} 0.699 & -0.108 & 0.065 & 0.704 \end{pmatrix}, \quad \lambda_2 = -0.435$$
$$v^3 = \begin{pmatrix} 0.699 & -0.108 & -0.065 & -0.704 \end{pmatrix}, \quad \lambda_3 = 0.435$$
$$v^4 = \begin{pmatrix} 0.108 & 0.699 & 0.704 & -0.064 \end{pmatrix}, \quad \lambda_4 = 0.736.$$

These eigenvectors form a basis and hence α can be decomposed in this basis as

$$\alpha = \gamma_1 v^2 + \gamma_2 v^2 + \gamma_3 v^3 + \gamma_4 v^4,$$

with $\gamma_i = \alpha^T v^i$ for $i = 1, \ldots, 4$. Applying the Hamiltonian therefore leads to

$$\psi = e^{-iHt}\alpha = e^{-i\lambda_1 t}\gamma_1 v^1 + e^{-i\lambda_2 t}\gamma_2 v^2 + e^{-i\lambda_3 t}\gamma_3 v^3 + e^{-i\lambda_4 t}\gamma_4 v^4,$$

or, in Dirac notation,

$$\begin{aligned}|\psi\rangle = &(-0.108e^{-i\lambda_1 t}\gamma_1 + 0.699e^{-i\lambda_2 t}\gamma_2 + 0.699e^{-i\lambda_3 t}\gamma_3 + 0.108e^{-i\lambda_4 t}\gamma_4)|00\rangle \\ &+ (-0.699e^{-i\lambda_1 t}\gamma_1 - 0.108e^{-i\lambda_2 t}\gamma_2 - 0.108e^{-i\lambda_3 t}\gamma_3 + 0.699e^{-i\lambda_4 t}\gamma_4)|01\rangle \\ &+ (0.704e^{-i\lambda_1 t}\gamma_1 + 0.065e^{-i\lambda_2 t}\gamma_2 - 0.065e^{-i\lambda_3 t}\gamma_3 + 0.704e^{-i\lambda_4 t}\gamma_4)|10\rangle \\ &+ (-0.065e^{-i\lambda_1 t}\gamma_1 + 0.704e^{-i\lambda_2 t}\gamma_2 - 0.704e^{-i\lambda_3 t}\gamma_3 - 0.064e^{-i\lambda_4 t}\gamma_4)|10\rangle\end{aligned}$$

3.5 Important Quantum Routines

The quantum machine learning algorithms in later chapters will make use of some well-known (and a few less well-known) quantum routines that shall be introduced briefly here. Also, we will discuss the asymptotic complexity - the growth of the number of elementary operations needed with the input - of the routines, a concept which will be introduced more rigorously in Sect. 4.1.

3.5.1 Grover Search

Grover's algorithm (here used synonymously with the term *amplitude amplification*) is a quantum routine that finds one or multiple entries in an unstructured (i.e., arbitrarily sorted) database of N entries in basis encoding, a task that on classical computers takes N operations at worst and $N/2$ on average. More generally, it is a routine that given a quantum state in superposition increases the amplitude of some desired basis states, which is a crucial tool for quantum computing. To illustrate this, imagine one had a 3-qubit register in uniform superposition that serves as an index register, joint with a 'flag' ancilla qubit in the ground state as well as the database entries e_i in basis encoding,

$$\begin{aligned}|\psi\rangle = &\alpha_1|000\rangle|0\rangle|e_1\rangle \\ &+ \alpha_2|001\rangle|0\rangle|e_2\rangle \\ &+ \alpha_3|010\rangle|0\rangle|e_3\rangle \\ &+ \alpha_4|011\rangle|0\rangle|e_4\rangle \\ &+ \alpha_5|100\rangle|0\rangle|e_5\rangle \\ &+ \alpha_6|101\rangle|0\rangle|e_6\rangle \\ &+ \alpha_7|110\rangle|0\rangle|e_7\rangle \\ &+ \alpha_8|111\rangle|0\rangle|e_8\rangle.\end{aligned}$$

3.5 Important Quantum Routines

Further, assume that there was a known quantum algorithm that 'marks' the desired output $|011\rangle$ of the computation by setting the ancilla to 1. This could be a quantum version of a classical routine that analyses an entry and flags it if it is recognised as the correct one, with the typical quantum property of applying it in parallel to an exponential amount of entries. The result is state

$$\begin{aligned}|\psi'\rangle = &\, \alpha_1|000\rangle|0\rangle|e_1\rangle \\ &+ \alpha_2|001\rangle|0\rangle|e_2\rangle \\ &+ \alpha_3|010\rangle|0\rangle|e_3\rangle \\ &+ \alpha_4|011\rangle|\mathbf{1}\rangle|e_4\rangle \\ &+ \alpha_5|100\rangle|0\rangle|e_5\rangle \\ &+ \alpha_6|101\rangle|0\rangle|e_6\rangle \\ &+ \alpha_7|110\rangle|0\rangle|e_7\rangle \\ &+ \alpha_8|111\rangle|0\rangle|e_8\rangle.\end{aligned}$$

Grover search is an iterative quantum algorithm that increases the desired amplitude α_3 so that $|\alpha_3|^2 \approx 1$ and a measurement reveals the result of the computation. It turns out that in order to increase the amplitude, one requires $\sqrt{N}$ iterations (where $N = 2^n$ is the number of basis states or amplitudes over which to 'search') and that this is a lower bound for quantum algorithms for this kind of task [46, 47]. In other words, for search in unstructured databases (a very generic search and optimisation problem) we will not see an exponential speed-up of quantum computers as Grover search is optimal. This is not surprising; if amplitude amplification could be done exponentially faster, we could answer the question whether there is a right solution and solve NP-hard problems at a wimp.[11]

One iteration of the Grover routine consists of the following steps:

1. Mark the desired state using the oracle and multiply the amplitude by -1
2. Apply a Hadamard transform on the index qubits
3. Apply a phase shift of -1 on every computational basis state but the first one ($|0\ldots 0\rangle$)
4. Apply a Hadamard transform on the index qubits.

The third step implements an operator $2|0\rangle\langle 0| - \mathbb{1}$ and together with the Hadamards an 'inversion about the average' is performed. Steps 2–4 make up the so called 'Grover operator' which alters each amplitude according to

$$\alpha_i \to -\alpha_i + 2\bar{\alpha},$$

---

[11] The quadratic speed-up seems to be intrinsic in the structure of quantum probabilities, and derives from the fact that one has to square amplitudes in order to get probabilities.

where $\bar{\alpha} = \frac{1}{N}\sum_i \alpha_i$ is the average of all amplitudes. After a number of steps proportional to $\sqrt{N}$ the probability of measuring the state of the marked amplitude is maximised if only one state has been marked. For several marked states the final probability depends on their number B and the optimal result (probability of one to measure one of the marked states) will be achieved after a number of steps proportional to $\sqrt{N/B}$.

A helpful analysis and extension of Grover search for quantum superpositions that are not uniform is given in [48] and shows that the distance of unmarked states to the average of all unmarked states remains constant in every iteration, and the same is true for marked states. In other words, the average gets shifted periodically. Also, the study finds that although one can show that the optimal probability is obtained after $\sqrt{N/B}$ iterations, the value of that probability can vary considerably depending on the initial distribution of the amplitudes. One therefore needs to pay attention when working with non-uniform initial distributions. If the number of states that are searched for is not known one can use the technique of *quantum counting* to get an estimate [49].

Example 3.7 (*Grover search*) Let us have a look at Grover search for two qubits. We start with a uniform superposition of four states with the aim of finding the third one

$$0.5|00\rangle + 0.5|01\rangle + 0.5|10\rangle + 0.5|11\rangle.$$

Step 1 in the first iteration marks the basis state

$$0.5|00\rangle + 0.5|01\rangle - 0.5|10\rangle + 0.5|11\rangle.$$

Steps 2–4 in the first iteration are the reflection yielding

$$|10\rangle.$$

Steps 1–4 of the second iteration result in:

$$-0.5|00\rangle - 0.5|01\rangle + 0.5|10\rangle - 0.5|11\rangle,$$

and Steps 1–4 of the third iteration yield:

$$-0.5|00\rangle - 0.5|01\rangle - 0.5|10\rangle - 0.5|11\rangle,$$

from which the entire cycle starts up to a sign and ends in the uniform superposition in three further steps. One finds that it takes one iteration until measuring the desired state has unit probability, a situation which is revisited after 2 cycles. Amplitude amplification has periodic dynamics.

3.5.2 Quantum Phase Estimation

Quantum phase estimation is a routine that writes information encoded in the phase φ of an amplitude $\alpha = |\alpha|e^{i\varphi}$ into a basis state by making use of the quantum Fourier transform. It is used extensively in quantum machine learning algorithms to extract eigenvalues of operators that contain information about a training set.

3.5.2.1 Discrete Fourier Transform

The quantum Fourier transform implements a discrete Fourier transform on the values of the amplitudes. As a reminder, the classical version of the Fourier transform maps a real vector $x \in \mathbb{R}^{2^n}$ to another vector $y \in \mathbb{R}^{2^n}$ via

$$y_k = \frac{1}{\sqrt{2^n}} \sum_{j=1}^{2^n} e^{\frac{2\pi i j k}{2^n}} x_j \quad k = 1, \ldots, 2^n, \tag{3.30}$$

using $\mathcal{O}(n 2^n)$ steps. The quantum version maps a quantum state with amplitudes encoding the x_k to a quantum state whose amplitudes encode the y_k,

$$\sum_{j=1}^{2^n} x_j |j\rangle \to \sum_{k=1}^{2^n} \underbrace{\frac{1}{\sqrt{2^n}} \left[\sum_{j}^{2^n} e^{\frac{2\pi i j k}{2^n}} x_j \right]}_{y_k} |k\rangle,$$

in only $\mathcal{O}(n^2)$ steps. We need $n = m + \lceil \log(2 + \frac{1}{2\epsilon}) \rceil$ qubits to get the first m bits of the binary string encoding the eigenstate with probability $1 - \epsilon$. A quantum Fourier transform applied to a state that is only in one computational basis state $|j\rangle$ (which corresponds to the classical case in which all but one x_j are zero) makes the sum over the j vanish and leaves

$$|j\rangle \to \sum_{k=1}^{2^n} \underbrace{\frac{1}{\sqrt{2^n}} \left[\sum_{j=1}^{2^n} e^{\frac{2\pi i j k}{2^n}} x_j \right]}_{y_k} |k\rangle. \tag{3.31}$$

In the quantum phase estimation procedure, the reverse of this transformation will be used.

3.5.2.2 Estimating Phases

Given a unitary operator acting on n qubits, which, applied to $|\phi\rangle$ will reveal $U|\phi\rangle = e^{2\pi i \varphi}|\phi\rangle$ with $\varphi \in [0, 1)$. The goal is to get an estimate of φ [50].

In order to use the quantum Fourier transform, one needs a unitary transformation that can implement powers of U conditioned on another index register of ν qubits,

$$\sum_{k=1}^{2^\nu} |k\rangle|\phi\rangle \to \sum_{k=1}^{2^\nu} |k\rangle U^{(k-1)}|\phi\rangle.$$

The $(k-1)$ in the exponent derives from our convention to denote by $|k\rangle$ the kth computational basis state starting with $k=1$. Of course, this routine effectively applies U for $k-1$ times, and unless U reveals some special structure, to stay efficient, k always has to be of the order or smaller than $n+\nu$. The routine to implement powers of U always has to be given with the algorithm using quantum phase estimation. Altogether, the transformation reads

$$\frac{1}{\sqrt{\nu}}\sum_{k=1}^{2^\nu} |k\rangle|\phi\rangle \to \frac{1}{\sqrt{2^\nu}}\sum_{k=1}^{2^\nu} |k\rangle U^{(k-1)}|\phi\rangle,$$

$$= \frac{1}{\sqrt{2^\nu}}\sum_{k=1}^{2^\nu} |k\rangle (e^{2\pi i (k-1)\varphi}|\phi\rangle),$$

$$= \frac{1}{\sqrt{2^\nu}}\sum_{k=1}^{2^\nu} e^{2\pi i (k-1)\varphi}|k\rangle|\phi\rangle.$$

Note that the $|\phi\rangle$ register is not entangled with the rest of the state and can therefore be discarded without any effect. The next step is to apply the inverse quantum Fourier transform on the remaining index register. Consider first the case that $\varphi = \frac{j}{2^\nu}$ for an integer j, or in other words, j can exactly be represented by ν binary digits. The inverse quantum Fourier transform from Eq. (3.31) can be applied and leaves us with state $|j\rangle$. More precisely, $j = b_1 2^0 + \cdots + b_\nu \frac{1}{2^{(\nu-1)}}$ has a ν-bit binary representation which is at the same time the binary fraction representation of $\varphi = b_1 \frac{1}{2^1} + \cdots + b_s \frac{1}{2^\nu}$ (see Eq. 3.28).

In general, $\varphi \neq \frac{j}{2^\nu}$, and the inverse quantum Fourier transform will result in

$$\frac{1}{\sqrt{2^\nu}}\sum_{k=1}^{2^\nu} e^{2\pi i (k-1)(\varphi - \frac{j}{2^\nu})}|i\rangle|\phi\rangle.$$

The probability distribution over the computational basis states

$$p_j = \left| e^{2\pi i (k-1)(\varphi - \frac{j}{2^\nu})} \right|^2,$$

depends on the difference between φ and the binary fraction representation of integer j. The more accurate or 'close' the binary fraction representation corresponding to integer j is, the higher the probability of measuring the basis state $|j\rangle$ that represents φ [50]. In a sense, the result is therefore a distribution over different binary representations of φ, and the smaller our error by representing it with n bits is, the narrower the variance of the distribution around the correct representation. This is what is meant when saying that the resulting state of the quantum phase estimation algorithm 'approximates' the phase. The resources needed for quantum phase estimation are the resources to implement powers of U as well as those to implement the quantum Fourier transform.

3.5.3 Matrix Multiplication and Inversion

Quantum phase estimation can be used to multiply a matrix $A \in \mathbb{R}^{N \times N}$ to a vector $x \in \mathbb{R}^N$ in amplitude encoding, and with a procedure to invert eigenvalues, this can be extended to matrix inversion. This is a rather technical quantum routine which is at the heart of one approach to quantum machine learning. To understand the basic idea, it is illuminating to write Ax in terms of A's eigenvalues λ_r and eigenvectors $v_r, r = 1, \ldots, R$. These fulfil the equations

$$A v_r = \lambda_r v_r.$$

The vector x can be written as a linear combination of the eigenvectors of A,

$$x = \sum_{r=1}^{R} (v_r^T x) v_r. \tag{3.32}$$

Applying A to x leads to

$$Ax = \sum_{r=1}^{R} \lambda_r (v_r^T x) v_r. \tag{3.33}$$

and applying A^{-1} yields

$$A^{-1} x = \sum_{r=1}^{R} \lambda_r^{-1} (v_r^T x) v_r. \tag{3.34}$$

Under certain conditions, quantum algorithms can find eigenvalues and eigenstates of unitary operators exponentially fast [51]. This promises to be a powerful tool, but one will see that it can only used for specific problems. We will introduce the matrix multiplication algorithm and then mention how to adapt it slightly in order to do matrix inversion as proposed in [45] (also called the *quantum linear systems of equations* routine or "HHL" after the seminal paper by Harrow, Hassidim and Lloyd [45]).

Consider a quantum state $|\psi_x\rangle$ that represents a normalised classical vector in amplitude encoding. The goal is to map this quantum state to a normalised representation of Ax with

$$|\psi_x\rangle = \sum_{r=1}^{R} \langle \psi_{v_r}|\psi_x\rangle |\psi_{v_r}\rangle \to |\psi_{Ax}\rangle = \sum_{r=1}^{R} \lambda_r \langle \psi_{v_r}|\psi_x\rangle |\psi_{v_r}\rangle.$$

This can be done in three steps. First, create a unitary operator $U = e^{2\pi i H_A}$ which encodes A in Hamiltonian encoding, and apply it to $|\psi_x\rangle$. Second, execute the phase estimation procedure to write the eigenvalues of A into a register in basis encoding. Third, use a small subroutine to translate the eigenvalues from basis encoding into amplitude encoding.

1. Simulating A

In the first step, we need a unitary operator U whose eigenvalue equations are given by $U|\psi_{v_r}\rangle = e^{2\pi i \lambda_r}|\psi_{v_r}\rangle$. If one can evolve a system with a Hamiltonian H_A encoding the matrix A, the operator U is given by $e^{2\pi i H_A}$. To resemble a Hamiltonian, A has to be Hermitian, but the trick from Eq. (3.29) circumvents this requirement by doubling the Hilbert with one additional qubit. Techniques to implement general H are called *Hamiltonian simulation* and are discussed in Sect. 5.4.

The first step prepares a quantum state of the form

$$\frac{1}{\sqrt{K}} \sum_{k=1}^{K} |k\rangle e^{-2\pi i (k-1) H_A} |\psi_x\rangle = \frac{1}{\sqrt{K}} \sum_{k=1}^{K} |k\rangle \sum_{r=1}^{R} e^{2\pi i (k-1) \lambda_r} \langle \psi_{v_r}|\psi_x\rangle |\psi_{v_r}\rangle,$$

where on the right side, $|\psi_x\rangle$ was simply expressed in A's basis as defined in Eq. (3.32), and $K = 2^\nu$ where ν is again the number of qubits in the index register. This is a slight simplification of the original proposal, in which for example the index register $|k\rangle$ is not in a uniform superposition to exploit some further advantages, but the principle remains the same.

2. Extracting the eigenvalues

In the second step, the quantum phase estimation routine is applied to the index register $|k\rangle$ to 'reduce' its superposition to basis states encoding the eigenvalues,

$$\frac{1}{\sqrt{K}} \sum_{r=1}^{R} \sum_{k=1}^{K} \alpha_{k|r} \langle \psi_{v_r}|\psi_x\rangle |k\rangle |\psi_{v_r}\rangle. \tag{3.35}$$

As explained for the quantum phase estimation routine, the coefficients lead to a large probability $|\alpha_{k|r}|^2$ for computational basis states $|k\rangle$ that approximate the eigenvalues λ_r well. If enough qubits ν are given in the $|k\rangle$ register, one can assume that (3.35) is approximately

$$\sum_{r=1}^{R} \langle \psi_{v_r}|\psi_x\rangle |\lambda_r\rangle |\psi_{v_r}\rangle,$$

where $|\lambda_r\rangle$ basis encodes a ν qubit approximation of λ_r. The time needed to implement this step with an error ϵ in the final state is in $\mathcal{O}(\frac{1}{\epsilon})$ [45].

3.5 Important Quantum Routines

3. Adjusting the amplitudes

The third step 'writes' the eigenvalues into the amplitudes by using a technique we will present in more detail in Sect. 5.2.2.3. The idea is to apply a sequence of gates which rotates an additional ancilla qubit as

$$\sum_{r=1}^{R} \langle \psi_{v_r} | \psi_x \rangle | \lambda_r \rangle | \psi_{v_r} \rangle | 0 \rangle \to \sum_{r=1}^{R} \langle \psi_{v_r} | \psi_x \rangle | \lambda_r \rangle | \psi_{v_r} \rangle (\sqrt{1 - \lambda_r^2} | 0 \rangle + \lambda_r | 1 \rangle).$$

We then do a *postselective measurement*, which means that the ancilla is measured, and the algorithm only continues if we measure it in state $|1\rangle$. We will call this procedure *branch selection*. This makes sure we have the state

$$\sum_{r=1}^{R} \lambda_r \langle \psi_{v_r} | \psi_x \rangle | \lambda_r \rangle | \psi_{v_r} \rangle | 0 \rangle \to \sum_{r=1}^{R} \lambda_r \langle \psi_{v_r} | \psi_x \rangle | \psi_{v_r} \rangle,$$

where on the right side the eigenvalue register was 'uncomputed' with an inverse quantum phase estimation algorithm, and the ancilla discarded. Note that this is only possible because no terms in the sum interfere as a result (one could say that the superposition is 'kept intact' by the $|\psi_{v_r}\rangle$ state). The final state $|\psi_{Ax}\rangle$ corresponds to a normalised version of Ax in amplitude encoding.

For matrix inversion, the last step is slightly adjusted: When conditionally rotating the ancilla qubit, one writes the inverse of the eigenvalue $\frac{1}{\lambda_r}$ into the respective amplitude. Since there is no guarantee that these are smaller than 1, a normalisation constant C has to be introduced that is of the order of the smallest eigenvalue, and the conditional measurement yields up to normalisation factors

$$\sum_{r=1}^{R} \langle \psi_{v_r} | \psi_x \rangle | \tilde{\lambda}_r \rangle | \psi_{v_r} \rangle (\sqrt{1 - \frac{C^2}{\lambda_r^2}} | 0 \rangle + \frac{C}{\lambda_r} | 1 \rangle) \to C \sum_{r=1}^{R} \frac{1}{\lambda_r} \langle \psi_{v_r} | \psi_x \rangle | \psi_{v_r} \rangle.$$

The conditional measurement is a non-unitary operation and requires the routine to be repeated on average $\mathcal{O}(\frac{1}{p_{\text{acc}}})$ times until it succeeds. For matrix multiplication, the acceptance probability is given by $p_{\text{acc}} = \sum_r |\langle \psi_{v_r} | \psi_x \rangle|^2 \lambda_r^2$ while for the inversion technique $p_{\text{acc}} = \sum_r |\langle \psi_{v_r} | \psi_x \rangle|^2 \frac{C^2}{\lambda_r^2} \leq \kappa^{-2}$, where κ is the condition number of the matrix defined as the ratio of the largest and the smallest eigenvalue. The condition number is a measure of how 'invertible' or 'singular' a matrix is. Just like when considering numerical stability in classical inversion techniques, when the condition number is large, the quantum algorithm takes a long time to succeed on average. This can sometimes be circumvented with so called preconditioning techniques [52].

Example 3.8 (*Simulation of the quantum matrix inversion routine*) Since the matrix inversion routine is rather technical, this example shall illuminate how a quantum state gets successively manipulated by the routine. The matrix A and vector b considered here are given by

$$A = \begin{pmatrix} \frac{2}{3} & \frac{1}{3} \\ \frac{1}{3} & \frac{2}{3} \end{pmatrix}, \quad b = \begin{pmatrix} 0.275 \\ 0.966 \end{pmatrix}.$$

The eigenvalues of A are $\lambda_1 = 1$ and $\lambda_2 = 1/3$, with the corresponding eigenvectors $v_1 = (1/\sqrt{2}, 1/\sqrt{2})^T$ and $v_2 = (-1/\sqrt{2}, 1/\sqrt{2})^T$. The number of qubits for the eigenvalue register is chosen as $\tau = 10$, and the binary representation of the eigenvalues then becomes $\lambda_1 = 1111111111$ and $\lambda_2 = 0101010101$. The error of this representation is < 0.001. The constant C is chosen to be half the smallest eigenvalue, $C = 0.5\lambda_{\min} = \frac{1}{6}$.

The qubits are divided into three registers: The first qubit is the ancilla used for the postselection, the following ten qubits form the eigenvalue register, and the last qubit initially encodes the vector b, while after the routine it will encode the solution x.

Encoding b into the last qubit yields

$$|\psi\rangle = 0.275|0\ 0000000000\ 0\rangle + 0.966|0\ 0000000000\ 1\rangle.$$

After simulating A and writing the eigenvalues into the second register via phase estimation in Step 2 we get

$$\begin{aligned}|\psi\rangle = &-0.343|0\ 0101010101\ 0\rangle \\ &+ 0.343|0\ 0101010101\ 1\rangle \\ &+ 0.618|0\ 1111111111\ 0\rangle \\ &+ 0.618|0\ 1111111111\ 1\rangle.\end{aligned}$$

Rotating the ancilla conditioned on the eigenvalue register leads to

$$\begin{aligned}|\psi\rangle = &-0.297|0\ 0101010101\ 0\rangle \\ &+ 0.297|0\ 0101010101\ 1\rangle \\ &+ 0.609|0\ 1111111111\ 0\rangle \\ &+ 0.609|0\ 1111111111\ 1\rangle \\ &- 0.172|1\ 0101010101\ 0\rangle \\ &+ 0.172|1\ 0101010101\ 1\rangle \\ &+ 0.103|1\ 1111111111\ 0\rangle \\ &+ 0.103|1\ 1111111111\ 1\rangle,\end{aligned}$$

and uncomputing (reversing the computation) in the eigenvalue register prepares the state

3.5 Important Quantum Routines

$$|\psi\rangle = 0.312|0\ 0000000000\ 0\rangle$$
$$+ 0.907|0\ 0000000000\ 1\rangle$$
$$- 0.069|1\ 0000000000\ 0\rangle$$
$$+ 0.275|1\ 0000000000\ 1\rangle$$

After a successful conditional measurement of the ancilla in 1 we get

$$|\psi\rangle = -0.242|1\ 0000000000\ 0\rangle + 0.970|1\ 0000000000\ 1\rangle.$$

The probability of success is given by

$$p(1) = (-0.069)^2 + 0.275^2 = 0.080.$$

The amplitudes now encode the result of the computation, $A^{-1}b$, as a normalised vector. The final, renormalised result of the quantum algorithm can be extracted by taking the last amplitude vector $(-0.242, 0.970)^T$ and multiplying it with the renormalisation factor $C/\sqrt{p(1)} = 0.588$. Note that we can estimate $p(1)$ from the conditional measurement of the algorithm, and C is our choice, so that the 'classical' renormalisation can always be accomplished if needed. In applications we would be interested in extracting some compact information from the final quantum state, rather than measuring each amplitude.

The result after renormalisation is $x = (-0.412, 1.648)^T$. A quick calculation by hand or using a linear algebra library confirms that this is the correct outcome up to an error from the finite bit precision used in the calculation.

The Dirac notation shows beautifully how the routine starts with a small superposition (here of only two basis states) that gets 'blown up' and then again reduced to encode the two-dimensional output.

References

1. Dirac, P.A.M.: The Principles of Quantum Mechanics. Clarendon Press (1958)
2. Gerthsen, K., Vogel, H.: Physik. Springer (2013)
3. Aaronson, S.: Quantum Computing Since Democritus. Cambridge University Press (2013)
4. Leifer, M.S., Poulin, D.: Quantum graphical models and belief propagation. Ann. Phys. **323**(8), 1899–1946 (2008)
5. Landau, L., Lifshitz, E.: Quantum mechanics: non-relativistic theory. In: Course of Theoretical Physics, vol. 3. Butterworth-Heinemann (2005)
6. Whitaker, A.: Einstein, Bohr and the Quantum Dilemma: From Quantum Theory to Quantum Information. Cambridge University Press (2006)
7. Einstein, A.: Über die von der molekularkinetischen Theorie der Wärme geforderte Bewegung von in ruhenden Flüssigkeiten suspendierten Teilchen. Annalen der Physik **322**(8), 549–560 (1905)
8. Einstein, A.: Zur Elektrodynamik bewegter Körper. Annalen der Physik **322**(10), 891–921 (1905)

9. Wilce, A.: Quantum logic and probability theory. In: Zalta, E.N. (ed.), The Stanford Encyclopedia of Philosophy. Online encyclopedia (2012). http://plato.stanford.edu/archives/fall2012/entries/qt-quantlog/
10. Rédei, M., Summers, S.J.: Quantum probability theory. Studies in Hist. Phil. Sci. Part B Stud. Hist. Phil. Modern Phys. **38**(2), 390–417 (2007)
11. Breuer, H.-P., Petruccione, F.: The Theory of Open Quantum Systems. Oxford University Press (2002)
12. Denil, M., De Freitas, N.: Toward the implementation of a quantum RBM. In: NIPS 2011 Deep Learning and Unsupervised Feature Learning Workshop (2011)
13. Grover, L.K.: A fast quantum mechanical algorithm for database search. In: Proceedings of the Twenty-Eighth Annual ACM Symposium on Theory of Computing, pp. 212–219. ACM (1996)
14. Deutsch, D., Jozsa, R.: Rapid solution of problems by quantum computation. In: Proceedings of the Royal Society of London A: Mathematical, Physical and Engineering Sciences, vol. 439, pp. 553–558. The Royal Society (1992)
15. Peter, W.: Shor. Polynomial-time algorithms for prime factorization and discrete logarithms on a quantum computer. SIAM J. Comput. **26**(5), 1484–1509 (1997)
16. Deutsch, D.: Quantum theory, the Church-Turing principle and the universal quantum computer. In: Proceedings of the Royal Society of London A: Mathematical, Physical and Engineering Sciences, vol. 400, pp. 97–117. The Royal Society (1985)
17. Nielsen, M.A., Chuang, I.L.: Quantum Computation and Quantum Information. Cambridge University Press, Cambridge (2010)
18. Barenco, A., Bennett, C.H., Cleve, R., DiVincenzo, D.P., Margolus, N., Shor, P., Sleator, T., Smolin, J.A., Weinfurter, H.: Elementary gates for quantum computation. Phys. Rev. A, **52**(5), 34–57 (1995)
19. Clarke, J., Wilhelm, F.K.: Superconducting quantum bits. Nature **453**(7198), 1031–1042 (2008)
20. Kok, P., Munro, W.J., Nemoto, K., Ralph, T.C., Dowling, J.P., Milburn, G.J.: Linear optical quantum computing with photonic qubits. Rev. Modern Phys. **79**(1), 135 (2007)
21. Juan, I.: Cirac and Peter Zoller. Quantum computations with cold trapped ions. Phys. Rev. Lett. **74**(20), 4091 (1995)
22. Nayak, C., Simon, S.H., Stern, A., Freedman, M., Sarma, S.D.: Non-Abelian anyons and topological quantum computation. Rev. Modern Phys. **80**(3), 1083 (2008)
23. Farhi, E., Goldstone, J., Gutmann, S., Sipser, M.: Quantum computation by adiabatic evolution (2000). arXiv:quant-ph/0001106, MIT-CTP-2936
24. Das, A., Chakrabarti, B.K.: Colloquium: quantum annealing and analog quantum computation. Rev. Modern Phys. **80**(3), 1061 (2008)
25. Neven, H., Denchev, V.S., Rose, G., Macready, W.G.: Training a large scale classifier with the quantum adiabatic algorithm. arXiv:0912.0779 (2009)
26. Heim, B., Rønnow, T.F., Isakov, S.V., Troyer, M.: Quantum versus classical annealing of Ising spin glasses. Science **348**(6231), 215–217 (2015)
27. Santoro, G.E., Tosatti, E.: Optimization using quantum mechanics: quantum annealing through adiabatic evolution. J. Phys. A **39**(36), R393 (2006)
28. Briegel, H.J., Raussendorf, R.: Persistent entanglement in arrays of interacting particles. Phys. Rev. Lett. **86**(5), 910 (2001)
29. Nielsen, M.A.: Cluster-state quantum computation. Reports Math. Phys. **57**(1), 147–161 (2006)
30. Prevedel, R., Stefanov, A., Walther, P., Zeilinger, A.: Experimental realization of a quantum game on a one-way quantum computer. New J. Phys. **9**(6), 205 (2007)
31. Park, H.S., Cho, J., Kang, Y., Lee, J.Y., Kim, H., Lee, D.-H., Choi, S.-K.: Experimental realization of a four-photon seven-qubit graph state for one-way quantum computation. Opt. Exp. **20**(7), 6915–6926 (2012)
32. Tame, M.S., Prevedel, R., Paternostro, M., Böhi, P., Kim, M.S., Zeilinger, A.: Experimental realization of Deutsch's algorithm in a one-way quantum computer. Phys. Rev. Lett. **98**, 140–501 (2007)

33. Tame, M.S., Bell, B.A., Di Franco, C., Wadsworth, W.J., Rarity, J.G.: Experimental realization of a one-way quantum computer algorithm solving Simon's problem. Phys. Rev. Lett. **113**, 200–501 (2014)
34. Lloyd, S., Braunstein, S.L.: Quantum computation over continuous variables. In: Quantum Information with Continuous Variables, pp. 9–17. Springer (1999)
35. Weedbrook, C., Pirandola, S., García-Patrón, R., Cerf, N.J., Ralph, T.C., Shapiro, J.H., Lloyd, S.: Gaussian quantum information. Rev. Modern Phys. **84**(2), 621 (2012)
36. Lau, H.-K., Pooser, R., Siopsis, G., Weedbrook, C.: Quantum machine learning over infinite dimensions. Phys. Rev. Lett. **118**(8), 080501 (2017)
37. Chatterjee, R., Ting, Y.: Generalized coherent states, reproducing kernels, and quantum support vector machines. Quant. Inf. Commun. **17**(15, 16), 1292 (2017)
38. Schuld, M., Killoran, N.: Quantum machine learning in feature Hilbert spaces. arXiv:1803.07128v1 (2018)
39. Gisin, N.: Weinberg's non-linear quantum mechanics and supraluminal communications. Phys. Lett. A **143**(1), 1–2 (1990)
40. Polchinski, J.: Weinberg's nonlinear quantum mechanics and the Einstein-Podolsky-Rosen paradox. Phys. Rev. Lett. **66**, 397–400 (1991)
41. Daniel, S.: Abrams and Seth Lloyd. Nonlinear quantum mechanics implies polynomial-time solution for NP-complete and #P problems. Phys. Rev. Lett. **81**(18), 39–92 (1998)
42. Peres, A.: Nonlinear variants of Schrödinger's equation violate the second law of thermodynamics. Phys. Rev. Lett. **63**(10), 1114 (1989)
43. Yoder, T.L., Low, G.H., Chuang, I.L.: Quantum inference on Bayesian networks. Phys. Rev. A **89**, 062315 (2014)
44. Ozols, M., Roetteler, M., Roland, J.: Quantum rejection sampling. ACM Trans. Comput. Theory (TOCT) **5**(3), 11 (2013)
45. Aram, W., Harrow, A.H., Lloyd, S.: Quantum algorithm for linear systems of equations. Phys. Rev. Lett. **103**(15), 150–502 (2009)
46. Boyer, M., Brassard, G., Høyer, P., Tapp, A.: Tight bounds on quantum searching. Fortsch. Phys. **46**, 493–506 (1998)
47. Ambainis, A.: Quantum lower bounds by quantum arguments. In: Proceedings of the Thirty-Second Annual ACM Symposium on Theory of Computing, pp. 636–643. ACM (2000)
48. Biham, E., Biham, O., Biron, D., Grassl, M., Lidar, D.A.: Grovers quantum search algorithm for an arbitrary initial amplitude distribution. Phys. Rev. A **60**(4), 27–42 (1999)
49. Brassard, G., Høyer, P., Mosca, M., Tapp, A.: Quantum amplitude amplification and estimation. Contemp. Math. **305**, 53–74 (2000)
50. Watrous, J.: Theory of Quantum Information. Cambridge University Press (2018)
51. Abrams, D.S., Lloyd, S.: Quantum algorithm providing exponential speed increase for finding eigenvalues and eigenvectors. Phys. Rev. Lett. **83**(24), 51–62 (1999)
52. Clader, D.B., Jacobs, B.C., Sprouse, C.R.: Preconditioned quantum linear system algorithm. Phys. Rev. Lett. **110**(25), 250–504 (2013)

Chapter 4
Quantum Advantages

Before coming to the design of quantum machine learning algorithms, this chapter is an interlude to discuss how quantum computing can actually assist machine learning. Although quantum computing researchers often focus on asymptotic computational speedups, there is more than one measure of merit when it comes to machine learning. We will discuss three dimensions here, namely the *computational complexity*, the *sample complexity* and the *model complexity*.[1] While the section on computational complexity allows us to establish the terminology already used in previous chapters with more care, the section on sample complexity ventures briefly into quantum extensions of statistical learning theory. The last section on model complexity provides arguments towards what has been called the *exploratory* approach to quantum machine learning, in which quantum physics is used as a resource to build new types of models altogether.

4.1 Computational Complexity of Learning

The concept of runtime complexity and speedups has already been used in the previous sections and is the most common figure of merit when accessing potential contributions of quantum computing to machine learning. Quantum machine learning inherited this focus on runtime speedups from quantum computing, where—for lack of practical experiments—researchers mostly resort to proving theoretical runtime bounds to advertise the power of their algorithms. Let us briefly introduce the basics of asymptotic computational complexity.

The runtime of an algorithm on a computer is the time it takes to execute the algorithm, in other words, the number of elementary operations multiplied by their respective execution times. In conventional computers, the number of elementary

---
[1] The three dimensions were first introduced by Peter Wittek and Vedran Dunjko.

operations could in theory still be established by choosing the fastest implementations for every subroutine and count the logic gates. However, with fast technological advancements in the IT industry it becomes obvious that a device-dependent runtime is not a suitable theoretical tool to measure the general speed of an algorithm. This is why computational complexity theory looks at the *asymptotic complexity* or the rate of growth of the runtime with the size n of the input. "Asymptotic" thereby refers to the fact that one is interested in laws for sufficiently large inputs only. If the resources needed to execute an algorithm or the number of elementary operations grow polynomially (that is, not more than a factor n^c for a constant c) with the size of the input n, it is *tractable* and the problem in theory *efficiently solvable*. Of course, if c is large, even polynomial runtime algorithms can take too long to make them feasible for certain applications. The argument is that if we just wait a reasonable amount of time, our technology could *in principle* master the task. Exponential growth makes an algorithm *intractable* and the problem *hard* because for large enough problem sizes, it quickly becomes impossible to solve the problem.

An illustrative example for an intractable problem is guessing the number combination for the security code of a safe, which without further structure requires a brute force search: While a code of two digits only requires 100 attempts in the worst case (and half of those guesses in the average case), a code of $n = 10$ digits requires ten billion guesses in the worst case, and a code of $n = 30$ digits has more possible configurations than our estimation for the number of stars in the universe. No advancement in computer technology could possibly crack this code.

When thinking about quantum computers, estimating the actual runtime of an algorithm is even more problematic. Not only do we not have a unique implementation of qubits yet that gives us a set of elementary operations to work with, but even if we agreed on a set of elementary gates as well as a technology, it is nontrivial to decompose quantum algorithms into this set. In many cases, we can claim that we know there is such a sequence of elementary gates, but it is by no means clear how to construct it. The vast majority of authors in quantum information processing are therefore interested in the asymptotic complexity of their routines, and how they compare to classical algorithms. In a qubit-based quantum computer, the input is considered to be the number of qubits n, which we already hinted at by the notation. To avoid confusion when looking at different concepts for the input, we will call polynomial algorithms regarding the number of qubits *qubit-efficient*. Algorithms which are efficient with respect to the number of amplitudes (see for example Grover search 3.5.1) will be referred to as *amplitude-efficient*.

The field of *quantum complexity theory* has been developed as an extension to classical complexity theory [1, 2] and is based on the question whether quantum computers are in principle able to solve computational problems faster in relation to the runtime complexity. In the following, complexity will always refer to the asymptotic runtime complexity unless stated otherwise. A runtime advantage in this context is called a *quantum enhancement*, *quantum advantage*, or simply a *quantum speedup*. Demonstrating an exponential speedup has been controversially referred to as *quantum supremacy*.

4.1 Computational Complexity of Learning

A collaboration of researchers at the forefront of benchmarking quantum computers came up with a useful typology for the term of 'quantum speedup' [3] which shows the nuances of the term:

1. A *provable quantum speedup* requires a proof that there can be no classical algorithm that performs as well or better than the quantum algorithm. Such a provable speedup has been demonstrated for Grover's algorithm, which scales quadratically better than classical [4] given that there is an oracle to mark the desired state [5].
2. A *strong quantum speedup* compares the quantum algorithm with the best known classical algorithm. The most famous example of a strong speedup is Shor's quantum algorithm to factorise prime numbers in time growing polynomially (instead of exponentially) with the number of digits of the prime number, which due to far-reaching consequences for cryptography systems gave rise to the first major investments into quantum computing.
3. If we relax the term 'best classical algorithm' (which is often not known) to the 'best available classical algorithm' we get the definition of a *common quantum speedup*.
4. The fourth category of *potential quantum speedup* relaxes the conditions further by comparing two specific algorithms and relating the speedup to this instance only.
5. Lastly, and useful when doing benchmarking with quantum annealers, is the *limited quantum speedup* that compares to "corresponding" algorithms such as quantum and classical annealing.

Although the holy grail of quantum computing remains to find a *provable exponential speedup*, the wider definition of a quantum advantage in terms of the asymptotic complexity opens a lot more avenues for realistic investigations. Two common pitfalls have to be avoided. Firstly, quantum algorithms often have to be compared with classical sampling, which is likewise non-deterministic and has close relations to quantum computing. Secondly, complexity can easily be hidden in spatial resources, and one could argue that quantum computers have to be compared to cluster computing, which will surely be its main competitor in years to come [3, 6].

We want to introduce some particulars of how to formulate the asymptotic complexity of a (quantum) machine learning algorithm. The 'size' of the input in the context of machine learning usually refers to the number of data points M as well as the number of features N. Since this differs from what the quantum computing community considers as an input, we will call algorithms that are efficient with regards to the data "data-efficient". When dealing with sparse inputs that can be represented more compactly, the number of features is replaced by the maximum number s of nonzero elements in any training input. Complexity analysis commonly considers other numbers of merit. The error ϵ of a mathematical object z is the precision to which z' calculated by the algorithm is correct, $\epsilon = ||z - z'||$ (with a suitable norm). When dealing with matrices, the condition number κ, which is the ratio of the largest and the smallest eigenvalue or singular value, is sometimes of interest to find an upper bound for the runtime (since κ gives us an idea of the eigenvalue spectrum of

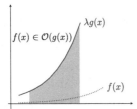 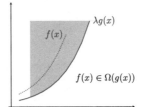 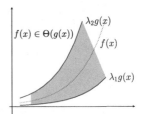

Fig. 4.1 Illustration of the big-O notation. If a function $f(x)$ (in this context f is the runtime and x the input) is 'in' $\mathcal{O}(g(x))$, there exists a $\lambda \in \mathbb{R}$ such that $|f(x)| \leq \lambda g(x)$ for large enough x. The Ω symbol stands for the inequality $|f(x)| \geq \lambda g(x)$, while the Θ symbol signifies that there are two $\lambda_1 < \lambda_2 \in \mathbb{R}$ such that $f(x)$ lies between the functions $\lambda_1 g(x)$ and $\lambda_2 g(x)$

the matrix). Many quantum machine learning algorithms have a chance of failure, for example because of a conditional measurement. In this case the average number of attempts required to execute it successfully needs to be taken into account as well. A success probability of p_s generally leads to a factor of $1/p_s$ in the runtime. For example, if the algorithm will only succeed in 1% of the cases, one has to repeat it on average for 100 times to encounter a successful run.

To express the asymptotic complexity, the runtime's dependency on these variables is given in the "big-O" notation (see Fig. 4.1):

- $\mathcal{O}(g(x))$ means that the true runtime f has an upper bound of $\lambda g(x)$ for some $\lambda \in \mathbb{R}$, a function g and the variable x.
- $\Omega(g(x))$ means that the runtime has a lower bound of $\lambda g(x)$ for some $\lambda \in \mathbb{R}$.
- $\Theta(g(x))$ means that the run time has a lower bound of $\lambda_1 g(x)$ and an upper bound of $\lambda_2 g(x)$ for some $\lambda_1, \lambda_2 \in \mathbb{R}$.

Having introduced time complexity, the question remains what speedups can be detected specifically in quantum machine learning. Of course, machine learning is based on common computational routines such as search or matrix inversion, and quantum machine learning derives its advantages from tools developed by the quantum information processing community. It is therefore no surprise that the speedups achieved in quantum machine learning are directly derived from the speedups in quantum computing. Roughly speaking (and with more details in the next chapters), three different types of speedups can be claimed:

1. Provable quadratic quantum speedups arise from variations of Grover's algorithm or quantum walks applied to search problems. Note that learning can always be understood as a search in a space of hypotheses [7].
2. Exponential speedups are naturally more tricky, even if we only look at the categories of either *strong* or *common speedups*. Within the discipline, they are usually only claimed by quantum machine learning algorithms that execute linear algebra computations in amplitude encoding. But these come with serious conditions, for example about the sparsity or redundancy in the dataset.

4.1 Computational Complexity of Learning

3. More specific comparisons occur when quantum annealing is used to solve optimisation problems derived from machine learning tasks [8], or to prepare a quantum system in a certain distribution [9]. It is not clear yet whether any provable speedups will occur, but benchmarking against specific classical algorithms show that there are problems where quantum annealing can be of advantage for sampling problems.

Altogether, computational complexity is of course important for machine learning, since it defines what is practically possible. However, limiting the focus only on this aspect severely and unnecessarily limits the scope of quantum machine learning. Very often, provably hard problems in machine learning can be solved for relevant special cases, sometimes without even a solid theoretical understanding why this is so. It is therefore important to look for other aspects in which quantum computing could contribute to machine learning.

4.2 Sample Complexity

Besides the asymptotic computational complexity, the so called *sample complexity* is an important figure of merit in classical machine learning, and offers a wealth of theoretical results regarding the subject of learning, summarised under the keyword *statistical learning theory* [10]. It also offers a door to explore more fundamental questions of quantum machine learning, for example what it means to learn in a quantum setting, or how we can formulate a quantum learning theory [11]. Although such questions are not the focus of this book, we use this opportunity for a small excursion into the more theoretical branch of quantum machine learning.

loosensee 1 Sample complexity refers to the number of samples needed to generalise from data. Samples can either be given as training instances drawn from a certain distribution (*examples*) or by computing outputs to specifically chosen inputs (*queries*). For the supervised pattern recognition problem we analysed so far, the dataset is given as examples and no queries can be made. One can easily imagine a slightly different setting where queries are possible, for example when a certain experiment results in input-output pairs, and we can choose the settings for the experiment to generate the data.

Considerations about sample complexity are usually based on binary functions $f : \{0, 1\}^N \to \{0, 1\}$. The sample complexity of a machine learning algorithm refers to the number of samples that are required to learn a *concept* from a given *concept class*. A concept is the rule f that divides the input space into subsets of the two class labels 0 and 1, in other words, it is the law that we want to recover with a model.

There are two important settings in which sample complexity is analysed.

1. In *exact learning from membership queries* [12], one learns the function f by querying a membership oracle with inputs x and receives the answer whether $f(x)$ evaluates to 1 or not.

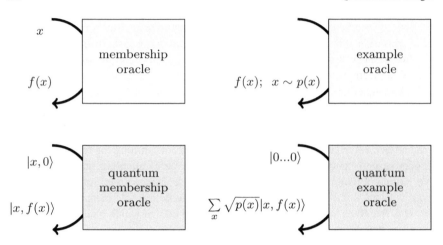

Fig. 4.2 Different types of oracles to determine the sample complexity of a learning algorithm. A membership oracle takes queries for a certain input x and returns the class $f(x)$, while an example oracle is activated and draws samples of x from a certain (usually unknown) distribution $p(x)$, returning the class of x. The quantum version of a membership oracle is a function evaluation on a register $|x, 0\rangle$, while a quantum example oracle has been defined as the qsample of the distribution $p(x)$

2. The framework of *Probably Approximately Correct (PAC)* learning was introduced by Valiant [13] and asks how many examples from the original concept are needed in the worst case to train a model so that the probability of an error ϵ (i.e., the probability of assigning the wrong label) is smaller than $1 - \delta$ for $0 \leq \delta \leq 1$. The examples are drawn from an arbitrary distribution via an example oracle. This framework is closely related to the Vapnik-Chervonenkis- or VC-dimension of a model (see Sect. 4.3).

To translate these two settings into a quantum framework (see Fig. 4.2), a quantum membership oracle as well as a quantum example oracle are introduced. They are in a sense parallelised versions of the classical sample generators, and with quantum interference of amplitudes we can hope to extract more information from the distribution than possible in the classical case. Rather surprisingly, it turns out that the classical and quantum sample complexity are polynomially equivalent, or as stated by Servedio and Gortler [14]:

> [F]or any learning problem, if there is a quantum learning algorithm which uses polynomially many [samples] then there must also exist a classical learning algorithm which uses polynomially many [samples].

Note that this only concerns the sample complexity; the same authors find an instance for a problem that is efficiently learnable by a quantum algorithm in terms of computational complexity, while the best classical algorithm is intractable.

4.2.1 Exact Learning from Membership Queries

Sample complexity in relation to queries is closely related to the concept of 'quantum query complexity' which is an important figure of merit in quantum computing in the oracular setting (for example to determine the runtime of Grover's algorithm). A quantum oracle can be described as a unitary operation

$$U : |x, 0\rangle \rightarrow |x, 0 \oplus f(x)\rangle,$$

where $x \in \{0, 1\}^n$ is encoded in the computational basis.

Two famous quantum algorithms that demonstrate how for specific types of problems only a single quantum query can be sufficient, are the Deutsch-Josza algorithm (see Sect. 3.3.1) as well as the famous Bernstein-Vazirani algorithm [1] that we did not discuss here. They are both based on the principle of applying the quantum oracle to a register in uniform superposition, thereby querying all possible inputs in parallel. Writing the outcome into the phase and interfering the amplitudes then reveals information on the concept, for example if it was a balanced (half of all inputs map to 1) or constant (all inputs map to 1) function. Note that this does not mean that the function itself is learnt (i.e., which inputs of the balanced function map to 0 or 1 respectively) and is therefore not sufficient as an example to prove theorems on general quantum learnability.

In 2003, Hunziker et al. [15] introduced a larger framework which they call "impatient learning" and proposed the following two conjectures on the actual number of samples required in the (asymptotic) quantum setting:

Conjecture 1: For any family of concept classes $C = \{C_i\}$ with $|C| \rightarrow \infty$, there exists a quantum learning algorithm with membership oracle query complexity $\mathcal{O}(\sqrt{|C|})$.

Conjecture 2: For any family of concept classes $C = \{C_i\}$ containing $|C| \rightarrow \infty$ concepts, there exists a quantum learning algorithm with membership oracle query complexity $\mathcal{O}(\frac{\log |C|}{\sqrt{\gamma}})$, where $\gamma \leq 1/3$ is a measure of how easy it is to distinguish between concepts, and small γ indicate a harder class to learn.

While the first conjecture was proven by Ambainis et al. [16], the second conjecture was resolved in a series of contributions [14, 17, 18]. The classical upper bound for exact learning from membership queries is given by $\mathcal{O}(\frac{\log |C|}{\gamma})$.

Servedio and Gortler [14] compared these results and found that if any class C of boolean functions $f : \{0, 1\}^n \rightarrow \{0, 1\}$ is learnable from Q quantum membership queries, it is then learnable by at most $\mathcal{O}(nQ^3)$ classical membership queries. This result shows that classical and quantum learnability have at most a polynomial overhead. It becomes apparent that no exponential speedup can be expected from quantum sample complexity for exact learning with membership queries.

4.2.2 PAC Learning from Examples

It is a well-established fact from classical learning theory that a (ϵ, δ)-PAC learning algorithm for a non-trivial concept class C of Vapnik-Chervonenkis-dimension d requires at least $\Omega(\frac{1}{\epsilon} \log \frac{1}{\delta} + \frac{d}{\epsilon})$ examples, but can be learnt with at most $\mathcal{O}(\frac{1}{\epsilon} \log \frac{1}{\delta} + \frac{d}{\epsilon} \log \frac{1}{\epsilon})$ examples [19, 20].

A first contribution to quantum PAC learning was made by Bshouty and Jackson in 1998 [21], who define the important notion of a quantum example oracle. The quantum example oracle is equivalent to what has been introduced earlier as a *qsample*,

$$\sum_i \sqrt{p(x)} |x, f(x)\rangle, \tag{4.1}$$

where the probabilities of measuring the basis states in the computational basis reflects the distribution $p(x)$ from which the examples are drawn. They show that a certain class of functions (so called "polynomial-size disjunctive normal form expressions" which are actively investigated in the PAC literature) are efficiently learnable by a quantum computer from 'quantum example oracles' which draw examples from a uniform distribution. This speedup is achieved by interfering the amplitudes of the qsample with a quantum Fourier transform.[2] However, the PAC model requires learnability under *any* distribution, and the results of Bshouty and Jackson do therefore not lead to a statement about the general quantum PAC case.

Servedio and Gortler [14] use the definition of a quantum example oracle from Eq. (4.1) to show that the equivalence of classical and quantum learning extends to the PAC framework. They prove that if any class C of boolean functions $f\{0, 1\}^n \rightarrow \{0, 1\}$ is learnable from Q evaluations of the quantum example oracle, it is then learnable by $\mathcal{O}(nQ)$ classical examples. Improvements in a later paper by Atici and Servedio [17] prove a lower bound on quantum learning that is close to the classical setting, i.e. that any (ϵ, δ)-PAC learning algorithm for a concept class of Vapnik-Chervonenkis-dimension d must make at least $\Omega(\frac{1}{\epsilon} \log \frac{1}{\delta} + d + \frac{\sqrt{d}}{\epsilon})$ calls to the quantum example oracle from Eq. (4.1). This was again improved by [22] to finally show that in the PAC setting, quantum and classical learners require the same amount of examples up to a constant factor. This also holds true for the *agnostic* learning framework, which loosens PAC learning by looking for a hypothesis that is sufficiently close to the best one [11].

As a summary, under the quantum formulation of classical learning theory based on the oracles from Fig. 4.2, we are not expecting any exponential advantages from quantum computing.

---

[2] Applying a quantum Fourier transform effectively changes the distribution from which to sample, which leaves some question whether the comparison to a static 'classical example generator' is fair. However, they show that while the quantum example oracle can be simulated by a membership query oracle, this is not true vice versa. It seems therefore that the quantum example oracle ranges somewhere between a query and an example oracle.

4.2.3 Introducing Noise

Even though the evidence suggests that classical and quantum sample complexity are similar up to at most polynomial factors, an interesting observation derives from the introduction of noise into the model. Noise refers to corrupted query results or examples for which the value $f(x)$ is flipped with probability μ. For the quantum example oracle, noise is introduced by replacing the oracle with a mixture of the original qsample and a corrupted qsample weighted by μ. In the PAC setting with a uniform distribution investigated by Bshouty and Jackson, the quantum algorithm can still learn the function by consuming more examples, while noise is believed to render the classical problem unlearnable [21]. A similar observation is made by Cross, Smith and Smolin [23], who consider the problem of learning n-bit parity functions by membership queries and examples, a task that is easy both classically and quantumly, and in the sample as well as time complexity sense. Parity functions evaluate to 1 if the the input string contains an odd number of 1's. However, Cross et al. find that in the presence of noise, the classical case becomes intractable while the quantum samples required only grow logarithmically. To ensure a fair comparison, the classical oracle is modelled as a dephased quantum channel of the membership oracle and the quantum example oracle respectively. As a toy model the authors consider a slight adaptation of the Bernstein-Vazirani algorithm. These observations might be evidence enough that a fourth category of potential quantum advantages, the robustness against noise, could be a fruitful avenue for further research.

4.3 Model Complexity

Lastly, we will come to the third potential advantage that quantum computing could offer machine learning. The term 'model complexity' is a wide concept that refers to the flexibility, capacity, richness or expressive power of a model. For example, a model giving rise to linear decision boundaries is less flexible than a model with more complex decision boundaries. A linear fit in regression is less flexible than a higher-order polynomial. Roughly speaking, more complex model families offer a larger space of possible trained models and therefore have a better chance to fit the training data.

Of course, this does not mean that more flexible models have a higher generalisation power. In fact, they are much more prone to overfitting, which is quantified by Vapnik's famous upper bound for the generalisation error ϵ_{gen}. With probability of at least $1 - \delta$ for some $\delta > 0$ the bound is given by [24]

$$\epsilon_{\text{gen}} \leq \epsilon_{\text{emp}} + \sqrt{\frac{1}{M}\left(d\left(\log\left(\frac{2M}{d}\right) + 1\right) + \log\left(\frac{4}{\delta}\right)\right)}. \qquad (4.2)$$

The generalisation error is bounded from above by the 'empirical error' ϵ_{emp} on the training set, plus an expression that depends on the so called Vapnik-Chervonenkis-dimension d. The VC-dimension of a model is closely related to the model complexity and is defined for binary models, $f : \mathcal{X} \to \{0, 1\}$. It refers to the maximum number of data points M assigned to two classes for which a hypothesis or 'trained version' of the model $f(x; \theta)$ makes no classification error, or *shatters* the data. For the next higher number of points $M + 1$ there is no dataset that can always be shattered, no matter how the labels are assigned. For example, a linear decision boundary can shatter three data points, but there are label assignments for which it cannot separate four data points. Inequality (4.2) states that we are interested in a low training error *as well as* a low flexibility of the model, as previously remarked in Sect. 2.2.2 in the context of overfitting and regularisation. This means that we are interested in the 'slimmest' model possible that can still capture the pattern which has to be learnt.

To investigate how quantum models can help machine learning is very much in the spirit of the 'exploratory approach' to quantum machine learning and will be subject of Chap. 8. By the term 'quantum model' we refer to a machine learning model that is derived from the dynamics of a quantum system and can therefore be implemented using a quantum device. For example, we will discuss Hopfield networks that are extended via the quantum Ising model, or a neural network where one layer is computed by a quantum circuit.

Asking how quantum models can offer advantages in terms of the model complexity has two dimensions. On the theoretical side, an analysis of the VC-dimension of quantum models could reveal interesting comparisons to equivalent classes of classical models. In the example above, one could compare a quantum and classical Hopfield network to find out what effect quantum extensions have on the VC-dimension. Given the challenge this poses already in classical machine learning, numerical studies may be crucial in answering this question.

On the more practical side, quantum models could prove to be a useful ansatz to capture patterns in certain datasets. An approach to find a slim but powerful model is to choose a model family which can express the natural dynamics of the system producing the data. For example, nonlinearities in the updating function of recurrent neural networks allow us to represent nonlinear dynamics of the real-world systems they try to recover. In probabilistic graphical models we assume that the system has inherent independence relations between variables. In short, an open question with regards to the model complexity is whether certain quantum models are better at capturing patterns and correlations in data, and if yes, which problems are suitable for quantum computing. One seemingly obvious case for the use of quantum models is when the system producing the data is a quantum system, which has been termed the QQ case in the introduction. Still, so far there are no studies that show how a quantum model is better in capturing quantum correlations. We see that the contribution of quantum computing to model complexity has many outstanding theoretical and practical research questions that the quantum machine learning community is only beginning to explore.

References

1. Bernstein, E., Vazirani, U.: Quantum complexity theory. SIAM J. Comput. **26**(5), 1411–1473 (1997)
2. Watrous, J.: Quantum computational complexity. In: Encyclopedia of Complexity and Systems Science, pp. 7174–7201. Springer (2009)
3. Rännow, T.F., Wang, Z., Job, J., Boixo, S., Isakov, S.V., Wecker, D., Martinis, J.M., Lidar, D.A., Troyer, M.: Defining and detecting quantum speedup. Science **345**, 420–424 (2014)
4. Grover, L.K.: A fast quantum mechanical algorithm for database search. In: Proceedings of the Twenty-Eighth Annual ACM Symposium on Theory of Computing, pp. 212–219. ACM (1996)
5. Bennett, C.H., Bernstein, E., Brassard, G., Vazirani, U.: Strengths and weaknesses of quantum computing. SIAM J. Comput. **26**(5), 1510–1523 (1997)
6. Steiger, D.S., Troyer, M.: Racing in parallel: quantum versus classical. In: Bulletin of the American Physical Society, vol. 61 (2016)
7. Sammut, C., Webb, G.I.: Encyclopedia of Machine Learning. Springer Science & Business Media (2011)
8. Neven, H., Denchev, V.S., Rose, G., Macready, W.G.: Training a large scale classifier with the quantum adiabatic algorithm. arXiv:0912.0779 (2009)
9. Amin, M.H.: Searching for quantum speedup in quasistatic quantum annealers. Phys. Rev. A **92**(5), 1–6 (2015)
10. Vapnik, V.N., Vapnik, V.: Statistical learning theory, vol. 1. Wiley, New York (1998)
11. Arunachalam, S., de Wolf, R.: Guest column: a survey of quantum learning theory. ACM SIGACT News **48**(2), 41–67 (2017)
12. Angluin, D.: Queries and concept learning. Mach. Learn. **2**(4), 319–342 (1988)
13. Valiant, L.G.: A theory of the learnable. Commun. ACM **27**(11), 1134–1142 (1984)
14. Rocco, A.S., Gortler, S.J.: Equivalences and separations between quantum and classical learnability. SIAM J. Comput. **33**(5), 1067–1092 (2004)
15. Hunziker, M., Meyer, D.A., Park, J., Pommersheim, J., Rothstein, M.: The geometry of quantum learning. Quant. Inf. Process. **9**(3), 321–341 (2010)
16. Ambainis, A., Iwama, K., Kawachi, A., Masuda, H., Putra, R.H., Yamashita, S.: Quantum identification of Boolean oracles. In: Annual Symposium on Theoretical Aspects of Computer Science, pp. 105–116. Springer (2004)
17. Atici, A., Servedio, R.A.: Improved bounds on quantum learning algorithms. Quant. Inf. Process. **4**(5), 355–386 (2005)
18. Kothari, R.: An optimal quantum algorithm for the oracle identification problem. In: Proceedings of the 31st International Symposium on Theoretical Aspects of Computer Science (STACS 2014), Leibniz International Proceedings in Informatics, vol. 25, pp. 482–493 (2014)
19. Blumer, A., Ehrenfeucht, A., Haussler, D., Warmuth, M.K.: Learnability and the Vapnik-Chervonenkis dimension. J. ACM (JACM) **36**(4), 929–965 (1989)
20. Hanneke, S.: The optimal sample complexity of PAC learning. J. Mach. Learn. Res. **17**(38), 1–15 (2016)
21. Bshouty, N.H., Jackson, J.C.: Learning DNF over the uniform distribution using a quantum example oracle. SIAM J. Comput. **28**(3), 1136–1153 (1998)
22. Arunachalam, S., de Wolf, R.: Optimal quantum sample complexity of learning algorithms. arXiv:1607.00932 (2016)
23. Cross, A.W., Smith, G., Smolin, J.A.: Quantum learning robust against noise. Phys. Rev. A **92**(1), 012327 (2015)
24. Schölkopf, B., Herbrich, R., Smola, A.: A generalized representer theorem. In: Computational Learning Theory, pp. 416–426. Springer (2001)

Chapter 5
Information Encoding

If we want to use a quantum computer to learn from classical data—which was in the introduction referred to as the *CQ* case—we have to think about how to represent features by a quantum system. We furthermore have to design a recipe for "loading" data from a classical memory into the quantum computer. In quantum computing, this process is called *state preparation* um machine learning algorithm.

Classical machine learning textbooks rarely discuss matters of data representation and data transfer to the processing hardware (although considerations of memory access become important in big data applications). For quantum algorithms, these questions cannot be ignored (see Fig. 5.1). The strategy of how to represent information as a quantum state provides the context of how to design the quantum algorithm and what speedups one can hope to harvest. The actual procedure of encoding data into the quantum system is part of the algorithm and may account for a crucial part of the complexity.[1] Theoretical frameworks, software and hardware that address the interface between the classical memory and the quantum device are therefore central for technological implementations of quantum machine learning. Issues of efficiency, precision and noise play an important role in performance evaluation. This is even more true since most quantum machine learning algorithms deliver probabilistic results and the entire routine—including state preparation—may have to be repeated many times. These arguments call for a thorough discussion of "data encoding" approaches and routines, which is why we dedicate this entire chapter to questions of data representation with quantum states. We will systematically go through the four encoding methods distinguished in Sect. 3.4 and discuss state preparation routines and the consequences for algorithm design.

---

[1]This is not only true for quantum machine learning algorithms. For example, the classically hard *graph isomorphism* problem is efficiently solvable on a quantum computer if a superposition of isomorph graphs can be created efficiently [1].

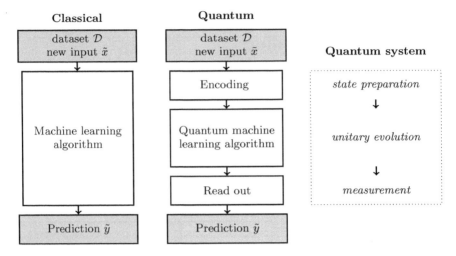

Fig. 5.1 In order to solve supervised machine learning tasks based on classical datasets, the quantum algorithm requires an information encoding and read out step that are in general highly non-trivial procedures, and it is important to consider them in the runtime. Adapted from [2]

Table 5.1 Comparison of the four encoding strategies for a dataset of M inputs with N features each. While basis, amplitude and Hamiltonian encoding aim at representing a full data set by the quantum system, qsample encoding works a little different in that it represents a probability distribution over random variables. It therefore does not have a dependency on the number of inputs M. *Only certain datasets or models can be encoded in this time. See text for details.

Encoding	Number of qubits	Runtime of state prep	Input features
Basis	N	$\mathcal{O}(MN)$	Binary
Amplitude	$\log N$	$\mathcal{O}(MN)/\mathcal{O}(\log(MN))$*	Continuous
Qsample	N	$\mathcal{O}(2^N)/\mathcal{O}(N)$*	Binary
Hamiltonian	$\log N$	$\mathcal{O}(MN)/\mathcal{O}(\log(MN))$*	Continuous

A central figure of merit for state preparation is the asymptotic runtime, and an overview of runtimes for the four encoding methods is provided in Table 5.1. In machine learning, the input of the algorithm is the data, and an efficient algorithm is efficient in the dimension of the data inputs N and the number of data points M. In quantum computing an efficient algorithm has a polynomial runtime with respect to the number of qubits. Since data can be encoded into qubits or amplitudes, the expression "efficient" can have different meanings in quantum machine learning, and this easily gets confusing. To facilitate the discussion, we will use the terms introduced in Sect. 4.1 and call an algorithm either *amplitude-efficient* or *qubit-efficient*, depending on what we consider as an input. It is obvious that if the data is encoded

5 Information Encoding

into the amplitudes or operators of a quantum system (as in amplitude and Hamiltonian encoding), amplitude-efficient state preparation routines are also efficient in terms of the data set. If we encode data into qubits, qubit-efficient state preparation is efficient in the data set size. We will see that there are some very interesting cases in which we encode data into amplitudes or Hamiltonians, but can guarantee qubit-efficient state preparation routines. In these cases, we prepare data in time which is logarithmic in the data size itself. Of course, this requires either a very specific access to or a very special structure of the data.

Before we start, a safe assumption when nothing more about the hardware is known is that we have a n-qubit system in the ground state $|0...0\rangle$ and that the data is accessible from a classical memory. In some cases we will also require some specific classical preprocessing. We consider data sets $\mathcal{D} = \{x^1, ..., x^M\}$ of N-dimensional real feature vectors. Note that many algorithms require the labels to be encoded in qubits entangled with the inputs, but for the sake of simplicity we will focus on unlabelled data in this chapter.

5.1 Basis Encoding

Assume we are given a binary dataset $\mathcal{D}$ where each pattern $x^m \in \mathcal{D}$ is a binary string of the form $x^m = (b_1^m, ..., b_N^m)$ with $b_i^m \in \{0, 1\}$ for $i = 1, ..., N$. We can prepare a superposition of basis states $|x^m\rangle$ that qubit-wise correspond to the binary input patterns,

$$|\mathcal{D}\rangle = \frac{1}{\sqrt{M}} \sum_{m=1}^{M} |x^m\rangle. \tag{5.1}$$

For example, given two binary inputs $x^1 = (01, 01)^T$, $x^2 = (11, 10)^T$, where features are encoded with a binary precision of $\tau = 2$, we can write them as binary patterns $x^1 = (0110)$, $x^2 = (1110)$. These patterns can be associated with basis states $|0110\rangle$, $|1110\rangle$, and the full data superposition reads

$$|\mathcal{D}\rangle = \frac{1}{\sqrt{2}}|0101\rangle + \frac{1}{\sqrt{2}}|1110\rangle. \tag{5.2}$$

The amplitude vector corresponding to State (5.1) has entries $\frac{1}{\sqrt{M}}$ for basis states that are associated with a binary pattern from the dataset, and zero entries otherwise. For Eq. (5.2), the amplitude vector is given by

$$\alpha = (0, 0, 0, 0, 0, \frac{1}{\sqrt{2}}, 0, 0, 0, 0, 0, 0, 0, 0, \frac{1}{\sqrt{2}}, 0)^T$$

Since—except in very low dimensions—the total number of amplitudes $2^{N\tau}$ is much larger than the number of nonzero amplitudes M, basis encoded datasets generally give rise to sparse amplitude vectors.

5.1.1 Preparing Superpositions of Inputs

An elegant way to construct such 'data superpositions' in time linear in M and N has been introduced by Ventura, Martinez and others [3, 4], and will be summarised here as an example of basis encoded state preparation. The circuit for one step in the routine is shown in Fig. 5.2. We will simplify things by considering binary inputs in which every bit represents one feature, or $\tau = 1$.

We require a quantum system

$$|l_1, ..., l_N; a_1, a_2; s_1, ..., s_N\rangle$$

with three registers: a *loading register* of N qubits $|l_1, ..., l_N\rangle$, the *ancilla register* $|a_1, a_2\rangle$ with two qubits and the N-qubit *storage register* $|s_1, ..., s_N\rangle$. We start in the ground state and apply a Hadamard to the second ancilla to get

$$\frac{1}{\sqrt{2}}|0, ..., 0; 0, 0; 0, ..., 0\rangle + \frac{1}{\sqrt{2}}|0, ..., 0; 0, 1; 0, ..., 0\rangle.$$

The left term, flagged with $a_2 = 0$, is called the *memory branch*, while the right term, flagged with $a_2 = 1$, is the *processing branch*. The algorithm iteratively loads patterns into the loading register and 'breaks away' the right size of terms from the processing

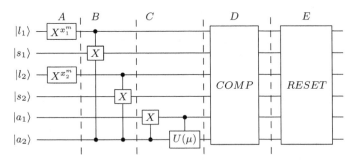

Fig. 5.2 Circuit for one step of Ventura and Martinez state preparation routine as described in the text for an example of 2-bit input patterns. The storage and loading qubits are in a slightly different order to facilitate the display. (A) The pattern is written into the loading register by NOT gates. Taking the gate to the power of the bit value x_i^m is a convenient way to apply the flip only when $x_i^m = 1$. (B) Transfer the pattern to the storage register of the processing branch. (C) Split the processing branch. (D) Flip the first ancilla back conditioned on a successful comparison of loading and storage branch. (E) Reset the registers to prepare for the next patterns

5.1 Basis Encoding

branch to add it to the memory branch (Fig. 5.3). This way the superposition of patterns is 'grown' step by step.

To explain how one iteration works, assume that the first m training vectors have already been encoded after iterations $1, ..., m$ of the algorithm. This leads to the state

$$|\psi^{(m)}\rangle = \frac{1}{\sqrt{M}} \sum_{k=1}^{m} |0, ..., 0; 00; x_1^k, ..., x_N^k\rangle + \sqrt{\frac{M-m}{M}} |0, ..., 0; 01; 0, ..., 0\rangle. \tag{5.3}$$

The memory branch stores the first m inputs in its storage register, while the storage register of the processing branch is in the ground state. In both branches the loading register is also in the ground state.

To execute the $(m+1)$th step of the algorithm, write the $(m+1)$th pattern $x^{m+1} = (x_1^{m+1}, ..., x_N^{m+1})$ into the qubits of the loading register (which will write it into both branches). This can be done by applying an X gate to all qubits that correspond to nonzero bits in the input pattern. Next, in the processing branch the pattern gets copied into the storage register using a CNOT gate on each of the N qubits. To limit the execution to the processing branch only we have to control the CNOTs with the second ancilla being in $a_2 = 1$. This leads to

$$\frac{1}{\sqrt{M}} \sum_{k=1}^{m} |x_1^{m+1}, ..., x_N^{m+1}; 00; x_1^k, ..., x_N^k\rangle$$

$$+ \sqrt{\frac{M-m}{M}} |x_1^{m+1}, ..., x_N^{m+1}; 01; x_1^{m+1}, ..., x_N^{m+1}\rangle.$$

Using a CNOT gate, we flip $a_1 = 1$ if $a_2 = 1$, which is only true for the processing branch. Afterwards apply the single qubit unitary

$$U_{a_2}(\mu) = \begin{pmatrix} \sqrt{\frac{\mu-1}{\mu}} & \frac{1}{\sqrt{\mu}} \\ \frac{-1}{\sqrt{\mu}} & \sqrt{\frac{\mu-1}{\mu}} \end{pmatrix}$$

with $\mu = M + 1 - (m+1)$ to qubit a_2 but controlled by a_1. On the full quantum state, this operation amounts to

$$\mathbb{1}_{\text{loading}} \otimes c_{a_1} U_{a_2}(\mu) \otimes \mathbb{1}_{\text{storage}}.$$

This splits the processing branch into two subbranches, one that can be "added" to the memory branch and one that will remain the processing branch for the next step,

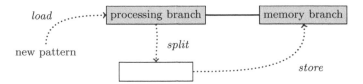

Fig. 5.3 Illustration of Ventura and Martinez state preparation routine [3]. The superposition is divided into a processing and memory term, flagged by an ancilla. New training input patterns get successively loaded into the processing branch, which gets 'split' by a Hadamard on an ancilla, and the pattern gets 'merged' into the memory term

$$\frac{1}{\sqrt{M}} \sum_{k=1}^{m} |x_1^{m+1}, ..., x_N^{m+1}; 00; x_1^k, ..., x_N^k\rangle$$
$$+ \frac{1}{\sqrt{M}} |x_1^{m+1}, ..., x_N^{m+1}; 10; x_1^{m+1}, ..., x_N^{m+1}\rangle$$
$$+ \frac{\sqrt{M-(m+1)}}{\sqrt{M}} |x_1^{m+1}, ..., x_N^{m+1}; 11; x_1^{m+1}, ..., x_N^{m+1}\rangle$$

To add the subbranch marked by $|a_1 a_2\rangle = |10\rangle$ to the memory branch, we have to flip a_1 back to 1. To confine this operation to the desired subbranch we can condition it on an operation that compares if the loading and storage register are in the same state (which is only true for the two processing subbranches), and that $a_2 = 1$ (which is only true for the desired subbranch). Also, the storage register of the processing branch as well as the loading register of both branches has to be reset to the ground state by reversing the previous operations, before the next iteration begins. After the $(m+1)$th iteration we start with a state similar to Eq. (5.3) but with $m \to m+1$. The routine requires $\mathcal{O}(MN)$ steps, and succeeds with certainty.

There are interesting alternative proposals for architectures of *quantum Random Access Memories*, devices that load patterns 'in parallel' into a quantum register. These devices are designed to query an index register $|m\rangle$ and load the mth binary pattern into a second register in basis encoding, $|m\rangle|0...0\rangle \to |m\rangle|x^m\rangle$. Most importantly, this operation can be executed in parallel. Given a superposition of the index register, the quantum random access memory is supposed to implement the operation

$$\frac{1}{\sqrt{M}} \sum_{m=0}^{M-1} |m\rangle|0...0\rangle \to \frac{1}{\sqrt{M}} \sum_{m=0}^{M-1} |m\rangle|x^m\rangle. \tag{5.4}$$

We will get back to this in the next section, where another step allows us to prepare amplitude encoded quantum states. Ideas for architectures which realise this kind of 'query access' in logarithmic time regarding the number of items to be loaded have been brought forward [5], but a hardware realising such an operation is still an open challenge [6, 7].

5.1.2 Computing in Basis Encoding

Acting on binary features encoded as qubits gives us the most computational freedom to design quantum algorithms. In principle, each operation on bits that we can execute on a classical computer can be done on a quantum computer as well. The argument is roughly the following: A Toffoli gate implements the logic operation

Input	Output
0 0 0	0 0 0
0 0 1	0 0 1
0 1 0	0 1 0
0 1 1	0 1 1
1 0 0	1 0 0
1 0 1	1 0 1
1 1 **0**	1 1 **1**
1 1 **1**	1 1 **0**

and is a universal gate for Boolean logic [8]. Universality implies that *any* binary function $f : \{0, 1\}^{\otimes N} \to \{0, 1\}^{\otimes D}$ can be implemented by a succession of Toffoli gates and possibly some 'garbage' bits. The special role of the Toffoli gate stems from the fact that it is reversible. If one only sees the state after the operation, one can deduce the exact state before the operation (i.e., if the first two bits are 11, flip the third one). No information is lost in the operation, which is in physical terms a non-dissipative operation. In mathematical terms the matrix representing the operation has an inverse. Reversible gates, and hence also the Toffoli gate, can be implemented on a quantum computer. In conclusion, if any classical algorithm can efficiently be formulated in terms of Toffoli gates, and these can always be implemented on a quantum computer, this means that any classical algorithm can efficiently be translated to a quantum algorithm. The reformulation of a classical algorithm with Toffoli gates may however have a large polynomial overhead and slow down the routines significantly.

Note that once encoded into a superposition, the data inputs can be processed in quantum parallel (see Sect. 3.2.4). For example, if a routine

$$\mathcal{A}(|x\rangle \otimes |0\rangle) \to |x\rangle \otimes |f(x)\rangle$$

is known to implement a machine learning model f and write the binary prediction $f(x)$ into the state of a qubit, we can perform inference in parallel on the data superposition,

$$\mathcal{A}\left(\frac{1}{\sqrt{M}}\sum_{m=1}^{M}|x^m\rangle \otimes |0\rangle\right) \to \frac{1}{\sqrt{M}}\sum_{m=1}^{M}|x^m\rangle \otimes |f(x^m)\rangle.$$

From this superposition one can extract statistical information, for example by measuring the last qubit. Such a measurement reveals the expectation value of the prediction of the entire dataset.

5.1.3 Sampling from a Qubit

As the previous example shows, the result of a quantum machine learning model can be basis encoded as well. For binary classification this only requires one qubit. If the qubit is in state $|f(x)\rangle = |1\rangle$ the prediction is 1 and if the qubit is in state $|f(x)\rangle = |0\rangle$ the prediction is 0. A superposition can be interpreted as a probabilistic output that provides information on the uncertainty of the result.

In order to read out the state of the qubit we have to measure it, and we want to briefly address how to obtain the prediction estimate from measurements, as well as what number of measurements are needed for a reliable prediction. The field of reconstructing a quantum state from measurements is called *quantum tomography*, and there are very sophisticated ways in which samples from these measurements can be used to estimate the density matrix that describe the state. Here we consider a much simpler problem, namely to estimate the probability of measuring basis state $|0\rangle$ or $|1\rangle$. In other words, we are just interested in the diagonal elements of the density matrix, not in the entire state. Estimating the ouptut of a quantum model in basis encoding is therefore a 'classical' rather than a 'quantum task'.

Let the final state of the output qubit be given by the density matrix

$$\rho = \begin{pmatrix} \rho_{00} & \rho_{01} \\ \rho_{10} & \rho_{11} \end{pmatrix}.$$

We need the density matrix formalism from Chap. 3 here because the quantum computer may be in a state where other qubits are entangled with the output qubit, and the single-qubit-state is therefore a mixed state. We assume that the quantum algorithm is error-free, so that repeating it always leads to precisely the same density matrix ρ to take measurements from. The diagonal elements ρ_{00} and ρ_{11} fulfil $\rho_{00} + \rho_{11} = 1$ and give us the probability of measuring the qubit in state $|0\rangle$ or $|1\rangle$ respectively. We associate ρ_{11} with the probabilistic output $f(x) = p$ which gives us the probability that model f predicts $y = 1$ for the input x.

5.1 Basis Encoding

To get an estimate $\hat{p}$ of p we have to repeat the entire algorithm S times and perform a computational basis measurement on the output qubit in each run. This produces a set of samples $\Omega = \{y_1, ..., y_S\}$ of outcomes, and we assume the samples stem from a distribution that returns 0 with probability $1 - p$ and 1 with probability p. Measuring a single qubit is therefore equivalent to sampling from a Bernoulli distribution, a problem widely investigated in statistics.[2] There are various strategies of how to get an estimate $\hat{p}$ from samples Ω. Most prominent, maximum likelihood estimation leads to the rather intuitive 'frequentist estimator' $\hat{p} = \bar{p} = \frac{1}{S}\sum_{i=1}^{S} y_i$ which is nothing else than the average over all outcomes.

An important question is how many samples from the single qubit measurement we need to estimate p with error ϵ. In physics language, we want an 'error bar' ϵ of our estimation $\hat{p} \pm \epsilon$, or the *confidence interval* $[\hat{p} - \epsilon, \hat{p} + \epsilon]$. A confidence interval is valid for a pre-defined confidence level, for example of 99%. The confidence level has the following meaning: If we have different sets of samples $S_1, ..., S_s$ and compute estimators and confidence intervals for each of them, $\hat{p}_{S_1} \pm \epsilon_{S_1}, ..., \hat{p}_{S_s} \pm \epsilon_{S_s}$, the confidence level is the proportion of sample sets for which the true value p lies within the confidence interval around $\hat{p}$. The confidence level is usually expressed by a so called z-value, for example, a z-value of 2.58 corresponds to a confidence of 99%. This correspondence can be looked up in tables.

Frequently used is the Wald interval which is suited for cases of large S and $p \approx 0.5$. The error ϵ can be calculated as

$$\epsilon = z\sqrt{\frac{\hat{p}(1-\hat{p})}{S}}.$$

This is maximised for $p = 0.5$, so that we can assume the overall error of our estimation ϵ to be at most

$$\epsilon \leq \frac{z}{2\sqrt{S}}$$

with a confidence level of z. In other words, for a given ϵ and z we need $\mathcal{O}(\epsilon^{-2})$ samples of the qubit measurement. If we want to have an error bar of at most $\epsilon = 0.1$ and a confidence level of 99% we need about 167 samples, and an error of $\epsilon = 0.01$ with confidence 99% requires at most 17,000 samples. This is a vast number, but only needed if the estimator is equal to 0.5, which is the worst case scenario of an undecided classifier (and we may not want to rely on the decision very much in any case). One can also see that the bound fails for $p \to 0, 1$ [9].

There are other estimates that also work when p is close to either zero or one. A more refined alternative is the Wilson score interval [10] with the following estimator for p,

$$\hat{p} = \frac{1}{1+\frac{z^2}{S}}\left(\bar{p} + \frac{z^2}{2S}\right),$$

---

[2] Bernoulli sampling is equivalent to a (biased) coin toss experiment: We flip a coin S times and want to estimate the bias p, i.e. with what probability the coin produces 'heads'.

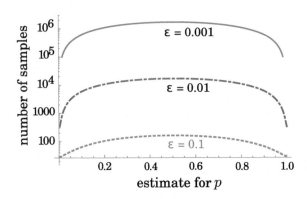

Fig. 5.4 Relationship between the sample size S and the mean value $\bar{p} = \frac{1}{S}\sum_{i=1}^{S} y_i$ for different errors ϵ for the Wilson score interval of a Bernoulli parameter estimation problem as described in the text

and the error

$$\epsilon = \frac{z}{1 + \frac{z^2}{S}}\left(\frac{\bar{p}(1-\bar{p})}{S} + \frac{z^2}{4S^2}\right)^{\frac{1}{2}},$$

with $\bar{p}$ being the average of all samples as defined above. Again this is maximised for $\bar{p} = 0.5$ and with a confidence level z we can state that the overall error of our estimation is bounded by

$$\epsilon \leq \sqrt{z^2 \frac{S + z^2}{4S^2}}.$$

This can be solved for S as

$$S \leq \frac{\epsilon^2 \sqrt{\frac{z^4(16\epsilon^2+1)}{\epsilon^4}} + z^2}{8\epsilon^2}.$$

With the Wilson score, a confidence level of 99% suggests that we need 173 single qubit measurements to guarantee an error of less than 0.1. However, now we can test the cases $\bar{p} = 0, 1$ for which $S = z^2(\frac{1}{2\epsilon} - 1)$. For $\epsilon = 0.1$ we only need about 27 measurements for the same confidence level (see Fig. 5.4).

5.2 Amplitude Encoding

An entire branch of quantum machine learning algorithms encode the dataset into the amplitudes of a quantum state

5.2 Amplitude Encoding

$$|\psi_D\rangle = \sum_{m=1}^{M}\sum_{i=1}^{N} x_i^m |i\rangle|m\rangle$$
$$= \sum_{m=1}^{M} |\psi_{x^m}\rangle|m\rangle. \tag{5.5}$$

This quantum state has an amplitude vector of dimension NM that is constructed by concatenating all training inputs, $\alpha = (x_1^1, ..., x_N^1, ..., x_1^M, ..., x_N^M)^T$. As discussed in Sect. 3.4.2, the dataset has to be normalised so that the absolute square of the amplitude vector is one, $|\alpha|^2 = 1$. The training outputs can either be basis encoded in an extra qubit $|y^m\rangle$ entangled with the $|m\rangle$ register, or encoded in the amplitudes of a separate quantum register,

$$|\psi_y\rangle = \sum_{m=1}^{M} y^m |m\rangle.$$

Amplitude encoding of datasets therefore requires the ability to prepare an arbitrary state

$$|\psi\rangle = \sum_i \alpha_i |i\rangle, \tag{5.6}$$

both efficiently and robustly.

The main advantage of amplitude encoding is that we only need $n = \log NM$ qubits to encode a dataset of M inputs with N features each. If the quantum machine learning algorithm is polynomial in n (or qubit-efficient), it has a logarithmic runtime dependency on the data set size. Promises of exponential speedups from qubit-efficient quantum machine learning algorithms sound strange to machine learning practitioners, because simply loading the NM features from the memory hardware takes time that is of course linear in NM. And indeed, the promises only hold if state preparation can also be done qubit-efficiently [11]. This is in fact possible in some cases that exhibit a lot of structure. As a trivial example, if the goal is to produce a vector of uniform entries we simply need to apply n Hadamard gates - a qubit-efficient state preparation time. Similarly, s-sparse vectors can be prepared with the routines from the previous section in time sn. On the other hand, there are subspaces in a Hilbert space that cannot be reached from a given initial state with qubit-efficient algorithms [12] (see Fig. 5.5). It is therefore an important and nontrivial open question which classes of relevant states for machine learning with amplitude encoding can be prepared qubit-efficiently.

Similar caution is necessary for the readout of all amplitudes. If the result of the computation is encoded in one specific amplitude α_i only, the number of measurements needed to retrieve it scale with the number of amplitudes (and to measure the

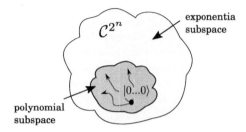

Fig. 5.5 Starting from the n-qubit ground state $|0...0\rangle$ some quantum states in $\mathbb{C}^{2^n}$ can be reached by a quantum algorithm that grows polynomially with n, while others require algorithms that grow exponentially with n. In amplitude encoding the number of features is 2^n, and state preparation routines that are polynomial in the number of qubits are therefore logarithmic in the number of features

state corresponding to that amplitude once requires on average $\mathcal{O}(1/|\alpha_i|^2)$ repetitions of the entire routine). This is why the result of quantum machine learning algorithms based on amplitude encoding is often designed in a different manner, for example through a single-qubit measurement as explained in the previous section,.

How can we prepare an arbitrary vector like in Eq. (5.6)? We will first discuss a scheme that is amplitude-efficient (and thereby efficient in N, M) and then look at qubit-efficient schemes and their limitations.

5.2.1 State Preparation in Linear Time

Given an n qubit quantum computer, the theoretical lower bound of the depth of an arbitrary state preparation circuit is known to be $\frac{1}{n}2^n$ [13–17]. Current algorithms perform slightly worse with slightly less than 2^n parallel operations, most of which are expensive 2-qubit gates. To illustrate one way of doing state preparation in linear time, let us have a look at the routine presented by Möttönen et al. [18]. They consider the reverse problem, namely to map an arbitrary state $|\psi\rangle$ to the ground state $|0...0\rangle$. In order to use this algorithm for our purpose we simply need to invert each and every operation and apply them in reverse order.

The basic idea is to control a rotation on qubit q_s by all possible states of the previous qubits $q_1, ..., q_{s-1}$, using sequences of so called *multi-controlled rotations*. In other words, we explicitly do a different rotation for each possible branch of the superposition, the one in which the previous qubits were in state $|0..0\rangle$ to the branch in which they are in $|1...1\rangle$. This is comparable to tossing $s - 1$ coins and manipulating the sth coin differently depending on the measurement outcome via a look-up table.

A full sequence of multi-controlled rotations around vectors v^i with angles β_i consists of the successive application of the 2^{s-1} gates

5.2 Amplitude Encoding

$$c_{q_1=0} \cdots c_{q_{s-1}=0} \, R_{q_s}(v^1, \beta_1) \, |q_1...q_{s-1}\rangle|q_s\rangle$$
$$c_{q_1=0} \cdots c_{q_{s-1}=1} \, R_{q_s}(v^2, \beta_2) \, |q_1...q_{s-1}\rangle|q_s\rangle$$
$$\vdots$$
$$c_{q_1=1} \cdots c_{q_{s-1}=1} \, R_{q_s}(v^{2^{s-1}}, \beta_{2^{s-1}}) \, |q_1...q_{s-1}\rangle|q_s\rangle.$$

For example, for $s = 3$ a full sequence consists of the gates

$$c_{q_1=0}c_{q_2=0} \, R_{q_3}(v^1, \beta_1) \, |q_1 q_2\rangle|q_3\rangle$$
$$c_{q_1=0}c_{q_2=1} \, R_{q_3}(v^2, \beta_2) \, |q_1 q_2\rangle|q_3\rangle$$
$$c_{q_1=1}c_{q_2=0} \, R_{q_3}(v^3, \beta_3) \, |q_1 q_2\rangle|q_3\rangle$$
$$c_{q_1=1}c_{q_2=1} \, R_{q_3}(v^4, \beta_4) \, |q_1 q_2\rangle|q_3\rangle,$$

which rotate q_3 in a different way for all four branches of the superposition of q_1, q_2. The circuit symbol of a single multi-controlled rotation is

$$
\begin{array}{r}
|q_1\rangle \\
\vdots \\
|q_{n-1}\rangle \\
|q_n\rangle
\end{array}
\quad \text{[circuit diagram]} \tag{5.7}
$$

where white circles indicate a control on qubit q being in state 1, or $c_{q=1}$, and black circles a control on qubit q being in state 0, $c_{q=0}$. The circuit diagram of a full sequence of multi-controlled rotations on the third of three qubits is

$$
\begin{array}{r}
|q_1\rangle \\
|q_2\rangle \\
|q_3\rangle
\end{array}
\quad R(v_{[1]}, \beta_1) \, R(v_{[2]}, \beta_2) \, R(v_{[3]}, \beta_3) \, R(v_{[4]}, \beta_4) \tag{5.8}
$$

Of course, we need a prescription to decompose sequences of multi-controlled rotations into elementary gates. If there are $s - 1$ control qubits, this is possible with 2^s CNOTs and 2^s single qubit gates [14, 18]. For example, a multi-controlled rotation applied to the third of three qubits would have the decomposition

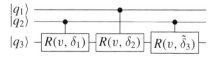

into three single-controlled rotations.

For a general quantum state one has to apply two *cascades* of such operations, where each cascade is a sequences of multi-controlled rotations that run trough all qubits q_1 to q_n. The first cascade uses R_z-rotations (which fixes the rotation axis v) and has the effect of 'equalising' the phases until the state only has one global phase which we can ignore (remember, we are doing reverse state preparation). The second cascade applies R_y rotations and has the effect of 'equalising' all amplitudes to result in the ground state. Since we are usually interested in real amplitude vectors, we can neglect the first cascade and end up with the circuit

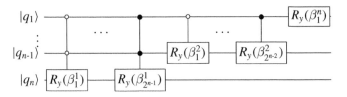

The choice of the rotation angles β is related to the amplitudes of the original state as follows:

$$\beta_j^s = 2\arcsin\left(\frac{\sqrt{\sum_{l=1}^{2^{s-1}}|\alpha_{(2j-1)2^{s-1}+l}|^2}}{\sqrt{\sum_{l=1}^{2^s}|\alpha_{(j-1)2^s+l}|^2}}\right) \quad (5.9)$$

Example 5.1 We want to prepare the state

$$|\psi\rangle = \sqrt{0.2}|000\rangle + \sqrt{0.5}|010\rangle + \sqrt{0.2}|110\rangle + \sqrt{0.1}|111\rangle,$$

with the amplitudes $a_0 = \sqrt{0.2}$, $a_2 = \sqrt{0.5}$, $a_6 = \sqrt{0.2}$, $a_7 = \sqrt{0.1}$. The circuit requires seven multi-controlled y-rotations

$$c_{q_1=0}c_{q_2=0}\, R_{y,q_3}(\beta_1^1), \quad \beta_1^1 = 0$$
$$c_{q_1=0}c_{q_2=1}\, R_{y,q_3}(\beta_2^1), \quad \beta_2^1 = 0$$
$$c_{q_1=1}c_{q_2=0}\, R_{y,q_3}(\beta_3^1), \quad \beta_3^1 = 0$$
$$c_{q_1=1}c_{q_2=1}\, R_{y,q_3}(\beta_4^1), \quad \beta_4^1 = 1.231..$$
$$c_{q_1=0}\, R_{y,q_2}(\beta_1^2), \quad \beta_1^2 = 2.014..$$
$$c_{q_1=1}\, R_{y,q_2}(\beta_2^2), \quad \beta_2^2 = 3.142..$$
$$R_{y,q_1}(\beta_1^3), \quad \beta_1^3 = 1.159..$$

5.2 Amplitude Encoding

We are left with only four gates to apply, and with the approximate values

$$\sin\left(\frac{1.231}{2}\right) = 0.577, \quad \cos\left(\frac{1.231}{2}\right) = 0.816$$

$$\sin\left(\frac{2.014}{2}\right) = 0.845, \quad \cos\left(\frac{2.014}{2}\right) = 0.534$$

$$\sin\left(\frac{3.142}{2}\right) = 1.000, \quad \cos\left(\frac{3.142}{2}\right) = 0$$

$$\sin\left(\frac{1.159}{2}\right) = 0.548, \quad \cos\left(\frac{1.159}{2}\right) = 0.837,$$

these gates can be written as

$$c_{q_1=1} c_{q_2=1} R_{y,q_3}(\beta_4^1) = \begin{pmatrix} 1 & 0 & 0 & 0 & 0 & 0 & 0 & 0 \\ 0 & 1 & 0 & 0 & 0 & 0 & 0 & 0 \\ 0 & 0 & 1 & 0 & 0 & 0 & 0 & 0 \\ 0 & 0 & 0 & 1 & 0 & 0 & 0 & 0 \\ 0 & 0 & 0 & 0 & 1 & 0 & 0 & 0 \\ 0 & 0 & 0 & 0 & 0 & 1 & 0 & 0 \\ 0 & 0 & 0 & 0 & 0 & 0 & 0.816 & 0.577 \\ 0 & 0 & 0 & 0 & 0 & 0 & -0.577 & 0.816 \end{pmatrix},$$

$$c_{q_1=0} R_{y,q_2}(\beta_1^2) = \begin{pmatrix} 0.534 & 0 & 0.845 & 0 & 0 & 0 & 0 & 0 \\ 0 & 0.534 & 0 & 0.845 & 0 & 0 & 0 & 0 \\ -0.845 & 0 & 0.534 & 0 & 0 & 0 & 0 & 0 \\ 0 & -0.845 & 0 & 0.534 & 0 & 0 & 0 & 0 \\ 0 & 0 & 0 & 0 & 1 & 0 & 0 & 0 \\ 0 & 0 & 0 & 0 & 0 & 1 & 0 & 0 \\ 0 & 0 & 0 & 0 & 0 & 0 & 1 & 0 \\ 0 & 0 & 0 & 0 & 0 & 0 & 0 & 1 \end{pmatrix},$$

$$c_{q_1=1} R_{y,q_2}(\beta_2^2) = \begin{pmatrix} 1 & 0 & 0 & 0 & 0 & 0 & 0 & 0 \\ 0 & 1 & 0 & 0 & 0 & 0 & 0 & 0 \\ 0 & 0 & 1 & 0 & 0 & 0 & 0 & 0 \\ 0 & 0 & 0 & 1 & 0 & 0 & 0 & 0 \\ 0 & 0 & 0 & 0 & 0 & 0 & 1 & 0 \\ 0 & 0 & 0 & 0 & 0 & 0 & 0 & 1 \\ 0 & 0 & 0 & 0 & -1 & 0 & 0 & 0 \\ 0 & 0 & 0 & 0 & 0 & -1 & 0 & 0 \end{pmatrix},$$

and

$$R_{y,q_1}(\beta_1^3) = \begin{pmatrix} 0.837 & 0 & 0 & 0 & 0.548 & 0 & 0 & 0 \\ 0 & 0.837 & 0 & 0 & 0 & 0.548 & 0 & 0 \\ 0 & 0 & 0.837 & 0 & 0 & 0 & 0.548 & 0 \\ 0 & 0 & 0 & 0.837 & 0 & 0 & 0 & 0.548 \\ -0.548 & 0 & 0 & 0 & 0.837 & 0 & 0 & 0 \\ 0 & -0.548 & 0 & 0 & 0 & 0.837 & 0 & 0 \\ 0 & 0 & -0.548 & 0 & 0 & 0 & 0.837 & 0 \\ 0 & 0 & 0 & -0.548 & 0 & 0 & 0 & 0.837 \end{pmatrix}.$$

Applying these gates to the amplitude vector $(\sqrt{0.2}, 0, \sqrt{0.5}, 0, 0, 0, \sqrt{0.2}, \sqrt{0.1})^T$, the effect of the first operation is to set the last entry to zero, the second operation sets the third entry to zero and the third operation swaps the 5th element with the 7th. The last operation finally cancels the 5th element and results in $(1, 0, 0, 0, 0, 0, 0, 0)^T$ as desired. State preparation means to apply the inverse gates in the inverse order.

5.2.2 Qubit-Efficient State Preparation

The above routine is always possible to use for amplitude-efficient state preparation, but requires an exponential number of operations regarding the number of qubits. In the gate model of quantum computing, a number of proposals have been brought forward to prepare specific classes of states qubit-efficiently, or in $\log(MN)$. We want to present some of those ideas and the conditions to which they apply.

5.2.2.1 Parallelism-Based Approach

Grover and Rudolph suggest a scheme that is linear in the number of qubits for the case that we know an efficiently integrable one-dimensional probability distribution $p(a)$ of which the state is a discrete representation [19]. More precisely, the desired state has to be of the form

$$|\psi\rangle = \sum_{i=1}^{2^n} \sqrt{p(i\Delta a)}|i\rangle = \sum_{i=1}^{2^n} \sqrt{p_i}|i\rangle,$$

with $\Delta a = \frac{1}{2^n}$. The desired quantum state is a coarse grained qsample (see Sect. 3.4.3) of a continuous distribution $p(a)$ (see Fig. 5.6). The "efficiently integrable" condition means that we need an algorithm on a classical computer that calculates definite integrals of the probability distribution efficiently, like for the Gaussian distribution.

5.2 Amplitude Encoding

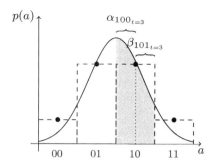

Fig. 5.6 Grover and Rudolph's state preparation algorithm for efficiently integrable probability distributions [19]. After step $t = 2$ the domain is divided into four regions $i = 00, 01, 10, 11$ of size $\Delta x = 1/2^2$, which defines a 2-qubit quantum state with amplitudes $p_{00}, p_{01}, p_{10}, p_{11}$ (indicated by the black dots). In the third step, each of the four regions (here demonstrated for $i = 10$) gets split into two. The parameters α and β are the probability to find a random variable in the left or right part of the region. This procedure successively prepares a finer and finer discretisation of the probability distribution

The idea of the scheme is to successively rotate each of the n qubits and get a finer and finer coarse graining of the distribution. Assume Step $(t - 1)$ prepared a qsample

$$|\psi^{(t-1)}\rangle = \sum_{i=1}^{2^{t-1}} \sqrt{p_i^{(t-1)}}|i\rangle|00...0\rangle.$$

The index register $|i\rangle$ contains the first $t - 1$ qubits. The next step t rotates the t'th qubit according to

$$|\psi^{(t)}\rangle = \sum_{i=1}^{2^t} \sqrt{p_i^{(t-1)}}|i\rangle(\sqrt{\alpha_i}|0\rangle + \sqrt{\beta_i}|1\rangle)|0...0\rangle,$$

such that α_i, $[\beta_i]$ are the probabilities for a random variable to lie in the left [right] half of the i'th interval of the probability distribution. In the tth step, the input domain is discretised into 2^t equidistant intervals which is visualised in Fig. 5.6. This defines a new qsample

$$|\psi^{(t)}\rangle = \sum_{i=1}^{2^t} \sqrt{p_i^{(t)}}|i\rangle|0...0\rangle,$$

where the index register now has t qubits and the remaining qubits in the ground state are reduced by one. With each step, this process prepares an increasingly fine discretisation of the probability distribution $p(a)$.

The Grover and Rudolph suggestion is in fact rather similar to the state preparation routine in the previous section, but this time we do not have to hardcode the 'lookup-table' for each state of the first $t-1$ qubits, but we can use quantum parallelism to compute the values α_i, β_i for all possible combinations. This more general version of the idea was proposed by Kaye and Mosca [20], who do not refer to probability distributions but demand that the conditional probability $p(q_k = 1|q_1...q_{k-1})$, which is the chance that given the state $q_1...q_{k-1}$ of the previous qubits, the k'th qubit is in state 1, is easy to compute. It is an interesting task to find classical datasets which allow for this trick.

5.2.2.2 Oracle-Based Approach

Soklakov and Schack [21] propose an alternative qubit-efficient scheme to approximately prepare quantum states whose amplitudes $\alpha_i = \sqrt{p_i}$ represent a discrete probability distribution p_i, $i = 1, ..., 2^n = N$, and all probabilities p_i are of the order of $1/\eta N$ for $0 < \eta < 1$. With this condition, they can use a Grover-type algorithm which with probability greater than $1 - \nu$ has an error smaller than ϵ in the result, and which is polynomial in η^{-1}, ϵ^{-1}, ν^{-1}. To sketch the basic idea (Fig. 5.7), a series of oracles is defined which successively marks basis states whose amplitudes are increased by amplitude amplification. The first oracle only marks basis states $|i\rangle$ for which the desired probability p_i is larger than a rather high threshold. With each step (defining a new oracle) the threshold is lowered by a constant amount. In the end, large amplitudes have been grown in more steps than small ones, and the final distribution is as fine as the number of steps allows. One pitfall is that we need to know the optimal number of Grover iterations for a given oracle. For this quantum counting can be applied to estimate the number of marked states in the current step. As with all oracle-based algorithms, the algorithm requires an implementation of the oracle to be used in practice.

5.2.2.3 Quantum Random Access Memory

Under the condition that the states to prepare are sufficiently uniform, a more generic approach is to refer to the quantum random access memory introduced in Eq. (5.4). The quantum random access memory allows access of classically stored information in superposition by querying an index register and is referred to by many authors in quantum machine learning [22–25]. In Sect. 5.1.1 this was used to prepare a dataset in basis encoding, but with one additional step we can extend its application to amplitude encoding as well. This step involves a conditional rotation and *branch selection* procedure that we already encountered in Step 3 of the quantum matrix inversion routine of Sect. 3.5.3. Let us assume the quantum random access memory prepared a state

5.2 Amplitude Encoding

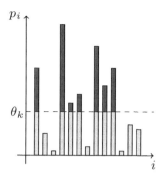

Fig. 5.7 In the k'th step of the routine of Soklakov and Schack [21], an oracle is applied to mark the states whose probabilities p_i are larger than a certain threshold θ_k (here in red). These states are amplified with the Grover iterator, resulting in a state that looks qualitatively like the red bars only. In the next step, the threshold is lowered to include the states with a slightly smaller probability in amplitude amplification, until the desired distribution of the p_i is prepared

$$\frac{1}{\sqrt{N}} \sum_{i=1}^{N} |i\rangle |x_i\rangle |0\rangle,$$

where $x_i \leq 1$ are the entries of a classical real vector we want to amplitude encode. Now rotate the ancilla qubit conditional on the $|x_i\rangle$ register to get

$$\frac{1}{\sqrt{N}} \sum_{i=1}^{N} |i\rangle |x_i\rangle (\sqrt{1 - |x_i|^2}|0\rangle + x_i |1\rangle). \quad (5.10)$$

The details of this step depend on how x_i is encoded into the register. If we assume binary fraction encoding, such a conditional rotation could be implemented by τ single-qubit conditional rotations on the ancilla that are controlled by the $q_1...q_\tau$ qubits of the second register. For qubit q_k the rotation would turn the amplitude closer to $|1\rangle$ by a value of $\frac{1}{2^k}$.

Now the ancilla has to be measured to 'select the desired branch' of the superposition (see Fig. 5.8). If the measurement results in $|1\rangle$ we know that the resulting state is in

$$\frac{1}{\sqrt{N p_{\text{acc}}}} \sum_{i=1}^{N} x_i |i\rangle |x_i\rangle |1\rangle,$$

where p_{acc} is the success or acceptance probability that renormalises the state. Discarding the last register and uncomputing the last but one (which is possible because the superposition is 'kept up' by the index register and no unwanted interference happens), we get a state that amplitude encodes a vector $x = (x_1, ..., x_N)^T$. Of course, one could also start with a non-uniform superposition and would get a product of the

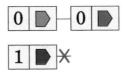

Fig. 5.8 Schematic illustration of the postselective measurement of the branch selection procedure. The two states 0 and 1 of the ancilla qubit are entangled with two different states marked in red and blue. A post-selective measurement selects only one branch of the superposition (here the 0-branch). The state entangled with this branch is formally renormalised

initial and new amplitude. If the measurement results in $|0\rangle$, we have to repeat the entire routine.

The routine for arbitrary state preparation with a quantum random access memory succeeds only with a probability of $p_{\text{acc}} = \sum_i |x_i|^2/N$ which obviously depends on the state to encode. A uniform distribution with $x_i = 1$ ensures that $p_{\text{acc}} = \sum_i \frac{1}{N} = 1$, and in the extreme case of only one non-zero amplitude of 1 we get $p_{\text{acc}} = \frac{1}{N}$, which is exponentially small in the number of qubits n. In other words, we would have to repeat the measurement $\mathcal{O}(N = 2^n)$ times on average to prepare the correct state. Luckily, very sparse states can be prepared by other methods. For example, for a one-sparse vector one can simply flip the qubit register from the ground state into a basis state representing the index i. Zhao et al. [26] therefore propose that in case of s sparse vectors, one does not apply the quantum random access memory to a uniform superposition, but a superposition

$$\frac{1}{\sqrt{s}} \sum_{i|x_i \neq 0} |i\rangle$$

of the basis states representing indices of non-zero amplitudes (Fig. 5.7).

There are many other possible ways to prepare quantum states beyond quantum circuits. An interesting perspective is offered by Aharanov et al. [1]. They present the framework of "adiabatic state generation" as a natural (and polynomially equivalent) alternative to state preparation in the gate model. The idea is to perform an adiabatic evolution of a quantum system that is initially in the ground state of a generic Hamiltonian, to the ground state of a final Hamiltonian. If the evolution is performed slow enough, we end up with the system in the ground state of the final Hamiltonian, which is the state we wish to prepare. This translates the question of which initial states can be easily prepared to the question of which ground states of Hamiltonians are in reach of adiabatic schemes, i.e. have small spectral gaps between the ground and the first excited state. A somewhat related idea is to use the unique stationary states of a dissipative process in an open quantum system.

In summary, quantum state preparation for amplitude encoded states relevant to machine learning algorithms is a topic that still requires a lot of attention, and claims of exponential speedups from qubit-efficient state preparation algorithms should therefore be viewed with a pinch of skepticism.

5.2 Amplitude Encoding

5.2.3 Computing with Amplitudes

In contrast to basis encoding, amplitude encoding radically limits the type of computations we can perform. As we saw in Chap. 3, there are two general operations for the manipulation of amplitudes in the formalism of quantum theory: unitary transformations and measurements.

As stated in detail in Chap. 3, unitary transformations are linear transformations of the amplitude vector that preserve its length. Even if we consider subsystems (where the evolution is not unitary any more), we still get linear dynamics, for example expressed by the so called Kraus formalism (see [27]). In fact, it has been shown that introducing a 'coherent' nonlinearity to quantum evolution implies the ability to solve NP-hard problems [28]. Essentially this is because we could increase exponentially small signals in a superposition to efficiently measurable information. With this, Grover search could be done exponentially faster. It has also been claimed that any nonlinear map would allow for super-luminal communication [29] and negate the laws of thermodynamics [30]. In short, a nonlinear version of quantum theory can be considered highly unlikely.[3]

A projective measurement is clearly a nonlinear operation, but its outcome is stochastic and if averaging over several runs it will reflect the distribution of amplitudes. But not all is lost: An important way to introduce nonlinearities on the level of amplitudes are post-selective measurements that we used in the branch selection scheme of the previous section. In a post-selective measurement of a single qubit, the state of the qubit is measured in the computational basis and the algorithm is only continued if this qubit is found in a particular state. This condition works like an "if" statement in programming. Post-selection has the effect of setting the amplitudes in a branch of superpositions to zero, namely the branch that is entangled with the qubit in the state of an unsuccessful measurement. Note that in most cases we can push the post-selection to the end of the algorithm and reject final measurement outcomes if the result of said qubit is in the desired state. A second, closely related idea is to use so called *repeat-until-success circuits* [32]. Here the unsuccessful measurement does not result in repeating the entire algorithm, but prescribes a small correction, which restores the state before the postselective measurement. Postselection can therefore be repeated until success is observed. As mentioned before, these procedures are non-unitary and for runtime considerations the likeliness of success has to be taken into account.

5.3 Qsample Encoding

Qsample encoding has been introduced to associate a real amplitude vector $(\sqrt{p_1}, ..., \sqrt{p_N})^T$ with a classical discrete probability distribution $p_1, ..., p_N$. This is in a sense a hybrid case of basis and amplitude encoding, since the information

---
[3]Note that there are *effective* nonlinear dynamics, see for example [31].

we are interested in is represented by amplitudes, but the N features are encoded in the qubits. An amplitude-efficient quantum algorithm is therefore exponential in the input dimension N, while a qubit-efficient algorithm is polynomial in the input.

For a given probability distribution, state preparation works the same way as in amplitude encoding. In some cases the distribution may be a discretisation of a parametrised continuous and efficiently integrable distribution $p(x)$, in which case the implicit Grover-Rudolph scheme can be applied.

The probabilistic nature of quantum systems lets us manipulate the classical distribution with the quantum system and implement some standard statistic computations generically. We can prepare joint distributions easily, marginalise over qubits 'for free', and we can perform a rejection sampling step via branch selection. These methods are nothing other than an application of basic quantum theory from Sects. 3.1.2.3 and 3.1.3.5, but can be useful when regarded from the perspective of manipulating distributions represented by a qsample.

5.3.1 Joining Distributions

When joining two quantum systems representing a qsample,

$$|\psi_1\rangle = \sum_{i=1}^{2^n} \sqrt{u_i} \, |i\rangle,$$

and

$$|\psi_2\rangle = \sum_{j=1}^{2^n} \sqrt{s_j} \, |j\rangle,$$

the joint state of the total system is described as a tensor product of the two states (see Sect. 3.1.2.3),

$$|\psi_1\rangle \otimes |\psi_2\rangle = \sum_{i,j}^{2^n} \sqrt{u_i s_j} \, |i\rangle |j\rangle.$$

Sampling the binary string ij from this state is observed with probability $u_i s_j$, which is a product of the two original probabilities. In other words, the qsample of two joint qsamples is a product distribution.

5.3.2 Marginalisation

Given a qsample,

$$|\psi\rangle = \sum_{i=1}^{2^n} \sqrt{p_i} \, |i\rangle.$$

5.3 Qsample Encoding

The mathematical operation of tracing over a qubit q_k corresponds to omitting the kth qubit from the description, leading to a statistical ensemble over all possible states of that qubit Sect. 3.1.2.3. The resulting state is in general (that is, unless the k'th qubit was unentangled) a mixed state.

To write down the effect of a trace operation is awkward and hides the simplicity of the concept, but we will attempt to do it anyways in order to show the equivalence to marginalisation. Remember that the ith computational basis state $|i\rangle$ represents a binary sequence, more precisely the binary representation of $i - 1$ (for example, $|1\rangle$ represents $|0...0\rangle$). Let us define the shorthand $i_{\neg k}$ for the binary sequence indicated by i, but without the kth bit, and let $i_{k=0,1}$ be the same sequence, but this time the kth bit is set to either 0 or 1. Also, let $\langle i|0_k\rangle$ and $\langle i|1_k\rangle$ be the inner product between $|i\rangle$ and a state $\mathbb{1} \otimes |0\rangle \otimes \mathbb{1}$ and $\mathbb{1} \otimes |1\rangle \otimes \mathbb{1}$, respectively, where $|0\rangle, |1\rangle$ sits at the kth position. With this notation, tracing out the kth qubit has the following effect,

$$\operatorname{tr}_k \{|\psi\rangle\langle\psi|\} = \operatorname{tr}_k \left\{ \sum_i \sum_{i'} \sqrt{p(i)p(i')}\, |i\rangle\langle i'| \right\}$$

$$= \sum_i \sum_{i'} \sqrt{p(i)p(i')} \left(\langle 0_k|i\rangle\langle i'|0_k\rangle + \langle 1_k|i\rangle\langle i'|1_k\rangle \right)$$

$$= \sum_{i_{\neg k}} \sum_{i'_{\neg k}} \left(\sqrt{p(i_{k=0})p(i'_{k=0})} + \sqrt{p(i_{k=1})p(i'_{k=1})} \right) |i_{\neg k}\rangle\langle i'_{\neg k}|.$$

The probability to measure the computational basis state $|i_{\neg k}\rangle$ is given by the corresponding diagonal element of the resulting density matrix, which is given by

$$\left(\sqrt{p(i_{k=0})p(i_{k=0})} + \sqrt{p(i_{k=1})p(i_{k=1})} \right) = \left(p(i_{k=0}) + p(i_{k=1}) \right).$$

This is the sum of the probability of finding k in 0 and the probability of finding k in 1, just as we would expect from common sense.

In the classical case, given a probability distribution $p(b_1...b_n)$ over n-bit strings $b_1...b_N$, marginalising over the state of the k'th bit will yield the marginal distribution

$$p(i_{\neg k}) = p(b_1..b_{k-1}b_{k+1}..b_n) \tag{5.11}$$
$$= p(b_1..b_k = 0..b_n) + p(b_1..b_k = 1..b_n). \tag{5.12}$$

This demonstrates that the tracing in the quantum formalism is a 'classical' statistical marginalisation. In other words, dropping the k'th qubit from our algorithm effectively implements a qsample representing a marginal distribution.

5.3.3 Rejection Sampling

There is a close relationship between branch selection or postselection and classical rejection sampling, which can help to understand, design and compare quantum algorithms. Rejection sampling is a method to draw samples from a distribution $p(x)$ that is difficult to sample from, but for which the probabilities at each point x are easy to evaluate. The idea is to use a distribution $q(x)$ from which it is easy to draw samples (i.e., a Gaussian or uniform distribution), and scale it such that $p(x) \leq Kq(x)$ for an integer $K < \infty$ (see Fig. 5.9). One then draws a sample x' from $q(x)$ as well as a random number u' from the interval $[0, 1]$. One accepts the sample if $u'Kq(x') < p(x')$ and rejects it otherwise. In other words, the chance to accept the sample from $q(x)$ depends on the ratio of the original and the alternative distribution.

Obviously, if $Kq(x')$ and $p(x')$ are almost equal, the acceptance probability is very high, but in regions where $p(x') << Kq(x')$ we will almost never accept. When repeating this procedure, the accepted samples x' are *effectively* drawn from $p(x)$. The general probability of acceptance can be calculated as

$$p_{\mathrm{acc}} = \left\langle p\left(u < \frac{p(x)}{Kq(x)}\right)\right\rangle$$
$$= \left\langle \frac{p(x)}{Kq(x)}\right\rangle$$
$$= \int \frac{p(x)}{Kq(x)} q(x) dx$$
$$= \frac{1}{K} \int p(x) dx = \frac{1}{K},$$

which shows that K is an important figure of merit for the runtime [33]. The larger it is, or the bigger the difference between the two distributions, the more samples are rejected.

In branch selection, we have a quantum state of the form

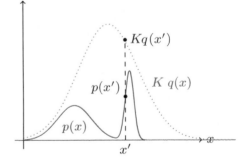

Fig. 5.9 Illustration of rejection sampling. In order to sample from the desired distribution $p(x)$ (smooth line) one samples x' instead from a distribution $K\,q(x)$ (dotted line) and accepts the sample with a probability depending on the proportion of the two distributions at that point

5.3 Qsample Encoding

$$\sum_{i=1}^{N} \sqrt{a_i}|i\rangle(\sqrt{1-b_i}|0\rangle + \sqrt{b_i}|1\rangle),$$

where from Eq. (5.10) the uniform coefficients $\frac{1}{\sqrt{N}}$ were replaced by more general amplitudes $\sqrt{a_1}, ..., \sqrt{a_N}$, and the value x_i was replaced by $\sqrt{b_i}$ to avoid confusion with the classical notation. For ease of comparison with classical probability theory, let us assume that the a_i and b_i are all real numbers in $[0, 1]$. By measuring the $|i\rangle$ register we sample from the distribution $\mathcal{A} = \{a_1, ..., a_N\}$, but our goal is to sample from the 'unknown' distribution $\mathcal{B} = \{a_1b_1, ..., a_Nb_N\}$. The conditional measurement on the ancilla therefore plays the role of the rejection step if we associate

$$q(x) \leftrightarrow \mathcal{A} \tag{5.13}$$
$$p(x) \leftrightarrow \mathcal{B}. \tag{5.14}$$

To see this, consider the probability b_i of measuring the ancilla in $|1\rangle$ for state $|i\rangle$ and compare it to the 'rejection sampling' probability of success for sample i. Using the replacement of Eqs. (5.14) we get $p_{acc}(i) = p(ua_i < a_ib_i) = p(u < b_i) = b_i$. The general probability of acceptance is

$$\begin{aligned} p_{acc} &= \langle p(u < b_i) \rangle \\ &= \langle b_i \rangle \\ &= \sum_i b_i a_i, \end{aligned}$$

where the expectation value is taken with respect to the distribution $\mathcal{A}$. This is equivalent to the probability of a successful branch selection.

5.4 Hamiltonian Encoding

While the approaches discussed so far encode features explicitly into quantum states, we can choose an implicit encoding which uses the features to define the evolution of the quantum system. This approach differs a bit from the previous ones. Instead of preparing a quantum state which contains the features or a distribution in its mathematical description, they now define an evolution applied to a state.

We mentioned in Sect. 3.4.4 that there are different ways to encode data into the dynamics of a quantum system, and focused on Hamiltonian encoding. Hamiltonian encoding associates the Hamiltonian of a system with a matrix that represents the data, such as the design matrix X containing the feature vectors as rows, or the Gram matrix $X^T X$. As discussed in Sect. 3.4.2, we may have to use preprocessing tricks that make this matrix Hermitian. Section 3.5.3 presented ways in which Hamiltonian encoding allows us to extract eigenvalues of matrices, or to multiply them to an amplitude vector. We therefore want to summarise some results on the resources needed to evolve a system by a given Hamiltonian. Similar to amplitude encoding, the general case will yield amplitude-efficient state preparation algorithms, while for some limited datasets we can get qubit-efficient schemes. Possibly not surprisingly, this is most prominently the case for sparse data.

To use Hamiltonian encoding we need to be able to implement an evolution

$$|\psi'\rangle = e^{-iH_A t}|\psi\rangle \tag{5.15}$$

on a quantum computer. The Hamiltonian H_A "encodes" a Hermitian matrix A of the same dimensions, which means that the matrix representation of H_A is entry-wise equivalent to A. The initial quantum state $|\psi\rangle$ describes a system of n qubits, and one can think of the final state $|\psi'\rangle$ as a quantum state that now "contains" the information encoded into the Hamiltonian—for example H's eigenvalues in the phase of the amplitudes.

The process of implementing a given Hamiltonian evolution on a quantum computer is called *Hamiltonian simulation* [34]. Since we only consider time-independent Hamiltonians, any unitary transformation can be written as $e^{-iH_A t}$. Hamiltonian simulation on a gate-based quantum computer[4] is therefore very closely related to the question of how to decompose a unitary into quantum gates. However, an operator-valued exponential function is not trivial to evaluate, and given H there are more practical ways than computing the unitary matrix $U = e^{-iH_A t}$ and decomposing it into gates.

The problem of (digital) Hamiltonian simulation can be formulated as follows: Given a Hamiltonian H, a quantum state $|\psi\rangle$, an error $\epsilon > 0$, the evolution time t (which can be imagined as a scaling factor to H), and an appropriate norm $||\cdot||$ that measures the distance between quantum states, find an algorithm which implements the evolution of Eq. (5.15) so that the final state of the algorithm, $|\tilde{\psi}\rangle$, is ϵ-close to the desired final state $|\psi'\rangle$,

$$|| |\psi'\rangle - |\tilde{\psi}\rangle || \leq \epsilon.$$

We want to summarise the basic ideas of Hamiltonian simulation and then state the results for qubit-efficient simulations.

5.4.1 Polynomial Time Hamiltonian Simulation

Consider a Hamiltonian H that can be decomposed into a sum of several "elementary" Hamiltonians $H = \sum_{k=1}^{K} H_k$ so that each H_k is easy to simulate. For non-commuting

---

[4] Hamiltonian simulation research can be distinguished into *analog* and *digital* approaches to simulation. Roughly speaking, analog simulation finds quantum systems that "naturally" simulate Hamiltonians, while digital simulation decomposes the time evolution into quantum gates, which is more relevant in the context of this book.

5.4 Hamiltonian Encoding

H_k, we cannot apply the factorisation rule for scalar exponentials, or

$$e^{-i \sum_k H_k t} \neq \prod_k e^{-iH_k t}.$$

An important idea introduced by Seth Lloyd in his seminal paper in 1996 [35] was to use the first-order Suzuki-Trotter formula instead,

$$e^{-i \sum_k H_k t} = \prod_k e^{-iH_k t} + \mathcal{O}(t^2). \tag{5.16}$$

The Trotter formula states that for small t the factorisation rule is approximately valid. This can be leveraged when we write the evolution of H for time t as a sequence of small time-steps of length Δt,

$$e^{-iHt} = (e^{-iH\Delta t})^{\frac{t}{\Delta t}}.$$

While for the left side, the Trotter formula has a large error, for each small time step Δt, the error in Eq. (5.16) becomes negligible. Of course, there is a trade-off: The smaller Δt, the more often the sequence has to be repeated. But overall, we have a way to simulate H by simulating the terms H_k.

This approach shows that if we know a decomposition of Hamiltonians into sums of elementary Hamiltonians that we know how to simulate, we can approximately evolve the system in the desired way. The case-specific decomposition may still be non-trivial, and some examples are summarised in [34]. One example for such a decomposition is a sum of Pauli operators. Every Hamiltonian can be written as

$$H = \sum_{k_1,\cdots,k_n=1,x,y,z} a_{k_1,\ldots,k_n}(\sigma^1_{k_1} \otimes \cdots \otimes \sigma^n_{k_n}). \tag{5.17}$$

The sum runs over all possible tensor products of Pauli operators applied to the n qubits, and the coefficients

$$a_{k_1,\cdots,k_n} = \frac{1}{2^n}\mathrm{tr}\{(\sigma^1_{k_1} \otimes \cdots \otimes \sigma^n_{k_n})H\},$$

define the entries of the Hamiltonian. For example, for two qubits we can write

$$H_2 = a_{1,1}(\sigma^1_1 \otimes \sigma^2_1) + a_{1,x}(\sigma^1_1 \otimes \sigma^2_x) + \cdots + a_{z,z}(\sigma^1_z \otimes \sigma^2_z).$$

This decomposition has $4^n = 2^n \times 2^n$ terms in general. If the evolution describes a physical problem we can hope that only local interactions—terms in which $\sigma^i = \mathbb{1}$ for all but a few neighbouring qubits—are involved. For machine learning this could also be interesting, when the features are generated by a "local" process and we can hope that correlations in the data reduce the number of terms in the sum of Eq. (5.17).

5.4.2 Qubit-Efficient Simulation of Hamiltonians

There are special classes of Hamiltonians that can be simulated in time that is logarithmic in their dimension. If the Hamiltonian encodes a data matrix, this means that the runtime is also logarithmic in the dimension of the dataset. The most prominent example are Hamiltonians that only act on a constant number of qubits, so called *strictly local Hamiltonians* [35]. This is not surprising when we remember that a local Hamiltonian can be expressed by a constant number of terms constructed from Pauli operators.

More generally, it has been shown that *sparse* Hamiltonians can be simulated qubit-efficiently [1, 36, 37]. An s-sparse Hamiltonian has at most s non-zero elements in each row and column. One usually assumes that the non-zero elements of are given by "oracular" access, which means that for integers $i, l \in 2^n \times [1...s]$, we have access to the lth non-zero element from the ith row. For example, the non-zero elements could be stored in a sparse array representation in a classical memory, which we can use to load them into a quantum register via the oracle call $|i, j, 0\rangle \to |i, j, H_{ij}\rangle$.

Recent proposals [37–39] manage to reduce the asymptotic runtime of former work considerably. They employ a combination of demanding techniques which go beyond the scope of this book, which is why we only state the result. To recapture, we want to simulate an s-sparse Hamiltonian H for time t and to error ϵ, and H's non-zero elements are accessible by an efficient oracle. Writing $\tau = s||H||_{\max}t$, the number of times we have to query the classical memory for elements of the Hamiltonian grows as

$$\mathcal{O}\left(\tau \frac{\log(\frac{\tau}{\epsilon})}{\log(\log(\frac{\tau}{\epsilon}))}\right)$$

and the number of 2-qubit gates we need grows with

$$\mathcal{O}\left(\tau \left(n + \log^{\frac{5}{2}}(\frac{\tau}{\epsilon})\right) \frac{\log(\frac{\tau}{\epsilon})}{\log(\log(\frac{\tau}{\epsilon}))}\right).$$

The runtime of the algorithm that simulates H is hence roughly linear in s and n, which is the number of qubits. This ensures a poly-logarithmic and therefore logarithmic dependency on the dimension of the matrix A we wish to encode into the Hamiltonian. It was also shown that this runtime is nearly optimal for Hamiltonian simulation of sparse Hamiltonians.

Note that there are other classes of qubit-efficiently simulable Hamiltonians. One class are Hamiltonians where the positions of the nonzero entries define a tree-like graph. For example, if a Hamiltonian has the non-zero elements at positions $(i, j) = \{(1, 2), (1, 3), (3, 6), (3, 7)\}$, we can read each index pair as an edge between nodes, and the graph structure is shown in Fig. 5.10. The proof draws on techniques of *quantum walks* [40]. How applicable such structures are to machine learning applications is an open question.

Fig. 5.10 Tree structure for the example of sparse Hamiltonian simulation in the text

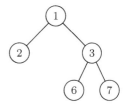

5.4.3 Density Matrix Exponentiation

There are special conditions under in which we can guarantee qubit-efficiency for densely populated Hamiltonians. The most important in the context of quantum machine learning is the technique of *density matrix exponentiation*. This technique can be used for more general amplitude-efficient simulations, but becomes qubit-efficient for low-rank Hamiltonians. Instead of querying an oracle, the data is initially encoded in a density matrix, for which we can apply techniques from previous sections. Although density matrix exponentiation relies on many technical details and its understanding requires a high level of quantum computing expertise, we want to try to briefly sketch the outline of the idea. This section is consequently suited for more advanced readers.

The goal of density matrix exponentiation is to simulate a Hamiltonian $e^{iH_\rho t}$ that encodes a non-sparse density matrix ρ and apply it to a quantum state σ. This is only possible because both operators are Hermitian, and every density matrix has an entry-wise equivalent Hamiltonian. It turns out that simulating H_ρ is approximately equivalent to simulating a swap operator S, applying it to the state $\rho \otimes \sigma$ and taking a trace operation. A swap operator is sparse and its simulation therefore efficient. It also means that whatever data we can encode in the entries of a density matrix, we can approximately encode it into a Hamiltonian.

The formal relation reads

$$\mathrm{tr}_2\{e^{-iS\Delta t}(\sigma \otimes \rho)e^{iS\Delta t}\} = \sigma - i\Delta t[\rho, \sigma] + \mathcal{O}(\Delta t^2) \quad (5.18)$$

$$\approx e^{-i\rho\Delta t}\sigma e^{i\rho\Delta t}. \quad (5.19)$$

Note that $\sigma - i\Delta t[\rho, \sigma]$ are the first terms of the exponential series $e^{-i\rho\Delta t}\sigma e^{i\rho\Delta t}$. In words, simulating the swap operator that 'exchanges' the state of the qubits of state ρ and σ for a short time Δt, and taking the trace over the second quantum system results in an effective dynamic as if exponentiating the second density matrix.

To prove the relation, consider two general mixed states $\rho = \sum_{i,i'=1}^{N} a_{ii'}|i\rangle\langle i'|$ and $\sigma = \sum_{jj'=1}^{N} b_{j,j'}|j\rangle\langle j'|$, where $\{|i\rangle\}, \{|j\rangle\}$ are the computational bases in the Hilbert spaces of the respective states, which have the same dimension. Now write the operator-valued exponential functions as a series

$$\text{tr}_2\{e^{-iS\Delta t}(\sigma \otimes \rho)e^{iS\Delta t}\} = \text{tr}_2\left\{\sum_{k,k'=0}^{\infty} \frac{(-i)^k \Delta t^k i^{k'} \Delta t^{k'}}{k!k'!} S^k(\sigma \otimes \rho)S^{k'}\right\}$$

and apply the swap operators. Since $S^2 = 1$, higher order terms in the series only differ in the prefactor, and for $\Delta t \ll 1$ the higher orders quickly vanish. We therefore only write the first few terms explicitly and summarise the vanishing tail of the series in a $\mathcal{O}(\Delta t^2)$ term,

$$\text{tr}_2\left\{\left(\sum_{ii'}\sum_{jj'} a_{ii'}b_{jj'}|j\rangle\langle j'| \otimes |i\rangle\langle i'|\right) + i\Delta t\left(\sum_{ii'}\sum_{jj'} a_{ii'}b_{jj'}|j\rangle\langle j'| \otimes |i\rangle\langle i'|\right)S\right.$$
$$\left. - i\Delta t S\left(\sum_{ii'}\sum_{jj'} a_{ii'}b_{jj'}|j\rangle\langle j'| \otimes |i\rangle\langle i'|\right) + \mathcal{O}(\Delta t^2)\right\}.$$

Next, apply the swap operator and the trace operation. Tracing out the second system means to 'sandwich' the entire expression by $\sum_k \langle k| \cdot |k\rangle$, where $\{|k\rangle\}$ is a full set of computational basis vectors in the Hilbert space of the second system. We get

$$\sum_{jj'} b_{jj'}|j\rangle\langle j'| + i\Delta t \sum_{ij}\sum_k a_{ki}b_{jk}|j\rangle\langle i| - i\Delta t \sum_{ij}\sum_k a_{ik}b_{kj}|i\rangle\langle j| + \mathcal{O}(\Delta t^2).$$

This is in fact the same as $\sigma - i\Delta t[\rho, \sigma] + \mathcal{O}(\Delta t^2)$, which can be shown by executing the commutator $[a, b] = ab - ba$,

$$\sigma + i\Delta t(\sigma\rho - \rho\sigma) + \mathcal{O}(\Delta t^2),$$

and inserting the expressions for ρ and σ,

$$\sigma + i\Delta t\left(\sum_{ii'}\sum_{jj'} a_{ii'}b_{jj'}|j\rangle\langle j'||i\rangle\langle i'| - \sum_{ii'}\sum_{jj'} a_{ii'}b_{jj'}|i\rangle\langle i'||j\rangle\langle j'|\right) + \mathcal{O}(\Delta t^2).$$

Note that the expressions $\rho\sigma$ and $\sigma\rho$ from the commutator are not abbreviated tensor products, but common matrix products, and we can use the relations $\langle j'|i\rangle = \delta_{j',i}$ and $\langle i'|j\rangle = \delta_{i',j}$ for orthonormal basis states. The error of the approximation is in $\mathcal{O}(\Delta t^2)$ which is negligible for sufficiently small simulation times Δt.

Density matrix exponentiation is often used in combination with phase estimation presented in Sects. 3.5.2 and 3.5.3. Once a density matrix containing data is exponentiated, we can 'apply' it to some amplitude encoded state and extract eigenvalues of ρ through a phase estimation routine using the the inverse quantum Fourier transform. However, to do so we need to be able to prepare powers of $(e^{-iH_\rho \Delta t})^k$ (compare to U^k in Sect. 3.5.2).

Lloyd et al. [41] show that this can be done by using of the order of $\mathcal{O}(\epsilon^{-3})$ copies of ρ joined with an index register of d qubits in superposition,

5.4 Hamiltonian Encoding

$$\sum_{k=1}^{2^d} |k\rangle\langle k| \otimes \sigma \otimes \rho^{(1)} \otimes \ldots \otimes \rho^{(2^d)}.$$

Instead of simulating a single swap operator, we now have to simulate a sequence of 2-qubit swap operators, each of which swaps the first state σ with the gth copy of ρ. The swap operator sequences are entangled with an index register, so that for index $|k\rangle$ the sequence of swap operators runs up to copy $\rho^{(k)}$,

$$\frac{1}{K}\sum_{k=1}^{K} |k\Delta t\rangle\langle k\Delta t| \otimes \prod_{g=1}^{k} e^{-iS_g \Delta t}$$

After taking the trace over all copies of ρ, this effectively implements the evolution

$$\sum_{k=1}^{K} |k\rangle\langle k| \otimes e^{-ikH_\rho \Delta t} \sigma\, e^{ikH_\rho \Delta t} + \mathcal{O}(\Delta t^2),$$

which is precisely in the form required to apply the quantum Fourier transform. For a proof, simply write out the expressions as seen above.

Density matrix exponentiation is qubit-efficient for low-rank matrices, given that the state ρ can be prepared qubit-efficiently. To see this, we first have to note that the desired accuracy is not necessarily a constant, but depends on the size of H_ρ, especially when we are interested in the eigenvalues of ρ. For example, consider the eigenvalues are approximately uniform. Since $\mathrm{tr}\rho = 1$, they are of the order of $\frac{1}{N}$ if N is the number of diagonal elements or the dimension of the Hilbert space of ρ. We certainly want the error to be smaller than the eigenvalues themselves, which means that $\epsilon < \frac{1}{N}$. The number of copies needed for the density matrix exponentiation in superposition hence grows with N^3, and since we have to apply swap operators to each copies, the runtime grows accordingly.

As a consequence, density matrix exponentiation for phase estimation is only polynomial in the number of qubits or qubit-efficient if the eigenvalues of H_ρ are dominated by a few large eigenvalues that do not require a small error to be resolved. In other words, the matrix has to be well approximable by a low-rank matrix. For design matrices containing the dataset, this means that the data is highly redundant.

References

1. Aharonov, D., Ta-Shma, A.: Adiabatic quantum state generation and statistical zero knowledge. In: Proceedings of the Thirty-Fifth annual ACM Symposium on Theory of Computing, pp. 20–29. ACM (2003)
2. Schuld, M., Petruccione, F.: Quantum machine learning. In: C. Sammut, G.I. Webb (eds.) Encyclopaedia of Machine Learning and Data Mining. Springer (2016)
3. Ventura, D., Martinez, T.: Quantum associative memory. Inf. Sci. **124**(1), 273–296 (2000)

4. Trugenberger, C.A.: Probabilistic quantum memories. Phys. Rev. Lett. **87**, 067901 (2001)
5. Giovannetti, V., Lloyd, S., Maccone, L.: Quantum random access memory. Phys. Rev. Lett. **100**(16), 160501 (2008)
6. Raoux, S., Burr, G.W., Breitwisch, M.J., Rettner, C.T., Chen, Y.-C., Shelby, R.M., Salinga, M., Krebs, D., Chen, S.-H., Lung, H.-L., et al.: Phase-change random access memory: a scalable technology. IBM J. Res. Dev. **52**(4.5):465–479 (2008)
7. Kyaw, T.H., Felicetti, S., Romero, G., Solano, E., Kwek, L.-C.: Scalable quantum memory in the ultrastrong coupling regime. Sci. Rep. **5**(8621) (2015)
8. Bennett, C.H.: Logical reversibility of computation. In: M. Demon (ed.) Entropy, Information, Computing, pp. 197–204 (1973)
9. Brown, L.D., Cai, T.T., DasGupta, A.: Interval estimation for a binomial proportion. Stat. Sci. 101–117 (2001)
10. Wilson, E.B.: Probable inference, the law of succession, and statistical inference. J. Am. Stat. Assoc. **22**(158), 209–212 (1927)
11. Aaronson, S.: Read the fine print. Nat. Phys. **11**(4), 291–293 (2015)
12. Kliesch, M., Barthel, T., Gogolin, C., Kastoryano, M., Eisert, J.: Dissipative quantum Church-Turing theorem. Phys. Rev. Lett. **107**(12), 120501 (2011)
13. Knill, E.: Approximation by quantum circuits (1995). arXiv:quant-ph/9508006
14. Mikko, M., Vartiainen, J.J., Bergholm, V., Salomaa, M.M.: Quantum circuits for general multiqubit gates. Phys. Rev. Lett. **93**(13), 130502 (2004)
15. Vartiainen, J.J., Möttönen, M., Salomaa, M.M.: Efficient decomposition of quantum gates. Phys. Rev. Lett. **92**(17), 177902 (2004)
16. Plesch, M., Brukner, Č.: Quantum-state preparation with universal gate decompositions. Phys. Rev. A **83**(3), 032302 (2011)
17. Iten, R., Colbeck, R., Kukuljan, I., Home, J., Christandl, M.: Quantum circuits for isometries. Phys. Rev. A **93**(3), 032318 (2016)
18. Möttönen, M., Vartiainen, J.J., Bergholm, V., Salomaa, M.M.: Transformation of quantum states using uniformly controlled rotations. Quantum Inf. Comput. **5**(467) (2005)
19. Grover, L., Rudolph,T.: Creating superpositions that correspond to efficiently integrable probability distributions (2002). arXiv:0208112v1
20. Kaye, P., Mosca, M.: Quantum networks for generating arbitrary quantum states. In: Proceedings of the International Conference on Quantum Information, OSA Technical Digest Series, pp. PB28. ICQI (2001). arXiv:quant-ph/0407102v1
21. Soklakov, A.N., Schack, R.: Efficient state preparation for a register of quantum bits. Phys. Rev. A **73**(1):012307 (2006)
22. Harrow, A.W., Hassidim, A., Lloyd, S.: Quantum algorithm for linear systems of equations. Phys. Rev. Lett. **103**(15), 150502 (2009)
23. Rebentrost, P., Mohseni, M., Lloyd, S.: Quantum support vector machine for big data classification. Phys. Rev. Lett. **113**, 130503 (2014)
24. Prakash, A.: Quantum Algorithms for Linear Algebra and Machine Learning. Ph.D thesis, EECS Department, University of California, Berkeley, Dec 2014
25. Wiebe, N., Braun, D., Lloyd, S.: Quantum algorithm for data fitting. Phys. Rev. Lett. **109**(5), 050505 (2012)
26. Zhao, Z., Fitzsimons, J.K., Fitzsimons, J.F.: Quantum assisted Gaussian process regression (2015). arXiv:1512.03929
27. Breuer, H.-P., Petruccione, F.: The Theory of Open Quantum Systems. Oxford University Press (2002)
28. Abrams, D.S., Lloyd, S.: Quantum algorithm providing exponential speed increase for finding eigenvalues and eigenvectors. Phys. Rev. Lett. **83**(24), 5162 (1999)
29. Gisin, N.: Weinberg's non-linear quantum mechanics and supraluminal communications. Phys. Lett. A **143**(1), 1–2 (1990)
30. Peres, A.: Nonlinear variants of Schrödinger's equation violate the second law of thermodynamics. Phys. Rev. Lett. **63**(10), 1114 (1989)

References

31. Meyer, D.A., Wong, T.G.: Nonlinear quantum search using the Gross-Pitaevskii equation. New J. Phys. **15**(6), 063014 (2013)
32. Paetznick, A., Svore, K.M.: Repeat-until-success: Non-deterministic decomposition of single-qubit unitaries. Quantum Inf. Comput. **14**, 1277–1301 (2013)
33. Andrieu, C., De Freitas, N., Doucet, A., Jordan, M.I.: An introduction to MCMC for machine learning. Mach. Learn. **50**(1–2):5–43 (2003)
34. Georgescu, I.M., Ashhab, S., Nori, F.: Quantum simulation. Rev. Mod. Phys. **86**, 153–185 (2014)
35. Lloyd, S.: Universal quantum simulators. Science **273**(5278), 1073 (1996)
36. Childs, A.M.: Quantum information processing in continuous time. Ph.D thesis, Massachusetts Institute of Technology (2004)
37. Berry, D.W., Ahokas, G., Cleve, R., Sanders, B.C.: Efficient quantum algorithms for simulating sparse Hamiltonians. Commun. Math. Phys. **270**(2), 359–371 (2007)
38. Berry, D.W., Childs, A.M., Cleve, R., Kothari, R., Somma, R.D.: Exponential improvement in precision for simulating sparse Hamiltonians. In: Proceedings of the 46th Annual ACM Symposium on Theory of Computing, pp. 283–292. ACM (2014)
39. Berry, D.W., Childs, A.M., Kothari, R.: Hamiltonian simulation with nearly optimal dependence on all parameters. In: IEEE 56th Annual Symposium on Foundations of Computer Science (FOCS), pp. 792–809. IEEE (2015)
40. Childs, A.M., Kothari, R.: Limitations on the simulation of non-sparse Hamiltonians. Quantum Inf. Comput. **10**(7), 669–684 (2010)
41. Lloyd, S., Mohseni, M., Rebentrost, P.: Quantum principal component analysis. Nat. Phys. **10**, 631–633 (2014)

Chapter 6
Quantum Computing for Inference

After the discussion of classical-quantum interfaces, we are now ready to dive into techniques that can bee used to construct quantum machine learning algorithms. As laid out in the introduction, there are two strategies to solve learning task with quantum computers. First, one can try to translate a classical machine learning method into the language of quantum computing. The challenge here is to combine quantum routines in a clever way so that the overall quantum algorithm reproduces the results of the classical model. The second strategy is more exploratory and creates new models that are tailor-made for the working principles of a quantum device. Here, the numerical analysis of the model performance can be much more important than theoretical promises of asymptotic speedups. In the remaining chapters we will look at methods that are useful for both approaches.

This chapter will focus on the first approach and present building blocks for quantum algorithms for *inference*, which we understand as algorithms that implement an input-output map $y = f(x)$ or a probability distribution $p(y|x)$ of a machine learning model. We will introduce a number of techniques, tricks, subroutines and concepts that are commonly used as building blocks in different branches of the quantum machine learning literature. The techniques rely very much on the encoding method, and the language developed in the previous chapter will play an important role.

The next Chap. 7 will look at *training*, or how to solve optimisation problems that typically occur in machine learning with a quantum computer. Chapter 8 considers the exploratory approach more closely and looks at genuine *quantum models* such as Ising models, generic quantum circuits or quantum walks, and shows some first attempts of how to turn these into machine learning algorithms.

6.1 Linear Models

A linear model, presented before in the contexts of linear regression (Sect. 2.4.1.1) and perceptrons (Sect. 2.4.2.1), is a parametrised model function mapping N-dimensional inputs $x = (x_1, ..., x_N)^T$ to K-dimensional outputs $y = (y_1, ..., y_K)^T$.

The function is linear in a set of weight parameters that can be written as a $K \times N$-dimensional weight matrix W,

$$f(x; W) = Wx. \tag{6.1}$$

The output can be the multi-dimensional result of a regression task, or it can be interpreted as a one-hot encoding to a multi-label classification task (see Sect. 2.1.2). Linear models also appear as *linear layers* of neural networks, i.e. a layer where the activation function is the identity.

When the output is a scalar, we have that $K = 1$ and the linear model becomes an inner product between the input and a weight vector $w = (w_1, ..., w_N)^T$,

$$f(x; w) = w^T x + w_0 = w_0 + w_1 x_1 + \cdots + w_N x_N. \tag{6.2}$$

As mentioned before, it is common to include the bias w_0 in the inner product by extending the parameter vector to $w = (w_0, w_1, ..., w_N)^T$ and considering the extended input $x = (1, x_1, ..., x_N)^T$, which is why we will mostly ignore it in the formalism.

In this section we present different strategies to implement a linear model with a quantum computer. First we look at models of the form (6.2) where the inputs are represented in amplitude encoding. The goal is to evaluate the inner product $\langle \psi_w | \psi_x \rangle$ if the weights are likewise amplitude encoded, or $\langle 0 | U(w) | \psi_x \rangle$ if they define a unitary evolution. Remember that we use $|\psi_a\rangle$ to denote the quantum state representing a real and normalised vector a in amplitude encoding, and that $U(w)$ is a parametrised unitary circuit. We will show how a general quantum evolution can be interpreted as a certain type of the general linear model (6.1), and how one can concatenate quantum evolutions as linear layers of a neural network. We then present strategies that encode the weights as a unitary evolution, $U(w)|x\rangle$, but where the inputs are now basis encoded. Lastly, we discuss how to add nonlinearities in the different encodings.

6.1.1 Inner Products with Interference Circuits

A natural way to represent a scalar-valued linear model on a quantum computer is to encode the inputs x and parameters w in quantum state vectors via amplitude encoding, $|\psi_x\rangle, |\psi_w\rangle$ (see Sects. 3.4.2, 5.2), and compute their inner product,

$$\langle \psi_w | \psi_x \rangle = f(x; w) = w^T x.$$

Inner products of quantum states are a central feature in quantum theory, where the terms 'ket' and 'bra' originate from "braket", $\langle \cdot | \cdot \rangle$, which is the mathematical symbol

6.1 Linear Models

of an inner product. However, while playing a prominent role in the mathematical description of quantum theory, it is actually not trivial to measure the inner product between two quantum states. There is a family of small quantum circuits that use interference between different branches of a superposition to fulfil this task [1]. The most well-known of the inner product interference routines is the so called *swap test* and returns the absolute value of the inner product of two separate quantum states.

6.1.1.1 The Swap Test

We consider two qubit registers in the respective quantum states $|\psi_a\rangle, |\psi_b\rangle$ that encode the real vectors a and b. The swap test is a common trick to 'write' the absolute square of their inner product, $|\langle\psi_a|\psi_b\rangle|^2$, into the probability of measuring an ancilla qubit in a certain state. To achieve this, one creates a superposition of the ancilla qubit and swaps the quantum states in one branch of the superposition, after which they are interfered (see Fig. 6.1).

Starting with a state $|0\rangle|\psi_a\rangle|\psi_b\rangle$, a Hadamard on the ancilla—the qubit in the first register—leads to

$$\frac{1}{\sqrt{2}}(|0\rangle + |1\rangle)|\psi_a\rangle|\psi_b\rangle.$$

We now apply a swap operator on the two registers $|\psi_a\rangle, |\psi_b\rangle$ which is conditioned on the ancilla being in state 1. This operation swaps the states $|\psi_a\rangle|\psi_b\rangle \to |\psi_b\rangle|\psi_a\rangle$ in the branch marked by the first qubit being in state 1,

$$\frac{1}{\sqrt{2}}(|0\rangle|\psi_a\rangle|\psi_b\rangle + |1\rangle|\psi_a\rangle|\psi_b\rangle).$$

Another Hadamard applied to the ancilla results in the state

$$|\psi\rangle = \frac{1}{2}|0\rangle \otimes (|\psi_a\rangle|\psi_b\rangle + |\psi_b\rangle|\psi_a\rangle) + \frac{1}{2}|1\rangle \otimes (|\psi_a\rangle|\psi_b\rangle - |\psi_b\rangle|\psi_a\rangle).$$

This computes two branches of a superposition, one containing a sum between the 'unswapped' and 'swapped' states of the two registers, and the other containing their

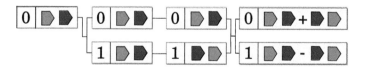

Fig. 6.1 Schematic illustration of the swap test routine. An ancilla qubit in state 0 is prepared together with two quantum states (turquois and red shapes). The ancilla is superposed and the two states are swapped in the branch marked by the ancilla's 1 state. The ancilla is then interfered, writing the sum and difference of the original and the swapped order into each branch

difference. The probability of measuring the ancilla qubit in state 0, the acceptance probability is given by

$$p_0 = |\langle 0|\psi\rangle|^2 = \frac{1}{2} - \frac{1}{2}|\langle\psi_a|\psi_b\rangle|^2, \quad (6.3)$$

and reveals the absolute value of the inner product through

$$|\langle\psi_a|\psi_b\rangle| = \sqrt{1 - 2p_0}. \quad (6.4)$$

In the more general case that the two input states are mixed states ρ_a and ρ_b, the same routine can be applied and the success probability of the postselective measurement is given by [2]

$$p_0 = \frac{1}{2} - \frac{1}{2}\text{tr}\{\rho_a \rho_b\}$$

Note that here the expression $\rho_a \rho_b$ is not an abbreviation of the tensor product, but a matrix product.

With a little trick in the way information is encoded into the amplitudes [3] it is possible to reveal the sign of the inner product using the swap test. One simply has to extend the vectors $a \to a'$ and $b \to b'$ by one extra dimension $N+1$, so that $a', b' \in \mathbb{R}^{N+1}$. The entry of the extra dimension is set to the constant value 1. To renormalise, we have to then multiply the entire $N+1$-dimensional vector with $\frac{1}{\sqrt{2}}$. This way, the amplitude vector $(\alpha_1, ..., \alpha_N)^T$ becomes $(\frac{1}{\sqrt{2}}\alpha_1, ..., \frac{1}{\sqrt{2}}\alpha_N, \frac{1}{\sqrt{2}})^T$. If part of the amplitude vector has been padded with zeros, this extension comes at no extra cost in the number of qubits. Only if we have already 2^n features (including the bias) to encode, this requires us to add one qubit to extend the dimensions of the Hilbert space.

With the extra constant dimension, the result of the swap test between $|\psi_{a'}\rangle, |\psi_{b'}\rangle$ will be

$$p(0) = \frac{1}{2} - \frac{1}{2}|\langle\psi_{a'}|\psi_{b'}\rangle|^2,$$
$$= \frac{1}{2} - \frac{1}{2}|\frac{1}{2}a_1 b_1 + \cdots + \frac{1}{2}a_N b_N + \frac{1}{2}|^2,$$
$$= \frac{1}{2} - \frac{1}{2}|\frac{1}{2}a^T b + \frac{1}{2}|^2.$$

Since $a^T b \in [-1, 1]$, the expression $|\frac{1}{2}a^T b + \frac{1}{2}|$ is guaranteed to lie in the positive interval $[0, 1]$. As opposed to Eq. (6.4) we therefore do not have to worry about only retrieving the absolute value. Hence, we can extract the inner product of the original vectors via

$$a^T b = 2\sqrt{1 - 2p_0} - 1,$$

and since $p_0 \in [0, \frac{1}{2}]$, this value does indeed lie in the interval $[-1, 1]$.

6.1 Linear Models

6.1.1.2 Interference Circuit

There is another, slightly more elegant routine to extract inner products, but it requires more sophisticated state preparation. First, let us note that for unit vectors there is a close relationship between the vector sum and the inner product: Given two real unit vectors a, b whose inner product one wishes to compute (with $|a|^2 = |b|^2 = 1$), then

$$
\begin{aligned}
(a+b)^T(a+b) &= \sum_i (a_i + b_i)^2 \\
&= \sum_i a_i^2 + \sum_i b_i^2 + 2\sum_i a_i b_i \\
&= 2 + 2a^T b.
\end{aligned}
$$

A geometric illustration is given in Fig. 6.2.

This fact has implicitly been used in the swap test routine, and helps to evaluate the inner product of two quantum states together with the correct sign without tricks like the constant shift introduced above. As a precondition, we need to be able to prepare the initial state

$$|\psi\rangle = \frac{1}{\sqrt{2}}\left(|0\rangle|\psi_a\rangle + |1\rangle|\psi_b\rangle\right). \tag{6.5}$$

Note that in comparison with the swap test routine, here there is a superposition of states $|\psi_a\rangle, |\psi_b\rangle$ in one register, as opposed to each state having its own register (see Fig. 6.3). In the standard computational basis, quantum state (6.5) corresponds to an amplitude vector

$$\alpha = (a_1, \ldots, a_N, b_1, \ldots, b_N)^T.$$

If we have a routine $\mathcal{A}$ to prepare $|\psi_a\rangle$ and another routine $\mathcal{B}$ to prepare $|\psi_b\rangle$, one has to implement these routines conditioned on the respective states of the ancilla qubit prepared in $|+\rangle = \frac{1}{\sqrt{2}}(|0\rangle + |1\rangle)$.

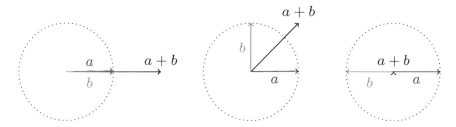

Fig. 6.2 Geometric illustration showing the relation between inner products of two normalised vectors a and b with their sum. The sum of parallel normalised vectors is at the maximum value, while the sum of antiparallel vectors is zero. This is exploited in the interference circuits introduces in this section

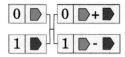

Fig. 6.3 Schematic illustration of the interference routine for the calculation of inner products. The two states (blue and red shape) are initially entangled with the 0 and 1 state of an ancilla. Interfering the two branches through a Hadamard gate applied to the ancilla writes the sum and difference of the two states into each branch

Once state (6.5) is prepared, a Hadamard gate on the ancilla will result in

$$|\psi\rangle = \frac{1}{2}|0\rangle \otimes (|\psi_a\rangle + |\psi_b\rangle) + \frac{1}{2}|1\rangle \otimes (|\psi_a\rangle - |\psi_b\rangle).$$

Formally speaking, the ancilla state $|0\rangle$ is entangled to an unnormalised quantum state $|\psi_{a+b}\rangle$ that encodes the sum of a and b via amplitude encoding. The acceptance probability $p(0) = |\langle 0|\psi\rangle|^2$ of the ancilla being measured in state 0 is given by

$$p(0) = \frac{1}{4}\langle\psi_{a+b}|\psi_{a+b}\rangle,$$
$$= \frac{1}{4}(2 + a^T b),$$
$$= \frac{1}{2} + \frac{1}{2}a^T b.$$

This time, there is no absolute value that obscures the sign of the result.

If a and b are complex vectors (for example, the result of a unitary evolution), the interference routine reveals only the real part of the inner product,

$$p(0) = \frac{1}{4}(2 + \sum_i a_i^* b_i + \sum_i a_i^* b_i),$$
$$= \frac{1}{2} + \frac{1}{2}\operatorname{Re}\{a^\dagger b\}.$$

This is not surprising, because a measurement always results in a real value, otherwise it would be unphysical.

Besides state preparation, evaluating the inner product on a quantum computer requires only one additional qubit and a number of gates that is at most linear in the number of qubits (and hence qubit-efficient, or logarithmic in the feature space dimension). The number of measurements grows with the desired precision (see Sect. 5.1.3). If weights and inputs are encoded in the amplitudes of the same register, we can even reduce this to a single Hadamard operation plus measurements, without the need for an extra qubit.

6.1.2 A Quantum Circuit as a Linear Model

We still assume that the inputs are given in amplitude encoding, $|\psi_x\rangle$, but now look at the case where the model parameters w are "dynamically encoded" in a unitary, or more general, in a quantum circuit.

6.1.2.1 Unitary Linear Model

Rather trivially, a linear unitary transformation can be interpreted as a special linear model,

$$|\psi_{x'}\rangle = U|\psi_x\rangle,$$

that maps a N-dimensional complex vector onto a N-dimensional complex vector. If we want to get a scalar output, we still need to sum up the elements of x'. An elegant way to do this is to compute the inner product of the final state with a uniform distribution $|\psi_u\rangle = \frac{1}{\sqrt{N}}\sum_i |i\rangle$, which would effectively implement the model

$$\begin{aligned} f(x;\theta) &= \langle \psi_u|U|\psi_x\rangle, \\ &= \underbrace{\frac{1}{\sqrt{N}}(u_{11}+\cdots+u_{N1})}_{w_1} x_1 + \cdots + \underbrace{\frac{1}{\sqrt{N}}(u_{N1}+\cdots+u_{NN})}_{w_N} x_N, \\ &= w_1 x_1 + \cdots + w_N x_N. \end{aligned}$$

If we summarise the coefficients $\frac{1}{\sqrt{N}}(u_{1i}+\cdots+u_{Ni})$ as w_i, $i = 1, ..., N$, like shown above, this is the original linear function from Eq. (6.2). From a geometrical perspective, the output is represented by a measure of 'uniformness' of the state $U|\psi_x\rangle$, or the overlap with the uniform superposition. Of course, this is just another way to formulate the idea in the previous section: If we write $|\psi_w\rangle = U^\dagger|\psi_u\rangle$ we can understand the unitary as a state preparation routine for the parameter quantum state from before. Training the model with data means to find a circuit or unitary evolution U that captures the input-output relation from the data.

6.1.2.2 Non-unitary Linear Model

An interesting twist appears when we look at linear models that map from $\mathbb{R}^N$ to a lower-dimensional vector space $\mathbb{R}^K$, $K < N$ to implement the layer of a neural net as in Eq. (6.1). If we want to allow for arbitrary concatenation of linear layers, the input encoding of the quantum algorithm has to be of the same type as the output encoding, which is not the case for the interference routines above. If we also want to reduce the dimensionality in the map, we have to consider a subsystem of the qubits as the output of the evolution (see also [4]). This forces us to use the language of density matrices which was introduced in Sect. 3.1.2.3.

Consider a density matrix ρ_x that encodes an input x on its diagonal, i.e. $\text{diag}\{\rho_x\} = (x_1, \ldots, x_N)^T$. For example, we could encode the square roots of the features x_i, $i = 1, \ldots, N$ in amplitude encoding in a $n = \log N$ qubit quantum system, $|\psi_{\sqrt{x}}\rangle = \sum_i \sqrt{x_i}|i\rangle$, and computing the density matrix of this pure state, $\rho_x = |\psi_{\sqrt{x}}\rangle\langle\psi_{\sqrt{x}}|$ results in the desired encoding. A unitary evolution in this language is formally written as

$$\rho_{x'} = U\rho_x U^\dagger.$$

From here, we trace out the first $1, \ldots, n-k$ of the n qubits so that we end up with only the last k qubits, reducing the Hilbert space from $N = \log n$ dimensions to $K = \log k$ dimensions. This yields the "hidden layer density matrix"

$$\rho_h = \text{tr}_{1,\ldots,n-k}\{\rho_{x'}\}.$$

We are not restricted to reducing the dimension: Adding qubits $\rho_h \otimes |0\ldots0\rangle\langle0\ldots0|$, and extending the further evolution to these new qubits can increase the dimension of the Hilbert space again.

Figure 6.4 shows an example of unitary evolutions, system reductions and extensions in the graphical representation of neural networks, where the diagonal of the density matrix at any given point of the evolution is interpreted as a layer of 'neurons'. Initially the quantum system consists of three qubits

$$|\text{in}\rangle\langle\text{in}| = |q_1, q_2, q_3\rangle\langle q_1, q_2, q_3|.$$

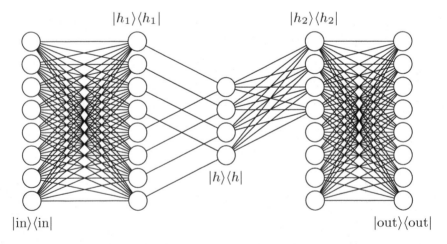

Fig. 6.4 Formally, performing unitary transformations, tracing qubits out and joining qubits corresponds to a neural network structure. The layers correspond to the diagonal of the according density matrix. Shown here is the evolution described in the text

6.1 Linear Models

A unitary evolution applies a linear transformation that leads to an intermediate layer,

$$|h_1\rangle\langle h_1| = U_1|\text{in}\rangle\langle\text{in}|U_1^\dagger.$$

Tracing out the first qubit yields the hidden layer,

$$|h\rangle\langle h| = \text{tr}_1\{|h_1\rangle\langle h_1|\},$$

and joining a qubit in the ground state,

$$|h_2\rangle\langle h_2| = |h_2\rangle\langle h_2| \otimes |q_4=0\rangle\langle q_4=0|.$$

creates a quantum state whose density matrix has again 8 diagonal elements, but 4 of them have a probability zero (i.e., those corresponding with the last qubit being in state 1, which it is obviously not). This density matrix can be interpreted as a second intermediate layer. Another full unitary on qubits q_2, q_3, q_4 generates a quantum state that represents the 'output layer',

$$|\text{out}\rangle\langle\text{out}| = U_2|h_2\rangle\langle h_2|U_2^\dagger.$$

The decision which qubit is traced out defines the connectivity between a bigger and a smaller layer.

Example 6.1 For $n=2$ and the 'input' density matrix

$$\rho_x = \begin{pmatrix} a_{11} & a_{12} & a_{13} & a_{14} \\ a_{21} & a_{22} & a_{23} & a_{24} \\ a_{31} & a_{32} & a_{33} & a_{34} \\ a_{41} & a_{42} & a_{43} & a_{44} \end{pmatrix},$$

tracing out the first or second qubit results respectively in

$$\text{tr}_1\rho_x = \begin{pmatrix} a_{11}+a_{33} & a_{12}+a_{34} \\ a_{21}+a_{43} & a_{22}+a_{44} \end{pmatrix}, \quad \text{tr}_2\rho_x = \begin{pmatrix} a_{11}+a_{22} & a_{13}+a_{24} \\ a_{31}+a_{42} & a_{33}+a_{44} \end{pmatrix}.$$

Tracing out the first qubit consequently adds inputs 1, 3 and 2, 4 to define the two units of the following layer, while tracing out the second qubit adds inputs 1, 2 and 3, 4.

Note that in general the first trace operation turns the pure state (full information on the state) into a mixed state (we lost the information about some qubits). The trace operation maintains the linearity, but now the full evolution of a linear layer is no longer described by a unitary operator.

6.1.3 Linear Models in Basis Encoding

We now tend to the case that inputs and outputs of the linear model are basis encoded. Implementing the model requires us to find a quantum circuit $U(w)$ that takes the state $|x\rangle|0...0\rangle$ to $|x\rangle|w^T x\rangle$, where the result of the linear model is written into the qubits of the second register.

There are many different ways to achieve this transformation, and as an example we will present an idea based on the quantum phase estimation algorithm from Sect. 3.5.2. It first constructs an evolution $U(w)|x\rangle|j\rangle = e^{2\pi i(j-1)w^T x}|x\rangle|j\rangle$, from which quantum phase estimation can write the phase $w^T x$ into the second quantum register [5].

Consider a state

$$|x_1...x_N\rangle|0...0\rangle, \tag{6.6}$$

where the second register contains ν qubits in the ground state (ν is a hyperparameter that determines the precision of the output). The first step is to put the second register into a uniform superposition by using Hadamards on each of its qubits, leading to $\frac{1}{\sqrt{2^\nu}}\sum_{j=1}^{2^\nu}|x_1,...,x_N\rangle|j\rangle$. The circuit with the desired eigenvalue for eigenstate $|x_1,...,x_N\rangle$ can be constructed by using a sequence $k=1,...,N$ of parametrised single qubit gates $U_k(w_k)$ acting on the kth qubit of the input register. The gates have the form of an S gate from Table 3.3 but with a variable phase defined by w_k,

$$U_k(w_k) = \begin{pmatrix} 1 & 0 \\ 0 & e^{2\pi i w_k} \end{pmatrix}.$$

The full circuit reads

$$U(w) = U_n(w_n)...U_2(w_2)U_1(w_1)U_0(w_0).$$

The "bias gate" U_0 adds a global phase of $2\pi i w_0$, where w_0 is a real bias parameter in $[0, 1)$. Altogether, applying this circuit to the initial state results in

$$U(w)|x_1,...,x_N\rangle|0...0\rangle = e^{2\pi i w^T x}|x_1,...,x_N\rangle|0...0\rangle.$$

For phase estimation, we need an oracle $\mathcal{O}$ that can apply powers of $U(w)$,

$$|x\rangle|j\rangle \xrightarrow{\mathcal{O}} U(w)^{j-1}|x\rangle|j\rangle,$$

for $j = 1...2^\nu$. Remember that we defined $|j\rangle$ to be the jth computational basis state, and since we want $|0...0\rangle$ to produce $U(w)^0$, we have to use $j-1$ in the exponent. For sufficiently small ν, this can always be constructed by applying $U(w)$ conditioned on the qubits of the index register. For example, if the last qubit is in 1, apply $U(w)$ once, if the second last qubit is also in 1, apply $U(w)$ twice more, and if the kth last qubit is in 1 apply $U(w)$ 2^{k-1} times more.

6.1 Linear Models

With help of such an oracle we can prepare a state

$$\frac{1}{\sqrt{2^\nu}} \sum_{j=1}^{2^\nu} |x_1, ..., x_N\rangle e^{2\pi i (j-1) w^T x} |j\rangle.$$

The quantum phase estimation routine (introduced in Sect. 3.5.2) applies an inverse quantum Fourier transform, and if $w^T x \approx (j-1)/2^\nu$ for a particular integer j, this approximately results in a quantum state $|x\rangle|w^T x\rangle$ that represents the "phase" $w^T x$ in basis encoding. For cases $w^T x \neq (j-1)/2^\nu$ it can be shown that in order to obtain $w^T x$ accurately up to ν_{acc} bits of precision with a success probability of $1 - \epsilon$, one has to choose $\nu = \nu_{\text{acc}} + \lceil \log(2 + \frac{1}{2\epsilon}) \rceil$ [6]. Since $w^T x \in [0, 1)$, The qubits $q_1, ..., q_\nu$ of register $|w^T x\rangle$ represent the phase via the relationship

$$w^T x \approx q_1 \frac{1}{2^0} + ... + q_\nu \frac{1}{2^{\nu-1}}.$$

The computational complexity of this routine with regards to N (the number of features) and τ (the number of bits to represent each feature) is in $\mathcal{O}(N\tau)$, comparable to the classical linear model. The dependency on the number of output qubits or precision hyperparameter ν is slightly more difficult. After ν Hadamards, the oracle to implement powers of $U(w)$ in the most naive form presented above needs to be called $\mathcal{O}(\sum_{i=0}^{2^\nu - 1} i)$ times. The inverse quantum Fourier transform requires $\frac{\nu(\nu+1)}{2} + 3\frac{\nu}{2}$ gates [6]. We will see below that when implementing a step function on the result to extend this routine to a perceptron model, we are only interested in the value of the first qubit, and ν is therefore sufficiently small.

Linear models in basis encoding such as the example presented here open up a couple of interesting avenues. For example, one can process training sets in superposition by starting with the state $\sum_{m=1}^{M} |x^m\rangle$ instead of $|x\rangle$, so that the entire procedure gets applied in parallel. The result would likewise be a superposition of outputs. From measurements on the output register one can retrieve the accuracy or success rate on an entire dataset in parallel.

6.1.4 Nonlinear Activations

A linear model is not very powerful when it comes to data that is not linearly separable. We therefore need nonlinearities in the model function. To focus the discussion, we will look at how to introduce nonlinearities from the perspective of neural networks where they enter as nonlinear activation functions φ mapping a net input v to $\varphi(v)$. However, the observations can be used in other contexts as well.

6.1.4.1 Amplitude Encoding

In terms of amplitude encoding, applying an activation function to a layer means to implement the map $|\psi_v\rangle \to |\psi_{\varphi(v)}\rangle$ for a vector or scalar v, an operation that cannot be achieved by a standard unitary quantum evolution. This has been discussed in Sect. 5.2, where it was illustrated why nonlinear transformations of amplitudes are very likely to be unphysical. However, as we mentioned there is one element in quantum theory that allows for nonlinear updates of the amplitudes: measurement. Measurement outcomes are non-deterministic, and we have seen in Sect. 5.2.2.3 how to make them deterministic by using branch selection. For example, we could store the net input v temporarily in basis encoding and use the trick of conditional rotation of an ancilla to prepare a state

$$|\psi\rangle|0\rangle|v\rangle \to |\psi\rangle \left(\varphi(v)|0\rangle + \sqrt{1-\varphi(v)^2}|1\rangle\right)|v\rangle,$$

where during the conditional rotation of the ancilla we also applied a function φ. From here, branch selection would lead to a state proportional to $\varphi(v)|\psi\rangle|v\rangle$ where—depending on the implementation—we could try and invert the storage operation to get rid of the last register. This is a costly detour, and it is therefore questionable if amplitude encoding is suitable for the implementation of feed-forward neural networks. In the next section we will look at an alternative strategy to give amplitude encoding more power, namely through so called *kernel methods*.

6.1.4.2 Basis Encoding

In basis encoding, nonlinear transformations of the form $|v\rangle|0\rangle \to |v\rangle|\varphi(v)\rangle$ can be implemented deterministically. As mentioned in Sect. 5.1.2 one can always take a classical algorithm to compute φ on the level of logical gates, translate it into a reversible routine and use this as a quantum algorithm. Depending on the desired nonlinearity, there may be much more efficient options. As three illustrative examples, we consider popular nonlinear functions used as activations in neural networks, the *step function*, as well as *rectified linear units* and a *sigmoid function* (see Fig. 2.18).

For simplicity, we will fix the details of the binary representation and consider fixed point arithmetic. The real scalar v is represented by a $(1 + \tau_l + \tau_r)$-bit binary sequence

$$b_s b_{\tau_l - 1} \cdots b_1 b_0 . b_{-1} b_{-2} \cdots b_{-\tau_r}. \tag{6.7}$$

The first bit b_s indicates the sign, and τ_l, τ_r are the numbers of bits left and right of the decimal dot (called *integer* and *fractional* bits). A real number can be retrieved from the bit string via

$$v = (-1)^{b_s}(b_{\tau_l-1} 2^{\tau_l - 1} + \cdots + b_0 2^0 + b_{-1} 2^{-1} + b_{-2} 2^{-2} + \cdots + b_{-\tau_r} 2^{-\tau_r}).$$

6.1 Linear Models

For example, the binary sequence 111.101 with $\tau_l = 2$ and $\tau_r = 3$ corresponds to the real number $(-1)(2 + 1 + 0.5 + 0.125) = -3.625$.

A step function defined as

$$\varphi(v) = \begin{cases} 0, & \text{if } v \leq 0, \\ 1, & \text{else,} \end{cases}$$

is most trivial to implement in this representation [5], and one does not even require an extra output register to encode the result. Instead, the sign qubit of the input register can be interpreted as the output, and the step function is implemented by the 'effective map' $|v\rangle = |b_s b_{\tau_l - 1} ...\rangle \rightarrow |b_s\rangle$.

Also rectified linear units,

$$\varphi(v) = \begin{cases} 0, & \text{if } v \geq 0, \\ v, & \text{else,} \end{cases}$$

are simple to realise. Conditioned on the sign (qu)bit b_s, either copy the input register into the output register, or do nothing. This requires a sequence of NOT gates, each controlled by $|b_s\rangle$ as well as $|b_k\rangle$ for $k = \tau_l - 1, ..., 0, ..., -\tau_r$. The NOT is applied to the k'th qubit of the output register.

The sigmoid nonlinearity,

$$\varphi(v; \delta) = \frac{1}{1 + e^{-\delta v}},$$

poses a bigger challenge. An implementation via division and exponential function is rather demanding in terms of the number of elementary operations as well as the spacial resources. There are a number of more efficient approximation methods, such as piecewise linear approximation [7]. If the overall precision (i.e. the number of bits to represent a real number) is not very high, one can write down a boolean function that reproduces the sigmoid function for binary numbers explicitly. This means to define the binary representation of $\varphi(v)$ for every possible binary representation of the input v. For example, if the input is $v = -3.625$ with the above binary representation 111.101, the output of the sigmoid function, $\varphi(v, \delta = 0.1) = 0.410355...$, is approximated by the bitstring 000.011.

Of course, this approach would mean to store a look-up-table of possible inputs and their outputs, whose size grows exponentially with the precision. However, there are optimisation methods to reduce the size of the look-up table significantly, such as Quine-McCluskey [8, 9] methods, whose idea is sketched in the following example.

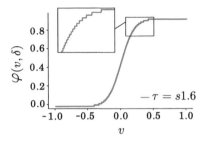

Fig. 6.5 Sigmoid function with $\delta = 10$ where the output has been approximated by a binary representation and reconverted into a real number. The precision $\tau = s1.6$ means that the binary representation has one sign bit, one integer bit and six fractional bits which is sufficient for a rather smooth approximation

Example 6.2 (Quine-McCluskey method) Consider an arbitrary boolean function:

x_1	x_2	x_3	y
0	0	0	0
0	0	1	1
0	1	0	0
0	1	1	1
1	0	0	1
1	0	1	0
1	1	0	1
1	1	1	0

We can see that the inputs {001, 011, 100, 110} belong to the output class 1. In a quantum algorithm one could for example apply four multi-controlled NOT gates, that together only flip the output qubit if the three qubits are in one of these states. But one can summarise this set significantly by recognising that the $y = 1$ cases all have the structure $\{0 \cdot 1, 1 \cdot 0\}$ (marked in the table above). Hence, we only have to control on 2 qubits instead of 3.

There are different options to compute the summarised boolean function, which is NP-hard in general, but only has to be done once. For low precisions the cost is acceptable. Figure 6.5 illustrates that one only needs 6 fractional bits for a rather smooth approximation of the sigmoid function that is scaled to the $[-1, 1]$ interval.

6.1.4.3 Angle Encoding

Before finishing this section, we want to have a look at a third type of encoding that has not been discussed before but may be interesting in the context of quantum neural networks. Assume that by some previous operations, the net input $\theta = w_0 +$

6.1 Linear Models

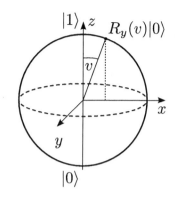

Fig. 6.6 A y-rotation $R_y(v)$ rotates the Bloch vector by an angle v around the y axis (that is, in the x-z-plane)

$w_1 x_1 + \ldots + w_N x_N$ is written into the angle of an ancilla or "net input qubit", which is entangled with an output qubit in some arbitrary state $|\psi\rangle$,

$$R_y(2v)|0\rangle \otimes |\psi\rangle_{\text{out}}.$$

As a reminder, the rotation around the y axis is defined as

$$R_y(2v) = \begin{pmatrix} \cos v & -\sin v \\ \sin v & \cos v \end{pmatrix}.$$

The net input qubit is hence rotated around the y axis by an angle v (see Fig. 6.6), and $R_y(2v)|0\rangle = \cos v |0\rangle + \sin v |1\rangle$.[1] The goal is now to rotate the output qubit by a nonlinear activation φ which depends on v, in other words, we want to prepare the output qubit in state $R_y(2\varphi(v))|\psi\rangle$.

An elegant way to accomplish this task has been introduced by Cao et al. [11] for the sigmoid-like function $\varphi(v) = \arctan(\tan^2(v))$ by using so called *repeat-until-success* circuits [12, 13]. As mentioned briefly in Sect. 5.2.3, repeat-until-success circuits apply an evolution to ancillas and an output qubit, and measuring the ancilla in state 0 leads to the desired state of the output qubit, while when measuring state 1, one can reset the circuit by a simple procedure and apply the evolution once more. This is repeated until finally the ancilla is measured in 0. In contrast to branch selection or postselective measurements, we do not have to repeat the entire routine up to the evolution in question when the undesired result was measured for the ancilla.

Let us have a closer look at the simplified circuit in Fig. 6.7 to understand the basic principle of the angle encoded activation function.[2] The first gate writes the net input into the angle of the 'net input qubit' as explained above. The second, conditional gate results in the state

---

[1] Such an *angle encoded* qubit has been called a *quron* [10] in the context of quantum neural networks.

[2] Thanks to Gian Giacomo Guerreschi for this simplified presentation.

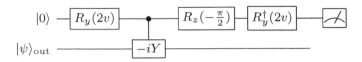

Fig. 6.7 Repeat-until-success circuit that turns the output qubit by a sigmoid-like activation function depending on the net input v encoded in the net qubit

$$R_y(2v)|0\rangle \otimes |\psi\rangle_{\text{out}} - \sin(v)|1\rangle \otimes (\mathbb{1} + iY)|\psi\rangle_{\text{out}}.$$

The third gate reverses the net input encoding step. Note that $R_y^\dagger(2v) = R_y(-2v)$. Due to the conditional evolution of the second gate, this leads to interference effects and yields

$$= |0\rangle|\psi\rangle_{\text{out}} - \sin(v)\left(\sin(v)|0\rangle + \cos(v)|1\rangle\right) \otimes (\mathbb{1} + iY)|\psi\rangle_{\text{out}}$$
$$= |0\rangle \left(\mathbb{1} - \sin^2(v)\mathbb{1} - i\sin^2(v)Y\right)|\psi\rangle_{\text{out}} - |1\rangle \sin(v)\cos(v)(\mathbb{1} + iY)|\psi\rangle_{\text{out}}$$
$$= |0\rangle \left(\cos^2(v)\mathbb{1} - i\sin^2(v)Y\right)|\psi\rangle_{\text{out}} - |1\rangle \sin(v)\cos(v)(\mathbb{1} + iY)|\psi\rangle_{\text{out}}.$$

If we measure the net input qubit in state $|0\rangle$ we have to renormalise by a factor $(\cos^4(v) + \sin^4(v))^{-\frac{1}{2}}$ and get the desired result

$$\frac{\cos^2(v)\mathbb{1} - i\sin^2(v)Y}{\sqrt{\cos^4(v) + \sin^4(v)}} = R_y(2\varphi(v)).$$

Else, if the measurement returns 1, we get the state $\frac{1}{\sqrt{2}}(\mathbb{1} + iY)|\psi\rangle_{\text{out}}$. Note that here the measurement induces a renormalisation factor $(\sin(v)\cos(v))^{-1}$ which cancels the dependence on v out. We therefore simply have to reverse the constant operation $\frac{1}{\sqrt{2}}(\mathbb{1} + iY)$ and can try again.

This basic idea can be extended to perceptron models and feed-forward neural networks for which the value of a neuron is encoded in the angles as described above [11]. Of course, the non-deterministic nature of repeat-until-success circuits implies that the runtime is on average slightly longer than the feed-forward pass in a classical neural network, since we have to repeat parts of the algorithm in case of unsuccessful measurement outcomes. Still, the proposal is an interesting realisation of a nonlinear activation function in the context of quantum computing.

6.2 Kernel Methods

Section 6.1.4 demonstrated that while there is a wealth of methods to implement linear models, nonlinearities can be a lot trickier, especially when it comes to amplitude encoding. In Sect. 2.2.4 of the introduction to machine learning another strategy has

6.2 Kernel Methods

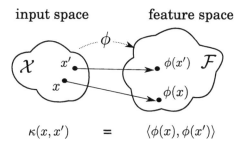

Fig. 6.8 Kernel methods compute the distance between data point x and x' through a kernel $\kappa(x, x')$. For every positive definite kernel there is a feature space F produced by mapping inputs with a nonlinear map $\varphi(x)$, so that the kernel of two inputs is equal to the inner product of their feature mapped versions, $\kappa(x, x') = \langle \varphi(x) | \varphi(x') \rangle$

been already revealed to give linear models more power: feature spaces, based on the theory of kernel methods. We will show in this section that feature spaces and quantum Hilbert spaces have a surprisingly similar mathematical formalism and how this can be used for quantum machine learning. More details can be found in [14].

To recapitulate the highlights from Sect. 2.2.4, one can understand a (positive semi-definite) kernel as a distance measure on the input space, and use it to compute the distance between each training input and the new input we aim to classify, and favour the class of "closer" training data when the decision for the prediction is taken. Such a kernel corresponds to an inner product of data points mapped to a higher dimensional *feature space* (see Fig. 6.8). The *representer theorem* shows how a large class of trained models can be expressed in terms of kernels. Finally, the *kernel trick* allows us to construct a variety of models that are formulated in terms of a kernel function by replacing it with another kernel function. The following Sect. 6.2.1 will review these concepts once more with slightly more mathematical foundation, after which we show how to understand state preparation and information encoding as a 'quantum feature map' (Sect. 6.2.3). Finally, we discuss how to compute kernels or distance measures on input space on a quantum computer (Sect. 6.2.4) and show how to construct density matrices that correspond to a kernel Gram matrix (Sect. 6.2.5).

6.2.1 Kernels and Feature Maps

Let us revisit some concepts from Sect. 2.2.4, following the textbook of Schölkopf and Smola [15]. Let $\mathcal{X}$ be a non-empty set of patterns or inputs. As a reminder, a (positive semi-definite) kernel was defined as a map $\kappa : \mathcal{X} \times \mathcal{X} \to \mathbb{C}$ where for any subset of the input set $x_1, \ldots, x_M \in \mathcal{X}$ with $M \geq 2$ the Gram matrix with entries

$$K_{ij} = \kappa(x_i, x_j)$$

is positive definite. Kernels fulfil $\kappa(x, x) \geq 0$ and $\kappa(x, x') = \kappa(x', x)^*$ where the star marks the complex conjugate.

We can always construct a feature map from a kernel. A feature map was introduced before as a map $\phi : \mathcal{X} \to \mathcal{F}$ from the input space into a feature space. Previously we looked at some examples where the feature space was a real vector space $\mathbb{R}^K$ with $K > N$, for example the map from $(x_1, x_2)^T$ to $(x_1, x_2, x_1^2 + x_2^2)^T$. It turns out a feature space can be formalised much more generally as we will see now.

The feature map that can always be constructed from a kernel κ (sometimes called the "canonical feature map") maps from the input set to complex-valued functions on the input set,

$$\phi : \mathcal{X} \to \mathbb{C}^\mathcal{X}, \quad x \to \kappa(\cdot, x). \tag{6.8}$$

Here, the feature vectors in $\mathbb{C}^\mathcal{X}$ are themselves functions that map from the input space to the space of complex numbers $\mathcal{X} \to \mathbb{C}$. The functions in feature space are kernels with one "open slot".

From this feature map we can construct a feature space that is a vector space with an inner product (see [15, 16]). The vector space contains linear combinations of $\{\kappa(\cdot, x_m)\}$, where $\{x^1, ..., x^M\} \subseteq \mathcal{X}$ is an arbitrary set of inputs,

$$f(\cdot) = \sum_{m=1}^{M} \nu_m \kappa(\cdot, x_m),$$

with coefficients $\nu_m \in \mathbb{R}$. Using a second vector of the same form,

$$g(\cdot) = \sum_{m'=1}^{M'} \mu_{m'} \kappa(\cdot, x_{m'}),$$

the inner product can be defined as

$$\langle f, g \rangle = \sum_{m=1}^{M} \sum_{m'=1}^{M'} \nu_m^* \mu_{m'} \kappa(x_m, x_{m'}).$$

With this inner product, the feature map has the property

$$\langle \phi(x) | \phi(x') \rangle = \langle \kappa(\cdot, x), \kappa(\cdot, x') \rangle = \kappa(x, x').$$

Read from right to left, this relationship is a core idea of the theory of kernel methods which we introduced before: A kernel of two inputs x and x' computes the inner product of the same inputs mapped to a feature space. As a mere technicality in this context, one can extend the inner product vector space by the norm $||f|| = \sqrt{\langle \cdot, \cdot \rangle}$ and the limit points of Cauchy series under this norm, and get a Hilbert space called the *Reproducing Kernel Hilbert Space*.

6.2 Kernel Methods

One can also argue vice versa and construct a kernel from a vector space with an inner product. It follows from the above that given a mapping $\phi : \mathcal{X} \to \mathcal{H}$, where $\mathcal{H}$ is a Hilbert space with vectors $\phi(x)$, one can define a kernel via

$$\kappa(x, x') = \langle \phi(x), \phi(x') \rangle.$$

To see why the resulting kernel is indeed positive semi-definite as requested in the definition, consider some patterns $x_1, \ldots, x_M \in \mathcal{X}$. Then for arbitrary $c_m, c_{m'} \in \mathbb{C}$,

$$\sum_{m,m'=1}^{M} c_m c_{m'}^* \kappa(x_m, x_{m'}) = \langle \sum_m c_m \phi(x_m), \sum_{m'} c_{m'} \phi(x_{m'}) \rangle$$
$$= || \sum_m c_m \phi(x_m) ||^2 \geq 0$$

The first equality sign is due to the bi-linearity of the inner product, while the inequality sign follows from the properties of a squared norm.

6.2.2 The Representer Theorem

Kernel methods use models for machine learning that are based on kernel functions that measure distances between data inputs. They have been popular in the 1990s, before neural networks became again the center of interest of mainstream machine learning research. However, they are not simply a specific ansatz for a model. The wide scope of kernel methods is revealed by the *representer theorem* introduced in Sect. 2.2.4. Also here we want to go into slightly more detail in order to motivate kernel methods as a rather general framework. This can also show a way of translating a range on models into a "kernelised" version.

Consider an input domain $\mathcal{X}$, a kernel $\kappa : \mathcal{X} \times \mathcal{X} \to \mathbb{R}$, a data set $\mathcal{D}$ consisting of data pairs $(x^m, y^m) \in \mathcal{X} \times \mathbb{R}$ and a class of model functions $f : \mathcal{X} \to \mathbb{R}$ that can be written in the form

$$f(x) = \sum_{l=1}^{\infty} \mu_l \kappa(x, x^l), \tag{6.9}$$

with $x, x^l \in \mathcal{X}$, $\mu_l \in \mathbb{R}$ and $||f|| < \infty$. (The norm is taken from the Reproducing Kernel Hilbert Space associated with κ). Furthermore, assume a cost function $C : \mathcal{D} \to \mathbb{R}$ that quantifies the quality of a model by comparing predicted outputs $f(x^m)$ with targets y^m, and which has a regularisation term of the form $g(||f||)$ where $g : [0, \infty) \to \mathbb{R}$ is a strictly monotonically increasing function. For example, C could be a least squares loss with a l_1 regularisation. The *representer theorem* says that any function f minimising the cost function C can be written as

$$f(x) = \sum_{m=1}^{M} \nu_m \kappa(x, x^m). \tag{6.10}$$

Note that the model function (6.9) from the feature space is expressed by an infinite sum, while expression (6.10) is formulated in terms of kernels over the (finite) training inputs x^m.[3] In short, a large class of typical machine learning problems can be solved by a model that is an expansion over kernels of training data. And although we derived it in a different fashion, the inference model for support vector machines in Eq. (2.39) had exactly this form (associating ν_m with the Lagrange parameters times the target outputs).

The following example shows how one can use the representer theorem to turn a simple linear model into its 'kernelised' version.

Example 6.3 (*Kernelised linear model*) Take a standard linear model function $f(x) = w^T x$. This is of the required form in Eq. (6.9) if we choose $\mathcal{X} = \mathbb{R}^N$, $\mu_l = w_l$ for $l = 1...N$ and $\mu_l = 0$ for $l > N$, as well as $x_l = \hat{e}_l$ to be the standard basis and κ to be an inner product kernel:

$$\sum_{l=1}^{\infty} \mu_l \kappa(x, x_i) = \sum_{i=1}^{N} w_i \, x^T \hat{e}_i = w^T x.$$

Considering a square loss objective with l_1 regularisation and $g(||f||) = ||f||$ with a standard norm, we can apply the representer theorem. According to the theorem, the optimal linear model with regards to the objective can be expressed as

$$f(x) = \sum_{m} \nu_m \kappa(x, x^m).$$

If we choose $\kappa(x, x^m) = x^T x^m$ (a linear kernel), this means that the weight vector has effectively been expanded in terms of the training inputs, $w = \sum_m \nu_m x^m$. In short, the optimal weight vector lies in the subspace spanned by the data.

The kernelised version of a model is now amenable to the kernel trick. By exchanging κ for another kernel κ', we can change the feature space in which the data gets effectively mapped.

---

[3] The form of Eq. (6.10) is a so called *non-parametric model*: The number of (potentially zero) parameters ν_m grows with the size M of the training set.

6.2.3 Quantum Kernels

Quantum states—just like feature vectors—also live in Hilbert spaces and allow for a very natural definition of a *quantum kernel*.[4] Input encoding (as discussed at length in Chap. 5) maps x to a vector in a Hilbert space. The main ingredient of a quantum kernel is simply to interpret the process of encoding an input $x \in \mathcal{X}$ into a quantum state $|\phi(x)\rangle$ as a feature map.[5]

If the quantum state $|\phi(x)\rangle$ is interpreted as a feature vector in amplitude encoding (which we call a *quantum feature state*), the inner product of two such quantum feature states can be associated with a quantum kernel

$$\kappa(x, x') = \langle \phi(x) | \phi(x') \rangle.$$

Hence, an inner product of quantum states produced by an input encoding routine is always a kernel. Any quantum computer that can compute such an inner product can therefore be used to estimate the kernel, and the estimates can be fed into a classical kernel method. If the result of this inner product is impossible to simulate on a classical computer, the kernel is classically intractable. A challenge is therefore to find quantum kernels for which we have a quantum advantage and which at the same time prove interesting for machine learning.

The quantum kernel itself depends solely on the input encoding routine, which is the state preparation circuit that 'writes' the input into the quantum state $|\phi(x)\rangle$. Different encoding strategies give rise to different kernels. To illustrate this, we look at five different input encoding strategies—some of which have been mentioned before—and present their associated quantum kernel (following [14]).

Basis encoding. If the input patterns x are binary strings of length n with integer representation i, and we chose basis encoding (see Sect. 5.1), the feature map maps the binary string to a computational basis state,

$$\phi : i \to |i\rangle.$$

The computational basis state corresponds to a standard basis vector in a 2^n-dimensional Hilbert space. The associated quantum kernel is given by the Kronecker delta

$$\kappa(i, j) = \langle i | j \rangle = \delta_{ij},$$

which is of course a very strict similarity measure on input space, since it only yields a nonzero value (indicating similarity) for inputs that are exactly the same.

---

[4] In fact, the Hilbert space of some quantum systems can easily be constructed as a reproducing kernel Hilbert space [14].

[5] Strictly speaking, according to the definition in Eq. (6.8) a feature map maps an input to a function, and not to a vector. However, a quantum state is also called a *wave function*, and a more general definition of a feature map is a map from $\mathcal{X}$ to a general Hilbert space. We can therefore overlook this subtlety here and refer to [14] for more details.

Amplitude encoding. If the input patterns $x \in \mathbb{R}^N$ are real vectors of length $N = 2^n$ and we chose amplitude encoding (see Sect. 5.2), we get the feature map

$$\phi : x \to |\psi_x\rangle.$$

If the input x is normalised to one, its dimension is a power of 2, $n = \log_2 N$, and we only use the real subspace of the complex vector space, this map is simply the identity. The kernel is the inner product

$$\kappa(x, x') = \langle \psi_x | \psi_{x'} \rangle = x^T x',$$

which is also known as a linear kernel.

Copies of quantum states. Using the trick from [17], we can map an input $x \in \mathbb{R}^N$ to copies of an amplitude encoded quantum state,

$$\phi : x \to |\psi_x\rangle \otimes \cdots \otimes |\psi_x\rangle$$

and, no surprise, get the homogeneous polynomial kernel

$$\kappa(x, x') = \langle \psi_x | \psi_{x'} \rangle \otimes \cdots \otimes \langle \psi_x | \psi_{x'} \rangle = (x^T x')^d.$$

If the original inputs are extended by some constant features, for example when padding the input to reach the next power of 2, the constant features can play a similar role to the offset c of a general polynomial kernel $\kappa(x, x') = (x^T x' + c)^d$ [18].

Angle encoding. Given $x = (x_1, .., x_N)^T \in \mathbb{R}^N$ once more, we can encode one feature per qubit like in Sect. 6.1.4.3. To recap, the feature x_i is encoded in a qubit as $|q(x_i)\rangle = \cos(x_i)|0\rangle + \sin(x_i)|1\rangle$ (see also [19]). We get the feature map

$$\phi : x \to \begin{pmatrix} \cos x_1 \\ \sin x_1 \end{pmatrix} \otimes \cdots \otimes \begin{pmatrix} \cos x_N \\ \sin x_N \end{pmatrix},$$

and the corresponding kernel is a cosine kernel:

$$\kappa(x, x') = \begin{pmatrix} \sin x_1 \\ \sin x_1 \end{pmatrix}^T \begin{pmatrix} \cos x'_1 \\ \sin x'_1 \end{pmatrix} \otimes \cdots \otimes \begin{pmatrix} \cos x_N \\ \sin x_N \end{pmatrix}^T \begin{pmatrix} \cos x'_N \\ \sin x'_N \end{pmatrix}$$

$$= \prod_{i=1}^{N} (\sin x_i \sin x'_i + \cos x_i \cos x'_i)$$

$$= \prod_{i=1}^{N} \cos(x_i - x'_i).$$

6.2 Kernel Methods

Coherent states. As remarked in Reference [20], *coherent states* can be used to explicitly compute so called *radial basis function kernels* (see Table 2.2),

$$\kappa(x, x') = e^{-\delta|x-x'|^2},$$

for the case $\delta = 1$.

Coherent states are known in the field of quantum optics as a description of light modes. Formally, they are superpositions of so called *Fock states*, which are basis states from an infinite-dimensional discrete basis $\{|0\rangle, |1\rangle, |2\rangle, ...\}$.

A coherent state has the form

$$|\alpha\rangle = e^{-\frac{|\alpha|^2}{2}} \sum_{n=0}^{\infty} \frac{\alpha^n}{\sqrt{n!}} |n\rangle,$$

for $\alpha \in \mathbb{C}$. Encoding a real scalar input $c \in \mathbb{R}$ into a coherent state $|\alpha = c\rangle = |\alpha_c\rangle$, induces a feature map to an infinite-dimensional space,

$$\phi : c \to |\alpha_c\rangle.$$

We can encode a real vector $x = (x_1, ..., x_N)^T$ in N joint coherent states,

$$|\alpha_x\rangle = |\alpha_{x_1}\rangle \otimes |\alpha_{x_2}\rangle \otimes \cdots \otimes |\alpha_{x_N}\rangle.$$

For two coherent states $|\alpha_x\rangle, |\alpha_{x'}\rangle$, the kernel corresponding to this feature map is

$$\kappa(x, x') = e^{-\left(\frac{|x|^2}{2} + \frac{|x'|^2}{2} - x^\dagger x'\right)},$$

and its absolute value is given by

$$|\kappa(x, x')| = |\langle \alpha_x | \alpha_{x'} \rangle| = e^{-\frac{1}{2}|x-x'|^2}.$$

Since products of kernels are also kernels, we can construct a new kernel

$$\tilde{\kappa}(x, x') = \kappa(x, x') \cdot \kappa(x, x')^*,$$

which is the desired radial basis function or Gaussian kernel.

According to the generalised definition of coherent states, one of their characteristics is that inner products of basis states are not orthogonal, which means that their inner product is not zero [21]. Generalised coherent states therefore allow the feature map to map inputs to basis states, while still producing more interesting kernels than a simple delta function as in the basis encoding example above [14, 20].

It is interesting to note that one cannot only use the quantum computer to estimate the kernel function, but apply a quantum circuit to state $|\phi(x)\rangle$ to process the input in the 'feature Hilbert space'. For example, a trainable or *variational circuit* (see

Sects. 7.3 and 8.2) can learn how to process the feature quantum state and compute a prediction from it. Such an approach has been termed the *explicit approach* to build a quantum classifier from feature maps, since it explicitly computes in feature space instead of implicitly using kernel functions [14].

6.2.4 Distance-Based Classifiers

The last section showed that inner products of quantum states $|\phi(x)\rangle$ compute a 'quantum' kernel function $\kappa(x, x') = \langle \phi(x)|\phi(x')\rangle$. Of course, we can also use quantum algorithms to implement kernelised classifiers directly, that is without preparing $|\phi(x)\rangle$. This is the strategy of classical kernel methods, which never visit the feature space, but perform computations on the inputs only.

In this section we illustrate with two examples how quantum algorithms can compute the prediction of a generic kernel-based model

$$f(x) = \sum_{m=1}^{M} \mu_m \kappa(x, x^m), \qquad (6.11)$$

with real weights μ_m and training inputs $\{x^m\}_{m=1}^{M}$ rather naturally. This is the same model for which the representer theorem promises wide applications (see Eq. 6.10). Remember that a kernel function is a distance measure between x and x', which is why such a model can also be called a distance-based classifier.

As so often in this book, the two following models are examples for basis and amplitude encoded inputs. The latter revisits the algorithm that has in a simplified example been introduced in the introduction.

6.2.4.1 Kernelised Classifier with Basis Encoded Inputs

To implement a kernelised binary classifier using basis encoding, we need a subroutine that calculates the distance between two vectors encoded into computational basis states. Obviously, the details of the binary representation as well as the desired measure dictate the design of such a routine. A refreshingly simple example has been proposed in [22] and was adapted to classification in [23] which we will extend here.

Assume that the inputs are encoded in a fixed point binary representation as introduced in Eq. (6.7). To work with a concrete case, we choose two digits left of the decimal point and four digits right of it. This means that every feature x_i is translated to a 6-bit binary sequence $b_s b_1 b_0 . b_{-1} b_{-2} b_{-3}$ with the relation

$$x_i = (-1)^{b_s}(b_1 2^1 + b_0 2^0 + b_{-1}\frac{1}{2^1} + b_{-2}\frac{1}{2^2} + b_{-3}\frac{1}{2^3}),$$

6.2 Kernel Methods

where the first bit b_s indicates the sign. Encoding all N features into the qubits of a $6N$-qubit register leads to the computational basis state

$$|x\rangle = |(b_s)_1, (b_1)_1, ..., (b_{-3})_1, \cdots, (b_s)_N, (b_1)_N, ..., (b_{-3})_N\rangle.$$

We want to compare two vectors x and x' through a quantum routine that takes the two states $|x\rangle|x'\rangle$ encoded in this fashion. Compare each pair of qubits $(b_j)_i$, $(b'_j)_i$ with an XOR gate that writes the result

$$(d_j)_i = \begin{cases} 0, & \text{if } (b_j)_i = (b'_j)_i \\ 1, & \text{else}, \end{cases}$$

into the corresponding qubit of the second register,

$$|(b_s)_1, \cdots, (b_{-3})_N\rangle|(d_s)_1...(d_{-3})_N\rangle.$$

The second register is now a 'distance' register $|d(x, x')\rangle$ which has ones at qubit positions where $|x\rangle$ and $|x'\rangle$ did not coincide. Formally, the distance can be read out from the distance register as

$$d(x, x') = \sum_{i=1}^{N} (-1)^{(d_s)_i} \sum_{j=1,0,...,-3} (d_j)_i \, 2^j.$$

We can extract this measure and write it into the phase of the quantum state with a suitable sequence of conditional phase-rotations. This ability to write the distance measure into the phase of a quantum state directly leads to a kernelised classifier. Assume we start with a basis encoded superposition of training inputs entangled with their targets and joined by the new input $\tilde{x}$, and in the fashion outlined above we turn the second register into a distance register $|d(x, x')\rangle$ that stores the distance bits. We also add a an ancilla in uniform superposition,

$$\frac{1}{\sqrt{M}} \sum_{m=1}^{M} |x^m, y^m\rangle |d(x^m, \tilde{x})\rangle \frac{1}{\sqrt{2}}(|0\rangle + |1\rangle).$$

We write the positive distance (that we assume here to lie in $[0, 1]$) into the phase of the branch entangled with the ancilla in 0, and the negative distance into the phase of the branch entangled with the ancilla in 1, getting

$$\frac{1}{\sqrt{2M}} \sum_{m=1}^{M} |x^m, y^m\rangle |d(x^m, x)\rangle (e^{i\frac{\pi}{2}d(x^m,\tilde{x})}|0\rangle + e^{-i\frac{\pi}{2}d(x^m,\tilde{x})}|1\rangle).$$

Now we interfere the two branches by a Hadamard gate applied to the ancilla qubit. This results in

$$\frac{1}{\sqrt{M}} \sum_{m=1}^{M} |x^m, y^m\rangle |d(x^m, \tilde{x})\rangle \left(\cos\left(\frac{\pi}{2} d(x^m, \tilde{x})\right)|0\rangle + i \sin\left(\frac{\pi}{2} d(x^m, \tilde{x})\right)|1\rangle \right).$$

A conditional measurement on the ancilla selects the cosine branch (flagged by $|0\rangle$) via postselection. The probability of acceptance is given by

$$p_{acc} = \frac{1}{M} \sum_{m=1}^{M} \cos^2\left(\frac{\pi}{2} d(x^m, \tilde{x})\right).$$

This probability is also a measure of how close the data is to the new input: If the distance $d(x^m, \tilde{x})$ is large, the sine branch of the superposition will have a larger probability to be measured. In the worst case scenario, all training vectors have a distance to $\tilde{x}$ that is close to 1 and the probability of the conditional measurement to succeed will be close to zero. However, in this case the data might not reveal a lot of information for the classification of the new input anyways, and the probability of acceptance can therefore be seen as a measure of how well-posed the classification problem is in the first place.

After a successful conditional measurement, the state becomes

$$\frac{1}{\sqrt{M p_{acc}}} \sum_{m=1}^{M} \cos\left(\frac{\pi}{2} d(x^m, \tilde{x})\right) |x^m; y^m\rangle,$$

where we ignored the distance register. The training inputs and targets are now weighed by the cosine of their distance to the new input. We can extract a classification from this state in two different manners. A measurement on the class qubit $|y^m\rangle$ will have a probability of

$$p(\tilde{y} = 0) = \frac{1}{M p_{acc}} \sum_{m|y^m=0} \cos^2\left(\frac{\pi}{2} d(x^m, \tilde{x})\right),$$

to predict class 0 and a complementary probability of

$$p(\tilde{y} = 1) = \frac{1}{M p_{acc}} \sum_{m|y^m=1} \cos^2\left(\frac{\pi}{2} d(x^m, \tilde{x})\right)$$

to predict class 1. Alternatively, when measuring the entire basis state $|x^m; y^m\rangle$, we have a probability of

$$p(m) = \frac{1}{M p_{acc}} \cos^2\left(\frac{\pi}{2} d(x^m, \tilde{x})\right),$$

6.2 Kernel Methods

to pick the mth training vector, and closer training vectors are thus more likely to be sampled by the measurement. Doing this repeatedly and recording the state of the class qubit of the samples, we can estimate whether label $y^m = 0$ or $y^m = 1$ is more present 'in the neighbourhood' of $\tilde{x}$.

6.2.4.2 Kernelised Classifier with Amplitude Encoded Inputs

We have already seen in Sect. 6.1 that interference can compute the Euclidean distance between amplitude vectors. This has also been used in the simplified example of a quantum machine learning algorithm in the introduction. Here we want to revisit this idea in more detail and present a slightly more rigorous description of a distance-based classifier via interference circuits following [24] (Fig. 6.9).

Consider four registers $|0\rangle_a|0\ldots0\rangle_m|0\ldots0\rangle_i|0\rangle_c$, that we will use to encode the ancilla, the m index over the training data, the i index over the features, and the class label, respectively. As an initial state, one requires the (normalised) training set $\{(x^m, y^m)\}_{m=1}^{M}$ and the new input $\tilde{x}$ to be encoded in the amplitudes of a quantum state of the form

$$\frac{1}{\sqrt{2M}} \sum_{m=1}^{M} (|0\rangle|\psi_{\tilde{x}}\rangle + |1\rangle|\psi_{x^m}\rangle) |y^m\rangle|m\rangle,$$

with

$$|\psi_{x^m}\rangle = \sum_{i=1}^{N} x_i^m |i\rangle, \quad |\psi_{\tilde{x}}\rangle = \sum_{i=1}^{N} \tilde{x}_i |i\rangle.$$

A Hadamard gate on the ancilla interferes the two states and results in

$$\frac{1}{2\sqrt{M}} \sum_{m=1}^{M} \left(|0\rangle\left[|\psi_{\tilde{x}}\rangle + |\psi_{x^m}\rangle\right] + |1\rangle\left[|\psi_{\tilde{x}}\rangle - |\psi_{x^m}\rangle\right]\right)|y^m\rangle|m\rangle.$$

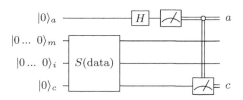

Fig. 6.9 Classification circuit of the distance-based classifier explained in the text. After state preparation S(data), the circuit only applies a Hadamard and two singe-qubit measurements to draw a sample of the prediction. Repeated applications of the circuit lead to an estimation for the prediction

A conditional measurement to find the ancilla in $|0\rangle$ selects the first branch with a success probability

$$p_{\text{acc}} = \frac{1}{4M} \sum_m \sum_i |\tilde{x}_i + x_i^m|^2,$$

which is equal to

$$1 - \frac{1}{4M} \sum_m \sum_i |x_i - x_i^m|^2.$$

As with the basis encoded kernelised classifier, the branch selection procedure is more likely to succeed if the collective Euclidean distance of the training set to the new input is small. In the worst case, $x \approx -x^m$ for all $m = 1...M$, and acceptance will be very unlikely. However, this means also here that the new input is 'far away' from the dataset, an indicator for the low expressive power of a classification algorithm based on distances.

If the conditional measurement was successful, the result is given by

$$\frac{1}{2\sqrt{Mp_{\text{acc}}}} \sum_{m=1}^{M} \sum_{i=1}^{N} (\tilde{x}_i + x_i^m) |i\rangle |y^m\rangle |m\rangle.$$

The probability of measuring the class qubit $|y^m\rangle$ in state 0 and predicting class 0 is given by

$$p(\tilde{y} = 0) = \frac{1}{4Mp_{\text{acc}}} \sum_{m|y^m=0} |\tilde{x} + x^m|^2,$$

which is the same as

$$p(\tilde{y} = 0) = 1 - \frac{1}{4Mp_{\text{acc}}} \sum_{m|y^m=0} |\tilde{x} - x^m|^2,$$

due to the normalisation of the inputs.

Expressing the probability to predict class 0 by the squared distance shows that it is higher the closer the class 0 training vectors are to the input. The kernel in Eq. (6.11) is hence given by $\kappa(x, x') = 1 - \frac{1}{c}|x - x'|^2$ where c is a constant. Figure 6.10 shows an example of how such a kernel weighs normalised 2-dimensional inputs.

Note that also in the amplitude encoded case, as an alternative the m-register could be measured to sample training inputs and their classes with a probability depending on their squared distance to the new input.

6.2 Kernel Methods

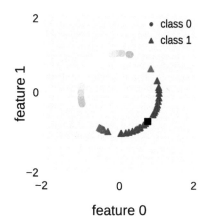

Fig. 6.10 Example of how the kernelised binary classifier in this section weighs neighbouring training inputs to come to a decision. The inputs are normalised and lie on a unit circle. The influence of a training input on the prediction of the new input (black square) is depicted by the colour scheme, and lighter dots have less influence

6.2.5 Density Gram Matrices

To conclude the discussion of kernel methods and quantum computing, we want to mention another close analogy that can be useful for the design of quantum machine learning algorithms, namely the formal equivalence between kernel Gram matrices and density matrices. Gram matrices are used in a number of models introduced in Chap. 2, for example the support vector machine or Gaussian processes.

As a reminder, given a training set of inputs $\{x^1, \cdots, x^M\}$ and a kernel $\kappa(x, x')$, the Gram matrix K is the matrix with entries

$$(K)_{m,m'} = \kappa(x^m, x^{m'}).$$

Gram matrices (of positive semi-definite kernels κ) are positive semi-definite and symmetric, and if we normalise it to unit trace,

$$K' = \frac{K}{\mathrm{tr}\{K\}},$$

it has the same mathematical form as a density matrix describing a mixed quantum state. One can therefore prepare a quantum state where the density matrix is entry-wise equivalent to a given Gram matrix. We call such a density matrix a *density Gram matrix*. This was first explicitly used for a quantum machine learning algorithm in [17] for a linear kernel,

$$\kappa(x, x') = x^T x'.$$

We want to sketch their idea of how to prepare the corresponding density Gram matrix.

We need to first prepare a quantum state where the entire dataset is amplitude encoded,

$$|\psi_D\rangle = \frac{1}{\sqrt{M}} \sum_{m=1}^{M} |m\rangle |\psi_{x^m}\rangle \tag{6.12}$$

with the normalised training inputs in amplitude encoding $|\psi_{x^m}\rangle = \sum_i x_i^m |i\rangle$. The corresponding pure density matrix reads

$$\rho_D = |\psi_D\rangle\langle\psi_D| = \frac{1}{M} \sum_{m,m'=1}^{M} |m\rangle\langle m'| \otimes |\psi_{x^m}\rangle\langle\psi_{x^{m'}}|. \tag{6.13}$$

Taking the partial trace over the register $|i\rangle$ in $|\psi_{x^m}\rangle$ computes the mixed quantum state of the m register only. Mathematically, we have to sum over a complete basis (i.e. the computational basis $\{|k\rangle\}$) in the N-dimensional Hilbert space of the qubits of register $|i\rangle$,

$$\mathrm{tr}_i\{\rho_D\} = \sum_{k=1}^{N} \langle k|\rho_D|k\rangle.$$

The computational basis states $\langle k|, |k\rangle$ do not act on the m register,

$$\mathrm{tr}_i\{\rho_D\} = \sum_{k=1}^{N} \langle k|\psi_D\rangle\langle\psi_D|k\rangle$$

$$= \frac{1}{M} \sum_{k=1}^{N} \sum_{m,m'=1}^{M} |m\rangle\langle m'| \ \langle k|\psi_{x^m}\rangle\langle\psi_{x^{m'}}|k\rangle$$

$$= \frac{1}{M} \sum_{m,m'=1}^{M} \left(\sum_{k=1}^{N} \langle k|\psi_{x^m}\rangle\langle\psi_{x^{m'}}|k\rangle \right) |m\rangle\langle m'|.$$

The expressions $\langle k|\psi_{x^m}\rangle$, $\langle\psi_{x^{m'}}|k\rangle$ are nothing else than the kth component of the mth and m'th training vector respectively. The expression in the bracket is therefore the inner product of these two training vectors. This becomes immediately clear when we turn around the inner products and remove the identity $\sum_k |k\rangle\langle k| = \mathbb{1}$,

$$\sum_{k=1}^{N} \langle k|\psi_{x^m}\rangle\langle\psi_{x^{m'}}|k\rangle = \sum_{k=1}^{N} \langle\psi_{x^{m'}}|k\rangle\langle k|\psi_{x^m}\rangle$$

$$= \langle\psi_{x^{m'}}|\psi_{x^m}\rangle.$$

We end up with an $M \times M$-dimensional density matrix that describes the state of the $|m\rangle$ register as a statistical mixture,

$$\mathrm{tr}_i\{\rho_D\} = \frac{1}{M} \sum_{m,m'=1}^{M} \langle\psi_{x^m}|\psi_{x^{m'}}\rangle \ |m\rangle\langle m'|.$$

6.2 Kernel Methods

This density matrix has entries

$$\rho_{m,m'} = \langle \psi_{x^m} | \psi_{x^{m'}} \rangle = (x^m)^T (x^{m'}) = \kappa(x^m, x^{m'}) \tag{6.14}$$

and is therefore identical to the normalised Gram matrix K' with a linear, i.e. inner product kernel.

As we have seen in Sect. 6.2.3 there is a simple way of generalising this result for polynomial kernels of order d,

$$\kappa(x, x') = \left(x^T x' + c\right)^d,$$

with $c = 0$. Instead of a single copy of $|\psi_{x^m}\rangle$ in Eq. (6.13), start with d copies of $|\psi_{x^m}\rangle$,

$$\frac{1}{\sqrt{M}} \sum_{m=1}^{M} |m\rangle \otimes |\psi_{x^m}\rangle \otimes \cdots \otimes |\psi_{x^m}\rangle.$$

Consequently, instead of using n qubits to encode 2^n-dimensional inputs, we need d registers of n qubits, or dn qubits altogether. The state preparation routine for the $|\psi_{x^m}\rangle$ has to be repeated (or applied in parallel) for each register. The density matrix corresponding to this state is

$$\rho_\mathcal{D} = \frac{1}{M} \sum_{m,m'=1}^{M} |m\rangle\langle m'| \otimes |\psi_{x^m}\rangle\langle \psi_{x^{m'}}| \otimes \cdots \otimes |\psi_{x^m}\rangle\langle \psi_{x^{m'}}|.$$

Tracing out all i registers now results in a reduced density matrix with entries

$$\rho_{m,m'} = \left((x^m)^T (x^{m'})\right)^d = \kappa(x^m, x^{m'}),$$

and corresponds to a normalised Gram matrix for a polynomial kernel with constant offset $c = 0$.

In order to include the constant we have to extend the amplitude encoded x^m by constant entries. Let $\sqrt{c} = \sqrt{c_1}, ..., \sqrt{c_N}$ be a N-dimensional vector of constants. One can add an ancilla to each $|i\rangle$ register in state $|\psi_{x^m}\rangle$ and prepare copies of

$$|0\rangle|\psi_{x^m}\rangle + |1\rangle|\psi_{\sqrt{c}}\rangle.$$

The inner products $\langle \psi_{x^m} | \psi_{x^{m'}} \rangle$ in Eq. (6.14) are now replaced by $\langle \psi_{x^m} | \psi_{x^{m'}} \rangle + \langle \psi_{\sqrt{c}} | \psi_{\sqrt{c}} \rangle$ and the resulting kernel has an offset c. With this trick, density Gram matrices of general polynomial kernels can be created.

6.3 Probabilistic Models

We have so far only dealt with deterministic models in this section. Quantum mechanics is a probabilistic theory, and it is therefore only natural to search for points of leverage for quantum computing with regards to probabilistic models.

From the perspective of supervised learning, a probabilistic model was defined as a generative probability distribution $p(x, y)$ or a discriminative probability distribution $p(y|x)$ over inputs x and labels y, which consist of binary random variables $x_1, \ldots, x_N, y_1, \ldots, y_K$. Probabilistic models often have a graphical representation, in which case they are referred to as *probabilistic graphical models*. Each variable is associated with a vertex of a graph. Edges indicate interdependence relations such as conditional independence or direct correlation.

6.3.1 Qsamples as Probabilistic Models

A quantum state can be associated with a probabilistic model by using a technique that has been introduced as *qsample* encoding in Sect. 5.3. A qubit is associated with the binary random variable, so that a basis state $|i\rangle = |x, y\rangle$ describes a specific sample of the random variables, which is a possible data point. The squared amplitude corresponding to this basis state is then interpreted as the probability of data point (x, y),

$$p(x, y) \Leftrightarrow |p(x, y)\rangle = \sum_{x,y} \sqrt{p(x, y)} |x, y\rangle.$$

One can now draw samples from the distribution by measuring the qubits in the computational basis. To take multiple samples, we have to re-prepare the qsample from scratch each time.

As mentioned in Sect. 5.3.2, using quantum states we get marginal distributions 'for free'. For example, given the qsample for $p(x, y)$, we get the qsample of the distribution over the labels only,

$$p(y) = \sum_x p(x, y) \Leftrightarrow \mathrm{tr}_x\{|p(x, y)\rangle\langle p(x, y)|\},$$

by simply tracing over (in other words, ignoring) the qubits that encode the inputs x. This is not surprising, because we are in a way "simulating" the probabilistic model by a probabilistic quantum system, where such operations come naturally. However, quantum theory does not seem to offer 'magic powers' for computing the discriminative distribution of a given observation $x = e$ for the input. In other words, given the qsample $|p(y|x)\rangle$ we cannot easily prepare the qsample $|p(y|e)\rangle$, at least not without further assumptions. This means that inference is still a hard problem.

6.3 Probabilistic Models

Most proposals combining quantum computing and probabilistic models use quantum algorithms to facilitate the preparation of (possibly discriminative) qsamples from which one can sample (x, y) or y for a given evidence e. Often, amplitude amplification offers the quantum advantage of a quadratic speedup. The ability of sampling can be used for inference as discussed in this chapter, but also for training, which is why we will revisit the idea of preparing qsamples with a quantum computer in the next chapter.

Here we will present two examples of classical models—Bayesian nets and Boltzmann machines—and illustrate how to prepare a qsample that corresponds to a probabilistic model. Each example reveals a different strategy of preparing the qsample, namely with regards to distributions where each variable is only conditionally dependent on a few parent variables, or distributions that have a good mean-field approximation.

6.3.2 Qsamples with Conditional Independence Relations

To recapitulate, Bayesian nets with vertices $s_1...s_G$ represent a probability distribution of the form

$$p(s_1, ..., s_G) = \prod_{i=1}^{G} p(s_i|\pi_i),$$

where π_i is the set of $|\pi_i|$ parent nodes to $s_i \in \{0, 1\}$. One can prepare a qsample $|p(s_1, ..., s_G)\rangle$ of this probability distribution in time $\mathcal{O}(n2^{|\pi|_{\max}})$ [25], where $|\pi|_{\max}$ is the largest number of parents any vertex has. For later use we denote this state preparation routine by $\mathcal{A}_S$.

The state preparation routine is similar to arbitrary state preparation we encountered in Sect. 5.2, but since the state of qubit i only depends on its parents, its rotation is only conditioned on the parent qubits. More precisely, we take a quantum state of G qubits corresponding to the random variables or vertices of the net, and rotate qubit q_i from $|0\rangle$ to $\sqrt{1 - p(s_i|\pi_i)}|0\rangle + \sqrt{p(s_i|\pi_i)}|1\rangle$, conditioned on the state of the qubits representing the parents π_i. Each possible state of the parents has to be considered with a separate conditional operation. The probability $p(s_i|\pi_i)$ can be read from the probability tables of the Bayesian net. The conditioning therefore introduces the exponential dependency on the number of parent nodes, since we need to cater for any of their possible binary values. For instance, if s_i has three parent nodes, then $p(s_i|\pi_i)$ describes a distribution over 2^3 possible values for the nodes in set π_i. We illustrate how to prepare a qsample for the Bayesian net shown in Fig. 2.25 in the following example.

Example 6.4 (Preparing a Bayesian net qsample) In the example Bayesian net from Sect. 2.4.3.1 which is reprinted in Fig. 6.11, one would require a register of three qubits $|0_R0_S0_G\rangle$ to represent the three binary variables 'Rain', 'Sprinkler' and

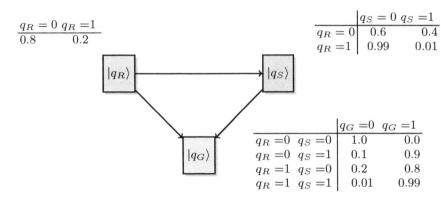

Fig. 6.11 The Bayesian network example from Sect. 2.4.3.1, which is used to demonstrate the preparation of a qsample to perform quantum inference on Bayesian networks. In the quantum model, each random variable corresponds to a qubit

'Grass'. The only node without a parent is 'Rain', and the rotation is therefore unconditional,

$$|0_R 0_S 0_G\rangle \rightarrow (\sqrt{0.8}|0_R\rangle + \sqrt{0.2}|1_R\rangle)|0_S 0_G\rangle.$$

Now rotate every successive node of the belief network conditioned on the state of its parents. This is possible because of the acyclic structure of the graph. For example, the second qubit will be rotated around the y-axis by $\sqrt{0.4}$ or $\sqrt{0.01}$ controlled by the first qubit being in $|0\rangle$ or $|1\rangle$,

$$\left[\sqrt{0.8}\,|0\rangle(\sqrt{0.6}\,|0\rangle + \sqrt{0.4}\,|1\rangle) + \sqrt{0.2}\,|1\rangle(\sqrt{0.99}\,|0\rangle + \sqrt{0.01}\,|1\rangle)\right]|0_G\rangle$$

Rotating the last qubit $|0_G\rangle$ requires four rotation gates, each controlled by two qubits, to finally obtain

$$|p(RSG)\rangle = \sqrt{0.48}\,|100\rangle + \sqrt{0.032}\,|101\rangle + \sqrt{0.00002}\,|110\rangle + \sqrt{0.00198}\,|111\rangle.$$

Obviously, for each qubit i one needs $2^{|\pi_i|}$ rotations. The resources for state preparation therefore grow with $\mathcal{O}(G2^{|\pi|_{\max}})$. We can therefore prepare quantum states for models with sparse dependence relations efficiently.

Once the qsample is prepared, the goal of inference in the context of supervised learning is to get the probability of a certain label given some input $x = \tilde{x}$. The new input plays the role of the observation or evidence e. For binary classification we can interpret the first node s_1 as the output and the remaining nodes $s_2, ..., s_G$ as the input to the model. The probability distribution of the Bayesian net can then be written as

$$p(s_1, ..., s_G) = p(y, x_1, ..., x_N) = p(x, y).$$

For inference, the nodes or qubits encoding x are replaced by the observed values or evidence $\tilde{x}$. Our goal is therefore to obtain the discriminative distribution $p(y|\tilde{x})$ that allows us to guess a label y. The most straight-forward approach is to draw samples (x, y) and reject every sample where $x \neq \tilde{x}$. The resulting samples are drawn from $p(y|\tilde{x})$.

With regards to the quantum model, inference requires us to prepare a qsample $|p(y|\tilde{x})\rangle$. Based on the ideas of quantum rejection sampling introduced in [26], Low and Chuang [25] show that amplitude amplification can make measurements of a qsample quadratically more efficient than classical rejection sampling. Given the original Bayesian net qsample $|p(x, y)\rangle$, one can formally separate the superposition into two subbranches, one with basis states that are consistent with the evidence $\tilde{x}$ and one with all other basis states x that do not confirm the evidence,

$$|p(x,y)\rangle = \sqrt{p(\tilde{x})}|y, \tilde{x}\rangle + \sqrt{1 - p(\tilde{x})}|s, x\rangle.$$

We can apply amplitude amplification to boost the probability of the left branch, the one containing the evidence. A part of the amplitude amplification routine is to apply the state preparation routine $\mathcal{A}_S$ for each iteration. The number of iterations of the algorithm (each applying $\mathcal{A}_S$) needed to fully amplify the 'evidence' branch—thereby making it certain to draw samples from qsample $|p(y|\tilde{x})\rangle$—is in $\mathcal{O}(p(\tilde{x})^{-\frac{1}{2}})$. Overall, one therefore achieves a runtime of $\mathcal{O}(n2^{|\pi|_{\max}} p(\tilde{x})^{-\frac{1}{2}})$, which is exponentially worse with respect to the maximum number of parents, but quadratically better with respect to the probability of observing the evidence than classical inference with rejection sampling.

6.3.3 Qsamples of Mean-Field Approximations

Bayesian nets allow for an elegant state preparation routine due to their intrinsic structure that factorises the distribution. In other cases one has to resort to approximations in order to use the favourable properties of factorisation. One idea is to start with a qsample of a mean-field approximation of the desired distribution (which has to be computed classically beforehand) and then use branch selection to refine the distribution. This has been introduced in the context of Boltzmann machines [27] which we will use as an example here. The advantage of these two steps is that a mean-field approximation is a product state, and we will see that it can be prepared by simply rotating each qubit in the qsample register successively. State preparation can—under certain circumstances—be therefore qubit-efficient.

To go into a bit more detail, remember that Boltzmann machines are probabilistic models defined by the model distribution

$$p(v, h) = \frac{e^{-E(v,h)}}{\sum_{v,h} e^{-E(v,h)}}$$

over the visible variables $v_1, ..., v_{N+K}$ and the hidden variables $h_1, ..., h_J$, with the learnable energy function $E(v, h) = E(v, h; \theta)$ (see also Eq. 2.26). As before, we summarise visible and hidden units with $s_1, ..., s_G$ where $G = N + K + J$.

Our goal is to prepare the qsample $|p(s)\rangle$. We start instead with preparing a qsample $|q(s)\rangle$ of the mean-field distribution $q(s)$. The mean-field distribution is a product distribution, which means it can be computed as the product of Bernoulli distributions over individual features,

$$q(s) = g(s_1; \mu_1)g(s_2; \mu_2)\cdots g(s_G; \mu_G).$$

The individual distributions have the form

$$g(s_i; \mu_i) = \begin{cases} \mu_i, & \text{if } s_i = 1, \\ 1 - \mu_i, & \text{else,} \end{cases}$$

for $0 \leq \mu_i \leq 1$ and $i = 1, ..., G$. The mean-field distribution has the property that it minimises the Kullback-Leibler divergence to the desired Boltzmann distribution $p(s, h)$ (which is a common measure of distance between distributions). The mean field parameters μ_i have to be obtained in a classical calculation. We assume furthermore that a real constant $k \geq 1$ is known such that $p(s) \leq kq(s)$.

To prepare a qsample of the form

$$\sum_s \sqrt{q(s)}|s\rangle,$$

where $|s\rangle = |s_1...s_G\rangle$ abbreviates a state of the network in basis encoding, one simply has to rotate qubit i around the y-axis to

$$|0\rangle_i \to \sqrt{1 - \mu_i}|0\rangle_i + \sqrt{\mu_i}|1\rangle_i,$$

for all qubits $i = 1, ..., G$ (similar to the Bayesian network state preparation routine, but without conditioning on the parents).

We have a qsample corresponding to the mean field approximation of the target distribution. The second step is based on branch selection presented in Sect. 5.2.2.3. First, add an extra register and load another distribution $\bar{p}(s)$ in basis encoding to obtain

$$\sum_s \sqrt{q(s)}|s\rangle|\bar{p}(s)\rangle.$$

This second distribution is constructed in such a way that $q(s)\bar{p}(s)$ is proportional to $p(s)$. Rotating an extra ancilla qubit,

$$\sum_s \sqrt{q(s)}|s\rangle|\bar{p}(s)\rangle \left(\sqrt{1 - \bar{p}(s)}|0\rangle + \sqrt{\bar{p}(s)}|1\rangle\right),$$

and selecting the branch in which the ancilla is in state $|1\rangle$ leads to the desired qsample $|p(s)\rangle$. The success probability is larger than $1/k$. If the approximation $q(s)$ was exact, $k = 1$, and the branch selection succeeds with certainty. The larger the divergence between $q(s)$ and $p(s)$, the smaller the success of the conditional measurement. This scheme is therefore useful for Boltzmann distributions that are close to a product distribution, or where the individual variables are not very correlated.

Boltzmann distributions play an important role in quantum machine learning due to their natural proximity to quantum mechanics. In the machine learning literature, Boltzmann distributions are frequently called 'Gibbs distribution'[6] and they do not only play a role in Boltzmann machines, but also in other models such as Markov logic networks [28]. In the next chapter we will discuss other ways of how to prepare Gibbs distribution qsamples with quantum annealing (Sect. 7.4), and variational circuits (Sect. 7.3). There are also conventional quantum circuits to prepare such states exactly [29, 30], but not qubit-efficiently.

In summary, interpreting a quantum state as a qsample corresponding to a probabilistic model is a fruitful way to combine machine learning and quantum computing, and a very natural intersection are Boltzmann or Gibbs distributions.

References

1. Cleve, R., Ekert, A., Macchiavello, C., Mosca, M.: Quantum algorithms revisited. In: Proceedings of the Royal Society of London A: Mathematical, Physical and Engineering Sciences, vol. 454, pp. 339–354. The Royal Society (1998)
2. Kobayashi, H., Matsumoto, K., Yamakami, T.: Quantum Merlin-Arthur proof systems: Are multiple Merlins more helpful to Arthur? In: Algorithms and Computation, pp. 189–198. Springer (2003)
3. Zhao, Z., Fitzsimons, J.K., Fitzsimons, J.F.: Quantum assisted Gaussian process regression (2015). arXiv:1512.03929
4. Romero, J., Olson, J.P., Aspuru-Guzik, A.: Quantum autoencoders for efficient compression of quantum data. Quantum Sci. Technol. **2**(4), 045001 (2017)
5. Schuld, M., Sinayskiy, I., Petruccione, F.: How to simulate a perceptron using quantum circuits. Phys. Lett. A **379**, 660–663 (2015)
6. Nielsen, M.A., Chuang, I.L.: Quantum computation and quantum information. Cambridge University Press, Cambridge (2010)
7. Tommiska, M.T.: Efficient digital implementation of the sigmoid function for reprogrammable logic. IEE Proc. Comput. Digit. Tech. **150**(6), 403–411 (2003)
8. Quine, V.W.: A way to simplify truth functions. Am. Math. Mon. **62**(9), 627–631 (1955)
9. McCluskey, J.E.: Minimization of boolean functions. Bell Labs Techn. J. **35**(6), 1417–1444 (1956)
10. Schuld, M., Sinayskiy, I., Petruccione, F.: The quest for a quantum neural network. Quantum Inf. Process. **13**(11), 2567–2586 (2014)
11. Cao, Y., Guerreschi, G.G., Aspuru-Guzik, A.: Quantum neuron: an elementary building block for machine learning on quantum computers (2017). arXiv:1711.11240
12. Paetznick, A., Svore, K.M.: Repeat-until-success: non-deterministic decomposition of single-qubit unitaries. Quantum Inf. Comput. **14**, 1277–1301 (2013)

---

[6]In statistical physics, a Gibbs distribution is a state of a system that does not change under the evolution of the system.

13. Wiebe, N., Kliuchnikov, V.: Floating point representations in quantum circuit synthesis. New J. Phys. **15**(9), 093041 (2013)
14. Schuld, M., Killoran, N.: Quantum machine learning in feature Hilbert spaces (2018). arXiv:1803.07128v1
15. Schölkopf, B., Herbrich, R., Smola, A.: A generalized representer theorem. In: Computational Learning Theory, pp. 416–426. Springer (2001)
16. Schölkopf, B., Smola, A.J.: Learning with kernels: support vector machines, regularization, optimization, and beyond. MIT Press (2002)
17. Rebentrost, P., Mohseni, M., Lloyd, S.: Quantum support vector machine for big data classification. Phys. Rev. Lett. **113**, 130503 (2014)
18. Schuld, M., Bocharov, A., Svore, K., Wiebe, N.: Circuit-centric quantum classifiers (2018). arXiv:1804.00633
19. Stoudenmire, E., Schwab, D.J.: Supervised learning with tensor networks. In: Advances in Neural Information Processing Systems, pp. 4799–4807 (2016)
20. Chatterjee, R., Ting, Y.: Generalized coherent states, reproducing kernels, and quantum support vector machines. Quantum Inf. Commun. **17**(15&16), 1292 (2017)
21. Klauder, J.R., Bo-Sture S.: Coherent States: Applications in Physics and Mathematical Physics. World Scientific (1985)
22. Trugenberger, C.A.: Quantum pattern recognition. Quantum Inf. Process. **1**(6), 471–493 (2002)
23. Schuld, M., Sinayskiy, I., Petruccione, F.: Quantum Computing for Pattern Classification. Lecture Notes in Computer Science, vol. 8862, pp. 208–220. Springer (2014)
24. Schuld, M., Fingerhuth, M., Petruccione, F.: Implementing a distance-based classifier with a quantum interference circuit. EPL (Europhys. Lett.) **119**(6), 60002 (2017)
25. Yoder, T.J., Low, G.H., Chuang, I.L.: Quantum inference on Bayesian networks. Phys. Rev. A **89**, 062315 (2014)
26. Ozols, M., Roetteler, M., Roland, J.: Quantum rejection sampling. ACM Trans. Comput. Theory (TOCT) **5**(3), 11 (2013)
27. Wiebe, N., Kapoor, A., Svore, K.M.: Quantum deep learning (2014). arXiv: 1412.3489v1
28. Wittek, P., Gogolin, C.: Quantum enhanced inference in markov logic networks. Sci. Rep. **7** (2017)
29. Brandão, F.G.S.L., Svore, K.M.: Quantum speed-ups for solving semidefinite programs. In: 2017 IEEE 58th Annual Symposium on Foundations of Computer Science (FOCS), pp. 415–426. IEEE (2017)
30. Poulin, D., Wocjan, P.: Sampling from the thermal quantum gibbs state and evaluating partition functions with a quantum computer. Phys. Rev. Lett. **103**(22), 220502 (2009)

Chapter 7
Quantum Computing for Training

The previous chapter looked into strategies of implementing inference algorithms on a quantum computer, or how to compute the prediction of a model using a quantum instead of a classical device. This chapter will be concerned with how to optimise models using quantum computers, a subject targeted by a large share of the quantum machine learning literature.

We will look at four different approaches to optimisation in quantum machine learning: linear algebra calculus, search, hybrid routines for gradient descent and adiabatic methods. The first two approaches are largely based on two famous quantum algorithms: the Harrow-Hassidim-Lloyd (HHL) algorithm for matrix inversion and Grover's routine for unstructured search, respectively. The third approach tackles gradient descent, one of the most frequently used optimisation methods in machine learning, which has no clear quantum equivalent. This is why hybrid or *variational* approaches are consulted, in which inference is made on a quantum device, but the optimisation is performed classically. Lastly, the adiabatic approach loosely summarises analog techniques where the result of an optimisation problem is encoded into the ground state of a quantum system, and the goal of optimisation is to generate this ground state, starting in a state that is relatively easy to prepare. Increasingly, this technique is also used to prepare qsamples which can be used as distributions for sampling-based classical training methods, a strategy that promises to be more fruitful in the face of noise-prone and sparsely connected early technologies.

The output of the quantum(-assisted) optimisation algorithm may be a quantum state representing the trained parameters, or it might be a classical description of the optimal parameters, or samples from a model distribution. While some of the routines presented here can be 'plugged' into classical machine learning algorithms, others are parts of quantum implementations of machine learning algorithms. Unsurprisingly, the choice of information encoding plays again an important role in this chapter.

7.1 Quantum Blas

The quantum computing community developed a rich collection of quantum algorithms for basic linear algebra subroutines, or *quantum blas* [1], in analogy to the linear algebra libraries of various programming platforms. Quantum blas include routines such as matrix multiplication, matrix inversion and singular value decomposition. These can be used to solve optimisation problems, and heavily rely on the idea of amplitude encoding. The quantum machine learning algorithms based on quantum blas are rather technical combinations of the subroutines introduced in previous chapters (see also Table 7.1), to which we will refer extensively. Amongst them are *state preparation for amplitude encoding* (Sect. 5.2), *Hamiltonian evolution* (Sect. 5.4), *density matrix exponentiation* (Sect. 5.4.3), *quantum matrix inversion* (Sect. 3.5.3), *quantum phase estimation* (Sect. 3.5.2), and *branch selection* (Sect. 5.2.3).

Quantum machine learning algorithms constructed from quantum blas inherit a runtime that is logarithmic in the input dimension N as well as the training set size M provided a certain structure in the inputs is given. They have different polynomial dependencies on the desired maximum error and/or the condition number of the design matrix constructed from the data. Our goal here is not to give an accurate runtime analysis, which can be found in the original references and is often paved with subtleties. Instead we want to focus on two main points: (1) How the machine learning problem translates to a linear algebra calculation, and (2) how to combine quantum subroutines to solve the task. Before coming to that part, the next section is an attempt to give readers with a less extensive background in quantum computing an idea of how the algorithms work.

7.1.1 Basic Idea

A number of learning algorithms contain linear algebra routines such as inverting a matrix or finding a matrix' eigenvalues and eigenvectors. The matrices are usually constructed from the training set (i.e., the design matrix that carries all training inputs as rows), and therefore grow with the dimension N and/or number of the training vectors M. The basic idea of the linear algebra approach in quantum machine learning is to use quantum systems for linear algebra calculus, where the design matrix is represented by the Hamiltonian of the system via dynamic encoding.

To illustrate this with an example, consider the linear algebra task of a multiplication of a vector and a unitary matrix, which we encountered in the discussion of linear models in amplitude encoding in the last chapter. The evolution of a quantum system can be mathematically expressed as $U\alpha$, where U is a unitary matrix and α is the complex vector of 2^n amplitudes. Performing this evolution effectively implements the unitary matrix multiplication, and the state of the quantum system after the evolution encodes the result of the multiplication. But one can do much more: Quantum systems are natural eigendecomposers in the sense that the results of measurements are eigenvalues of operators to certain eigenstates (see Sect. 3.1.3.4). One can say that the dynamics of the quantum system *emulates* a linear algebra calculation. This can be used for matrix inversion, as demonstrated in the algorithm for linear systems by Harrow, Hassidim and Lloyd [2] (see Sect. 3.5.3). In this algorithm one prepares a special quantum system whose evolution is defined by the matrix that is to be inverted, and the physical evolution allows us to invert the eigenvalues of that matrix and read out some desired information via measurements.

The crux of this idea is the same as in amplitude encoding. Many quantum blas require a number of manipulations that is polynomial in the number n of qubits, which means that they are qubit-efficient. Algorithms composed of such subroutines depend only logarithmic on the number $N = 2^n$ of amplitudes, and the same usually applies to the number of training vectors M. A logarithmic dependency is a significant speedup. For example, a training set of one billion (bn) vectors that each have dimension one million can be represented in a Hamiltonian (using the previous trick of extending the rectangular design matrix to a 2 bn × 2 bn Hermitian matrix) of a $n = 31$ qubit system.

Of course, there has to be a caveat: similar to state preparation, simulating such a Hamiltonian in general may take of the order of 4 bn operations. But we saw interesting exceptions in Sect. 5.4, for example where the Hamiltonian is sparse or low-rank. This obviously poses restrictions on the data one can deal with, and only little has been done to find out which datasets could fulfil the requirements of qubit-efficient processing. Still, the promises of simulating linear algebra calculus with quantum systems for big data applications are impressive, and therefore worth investigating.

7.1.2 Matrix Inversion for Training

This section presents some selected examples of classical machine learning algorithms that rely on eigenvalue decomposition or matrix inversion, and for which quantum algorithms have been proposed. The basic building blocks are summarised in Table 7.1.

7.1.2.1 Inverting Data Matrices

The first suggestion to use Harrow, Hassidim and Lloyd's (HHL's) quantum matrix inversion technique for statistical data analysis was proposed by Wiebe, Braun and Lloyd [3]. Closely related to machine learning, their goal was to 'quantise' linear regression for data fitting, in which the best model parameters of a linear function for some data points have to be found. In Sect. 2.4.1.1 it has been shown that basic linear regression (i.e., without regularisation) reduces to finding a solution to the equation

$$w = (X^T X)^{-1} X^T y, \tag{7.1}$$

Table 7.1 Simplified overview of the quantum blas based quantum machine learning algorithms presented in this chapter. The design matrix X is composed of the training inputs, and y is a vector of training outputs. u_r, v_r, σ_r are singular vectors/values of $X^T X$ and K is a kernel matrix. Abbreviations: HHL—quantum matrix inversion routine, HHL$^-$—quantum matrix multiplication routine (omitting the inversion of the eigenvalues), HHL^{DME}—quantum matrix inversion routine with density matrix exponentiation to simulate the required Hamiltonian, SWAP—interference circuit to evaluate inner products of quantum states.

Classical method	Computation	Strategy	Refs.
Data matrix inversion			
Linear regression (sparse)	$(X^\dagger X) X^\dagger y$	HHL$^-$, HHL	[3]
Linear regression (low-rank)	$\sum_{r=1}^{R} \sigma_r^{-1} u_r v_r^T y$	HHL^{DME}	[4]
Kernel matrix inversion			
Support vector machine	$K^{-1} y$	HHL^{DME}	[5]
Gaussian process	$\kappa^T K^{-1} y$ and $\kappa^T K^{-1} \kappa$	HHL + SWAP	[6]
Adjacency matrix inversion			
Hopfield network		HHL + HHL^{DME}	[1]

where w is a N-dimensional vector containing the model parameters, y is a M-dimensional vector containing the target outputs $y^1, \ldots, y^M$ from the dataset, and the rows of the *data matrix* X are the N-dimensional training inputs $x^1, \ldots, x^M$. The most demanding computational problem behind linear regression is the inversion of the square of the data matrix, $(X^T X)^{-1}$.

As a rough reminder, the HHL routine solves a linear system of equations $Az = b$ by amplitude encoding b into a quantum state and applying the evolution $e^{iH_A t}|\psi_b\rangle$, where A is encoded into the Hamiltonian H_A (where we can use the trick of Sect. 3.4.4 for non-Hermitian A and therefore assume without loss of generality that A is Hermitian). Quantum phase estimation can extract the eigenvalues of H_A and store them in basis encoding, and some quantum post-processing writes them as amplitudes and inverts them, which effectively computes the quantum state $|\psi_{A^{-1}b}\rangle$. Omitting the inversion step multiplies the matrix A as it is, $|\psi_{Ab}\rangle$.

From Eq. (7.1) we can see that in order to compute w, one has to perform a matrix multiplication as well as a multiplication with an inverted matrix. This can be done by applying HHL twice. The first step, $X^T y$, is computed following the HHL routine with $A \to X^T$ and $b \to y$, but without inverting X. In a second step, the original HHL routine for matrix inversion is used with $A^{-1} \to (X^T X)^{-1}$ applied to the result of the first step, $b \to X^T y$. The outputs of the quantum algorithm are the normalised trained model parameters encoded in the amplitudes of the quantum state $|\psi_w\rangle$. This quantum state can be used with a quantum inference algorithm from the last chapter, or by reading out the parameters via quantum tomography.

7.1 Quantum Blas

The runtime of the routine grows with the condition number κ of the data matrix X with $\mathcal{O}(\kappa^6)$, where the Hamiltonian simulation as well as the conditional measurement in the branch selection are each responsible for a term $\mathcal{O}(\kappa^3)$. Hamiltonian simulation furthermore contributes a linear dependency on the sparsity s of the data matrix with the most recent methods [7]. The runtime grows with the inverse error as $\mathcal{O}(1/\log \epsilon)$ when modern Hamiltonian simulation methods are used. Reading out the parameters through measurements requires a number of repetitions that is of the order of N.

A slight variation on the linear regression algorithm allows us to replace the sparsity condition with the requirement that the design matrix is of low rank, or close to a low rank matrix, and requires only one step of the matrix inversion technique [4, 8]. A low rank means that the data is highly redundant and reducible to only a few vectors. The basic idea is to express the design matrix as a singular value decomposition, which leads to a slightly different expression for the solution w in terms of the singular values σ_r and singular vectors u_r, v_r of the data design matrix,

$$w = \sum_{r=1}^{R} \sigma_r^{-1} u_r v_r^T y.$$

The singular vectors of X are the eigenvectors of $X^T X$, while the singular values of X are the square roots of the eigenvalues of $X^T X$. We therefore only need to perform one eigendecomposition of $X^T X$, thereby "saving" the matrix multiplication with X^T above. But there is another advantage to taking the route of the singular value decomposition. Since $X^T X$ is always a positive definite matrix, we can encode it in a density matrix $\rho_{X^T X}$ (as we have shown in Sect. 6.2.5) and use the technique of density matrix exponentiation to apply $e^{i\rho_{X^T X} t}$ for a time t. Density matrix exponentiation takes the role of Hamiltonian simulation in the original HHL routine. As discussed before, density matrix exponentiation can yield exponential speedups for eigenvalue extraction if $X^T X$ can be approximated by a low-rank matrix.

7.1.2.2 Inverting Kernel Matrices

The square of the design matrix, $X^T X$, is an example of a larger class of positive definite matrices, namely kernel Gram matrices that we discussed before. The idea to use density matrix exponentiation for optimisation was in fact first explored in the context of kernel methods by Rebentrost, Mohseni and Lloyd's quantum machine learning algorithm for support vector machines [5]. It was also the first proposal that applied quantum blas to machine learning in the stricter sense.

While support vector machines as presented in Sect. 2.4.4.3 do not directly lead to a matrix inversion problem, a version called *least-squares support vector machines* [9] turns the convex quadratic optimisation into least squares optimisation. In short, by replacing the inequality constraints in Eq. (2.36) with equalities, the support vector machine (with a linear kernel) becomes equivalent to a regression problem of the form

$$\begin{pmatrix} 0 & 1 \ldots 1 \\ 1 & \\ \vdots & K \\ 1 & \end{pmatrix} \begin{pmatrix} w_0 \\ \gamma \end{pmatrix} = \begin{pmatrix} 0 \\ y \end{pmatrix},$$

with the kernel Gram matrix K of entries $(K)_{mm'} = (x^m)^T x^{m'}$, the Lagrangian parameters $\gamma = (\gamma_1, \ldots, \gamma_M)^T$, and a scalar bias $w_0 \in \mathbb{R}$. To obtain the γ_i one has to invert the kernel matrix. Note that we simplified the formalism by ignoring so called 'slack parameters' that cater for non-linearly separable datasets.

After preparing the kernel matrix as a density matrix, one can use the density matrix exponentiation technique to simulate $e^{i\rho_K t}$ and proceed with the HHL algorithm to apply K^{-1} to a quantum state $|\psi_y\rangle$ to obtain $|\psi_{K^{-1}y}\rangle$. We call this version of HHL with density matrix exponentiation (DME) for the Hamiltonian simulation in short the HHL^{DME} routine.

The outcome of the algorithm is a quantum state that encodes the bias w_0 as well as the Lagrangian parameters $\gamma_1, \ldots, \gamma_M$ as

$$|\psi_{w_0, \gamma}\rangle = \frac{1}{w_0^2 + \sum_m \gamma_m^2} \left(w_0 |0..0\rangle + \sum_{m=1}^{M} \gamma_m |m+1\rangle \right).$$

With the help of an interference circuit, this state can be used to classify new inputs [5].

Another classical machine learning algorithm whose optimisation method requires a kernel matrix inversion are Gaussian processes (see Sect. 2.4.4.4). In Gaussian processes, the model distribution (see Eq. (2.42)) is given by

$$p(y|x, \mathcal{D}) = \mathcal{N}[y| \underbrace{\kappa^T K^{-1} y}_{\text{mean}}, \underbrace{\hat{\kappa} - \kappa^T K^{-1} \kappa}_{\text{covariance}}],$$

where $\hat{\kappa} = \kappa(\tilde{x}, \tilde{x})$ is the kernel function with the new input $\tilde{x}$ 'in both slots', κ describes the vector $(\kappa(\tilde{x}, x^1), \ldots, \kappa(\tilde{x}, x^M))^T$ which takes the new input and the respective training inputs, and K is the kernel Gram matrix for the training inputs with entries $\kappa(x^m, x^{m'})$ for $m, m' = 1 \ldots M$. Again the main computational task is the inversion of a $M \times M$ dimensional kernel matrix to compute the mean and variance at a new input $\tilde{x}$. In order to compute these values, Zhao et al. [6] essentially apply the HHL routine to prepare a quantum state representing $K^{-1}y$. For the inner product with κ^T one can use an interference circuit as the one shown in the previous chapter.

7.1.2.3 Inverting Adjacency Matrices

Amongst the problems that can be mapped to matrix inversion is (maybe surprisingly) also the Hopfield model. As presented in Sect. 2.4.2.3, the Hopfield model is a recurrent neural network with certain restrictions on the connectivity, which can be used for associative memory. To "train" such a model, one can set the weighing or adjacency matrix of the graph to

$$W = \frac{1}{MN} \sum_{m=1}^{M} x^m (x^m)^T - \frac{1}{N} \mathbb{1},$$

a technique known as the *Hebbian learning rule* [10]. This rule foresees that every weight w_{ij} connecting two units i, j is chosen as the average of the product $x_i x_j$ of features over the entire dataset. The famous thumb rule is "neurons that fire together, wire together".

Given a new pattern $\tilde{x}$ of which only some values are known (and the others are set to zero), the usual operation mode of a Hopfield network is to update randomly selected units according to a step activation rule (i.e., if the weighted sum of the adjacent neural values is larger than a threshold value, a neuron gets set to 1, else to 0). One can show that this decreases or maintains the energy $E = -\frac{1}{2} x^T W x + w^T x$ until the 'closest' memory pattern is found. The vector w contains the "constant fields" that weigh each feature.

The matrix inversion formulation of a Hopfield neural network can be achieved by noting that we want the known units of the new input to coincide with the solution, which means that

$$P x = \tilde{x}, \tag{7.2}$$

if P is a projector onto the known features' subspace. We can then construct the Langrangian for this optimisation problem to minimise E under the side constraint of Eq. (7.2),

$$\mathcal{L} = -\frac{1}{2} x^T W x + w^T x - \gamma^T (P x - \tilde{x}) + \frac{\lambda}{2} x^T x.$$

In this equation, the vector γ and the scalar λ are Lagrangian parameters to learn. The optimisation problem can be written as a system of linear equations of the form $Az = b$ with

$$A = \begin{pmatrix} W - \lambda \mathbb{1} & P \\ P & 0 \end{pmatrix}, \quad z = \begin{pmatrix} x \\ \gamma \end{pmatrix}, \quad b = \begin{pmatrix} w \\ \tilde{x} \end{pmatrix}.$$

Pattern retrieval or associative memory recall can be done by applying the pseudoinverse A^+ to a quantum state $|\psi_b\rangle$ that encodes the vector b. However, A cannot be assumed to be sparse. One can therefore use a mixed approach: Decompose A into the three matrices B, C, and D,

$$A = \begin{pmatrix} 0 & P \\ P & 0 \end{pmatrix} + \begin{pmatrix} -(\lambda + \frac{1}{N})\mathbb{1} & 0 \\ 0 & 0 \end{pmatrix} + \begin{pmatrix} \frac{1}{MN} \sum_m x^m (x^m)^T & 0 \\ 0 & 0 \end{pmatrix} = B + C + D.$$

According to the so called Suzuki-Trotter formula that we introduced in Eq. (5.16), instead of exponentiating or simulating e^{iAt}, one can simulate the three steps $e^{iBt}e^{iCt}e^{iDt}$, if we accept an error of $\mathcal{O}(t^2)$. While B and C are sparse and can be treated with Hamiltonian simulation, D is not sparse. However, D contains the sum of outer products of the training vectors which we can associate with the quantum state

$$\rho_W = \frac{1}{M} \sum_{m=1}^{M} |x^m\rangle\langle x^m|.$$

If we can prepare copies of ρ_W, we can use density matrix exponentiation to simulate e^{iDt}. The quantum algorithm for associative memory recall in Hopfield networks shows how to combine the preceding approaches to solve more and more complex problems.

7.1.3 Speedups and Further Applications

The preceding examples demonstrated how quantum blas can be combined and applied to machine learning optimisation problems whose solutions are formulated as a linear algebra computation. The subroutines are rather involved, and require a full-blown fault tolerant quantum computer that can execute a large number of gates coherently. As mentioned before, the promise of this approach does not lie in near-term applications, but in the longer-term potential for exponential speedups.

Overall, one can roughly summarise under which conditions quantum blas-based training algorithms are qubit-efficient, thereby bearing a super-polynomial speedup for machine learning tasks relative to the input dimension N and the data set size M (see also [11]):

1. The input matrix has a constant sparsity (when Hamiltonian simulation is used) or can be approximated by a constant rank matrix (when density matrix exponentiation is used). This means that the sparsity or the rank of the matrix do not grow with the data size and are reasonably small.
2. The input state which the (inverted) matrix is applied to can be prepared in logarithmic time in N, M.
3. The condition number of the matrix that has to be inverted depends at most poly-logarithmically on the data size and is reasonably small. Note that this can sometimes be ensured by clever preprocessing of the data [12].
4. We do not request the final state as classical information, which means that we do not have to read out every amplitude via repeated measurements. Authors refer to three possible scenarios here. We are a) only interested in some properties of

the solution, or we can b) feed the solution into an efficient quantum inference routine, or it is enough to c) sample from the solution.

Note that a question that is often left open is what these requirements posed on the data sets mean for machine learning, for example whether they still define a useful problem. Also reformulating a classical problem as a least-squares or quadratic optimisation problem might change the statistical properties of the model significantly and influence the quality of solutions.

Matrix inversion was used here as an illustrative example of how to exploit quantum blas for supervised machine learning. There are many other machine learning algorithms based on linear algebra routines. For example, the density matrix exponentiation routine introduced in Sect. 5.4.3 in the context of simulating Hamiltonians has originally been proposed as a principal component analysis technique [13], where the task is to identify the dominant eigenvalues of a data matrix. This scheme has been used for a quantum algorithm for dimensionality reduction and classification [14]. An application beyond supervised learning are recommendation systems, where the main task is *matrix completion* [15]. Recommendation systems are based on a preference matrix R that stores a rating of M users for N different items (an illustrative example is the user rating of movies) and which is incomplete. The learning task is to predict the unknown rating of a user for a specific item, which is given by an entry of the preference matrix. The complete preference matrix R is assumed to be of low rank, which can be interpreted as there are being only a few 'prototypes of taste' that allow us to deduce user rankings from others. Beyond these examples, the literature on quantum machine learning with quantum blas is continuously growing.

7.2 Search and Amplitude Amplification

The second line of approaches in using quantum computing for optimisation is based on Grover search and amplitude amplification, which has been introduced in Sect. 3.5.1. These algorithms typically promise a quadratic speedup compared to classical techniques. We illustrate the idea with three examples. First, we present the Dürr-Høyer algorithm which extends Grover's routine to solve an optimisation (rather than a search) problem. This technique has been proposed in the context of nearest neighbour [16] and clustering [17] methods to find closest data points. Second, we look at a slight modification [18] of Grover search to handle data superpositions, in which we want to maintain zero amplitude for data points that are not present in a given dataset to speed up the search. This has been applied to associative memory but can potentially be useful in other contexts as well. Third, we will look at an example that uses amplitude amplification to search for the best model, in this case amongst decision boundaries of a perceptron.

7.2.1 Finding Closest Neighbours

Dürr and Høyer [19] developed a quantum subroutine using Grover's algorithm to find the minimum of a function $C(x)$ over binary strings $x \in \{0, 1\}^n$. Let $\frac{1}{\sqrt{N}} \sum_x |x\rangle |x_{\text{curr}}\rangle$ be a superposition where we want to search through all possible basis states $|x\rangle$. In each step an oracle marks all states $|x\rangle$ that encode an input x so that $C(x)$ is smaller than $C(x_{\text{curr}})$ (see Fig. 7.1). To perform the comparison, x_{curr} is saved in an extra register. The marked amplitudes get amplified, and the $|x\rangle$ register measured to draw a sample x'. If the result x' is indeed smaller than x_{curr}, one replaces the current minimum by the newly found one, $x_{\text{curr}} = x'$. The routine gets repeated until the oracle runs empty, at which point x_{curr} is the desired minimum.

In principle, this routine could be used to find the set of model parameters $|w\rangle$ that minimises a given cost function, and brute force search could be improved by a quadratic speedup. However, it is not difficult to see that this search over all possible sets of parameters is hard in the first place. If we have D parameters and each is discretised with precision τ, we have to search over $2^{D\tau}$ binary strings, and even an improvement to $\sqrt{2^{D\tau}}$ would be hopeless.

However, the Dürr and Høyer routine can be applied to the task of finding the closest neighbour in clustering [17] and nearest neighbour methods [16]. For example, given a data superposition

$$|[\tilde{x}]\rangle \sum_{m=0}^{M-1} |m\rangle |[x^m]\rangle |0\ldots 0\rangle,$$

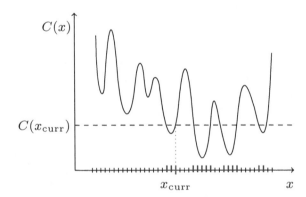

Fig. 7.1 Illustration of Dürr and Høyer's optimisation algorithm. A quantum register in uniform superposition represents the inputs x^m in basis encoding. In each iteration all states $|x^m\rangle$ with a lower cost than the current best solution, $C(x_{\text{curr}}) > C(x^m)$ (here shown with slightly longer blue ticks), are marked by an oracle and their amplitudes are amplified. The register is measured and if the result x' fulfills $C(x_{\text{curr}}) > C(x')$, the current candidate for the minimum (longest red tick) gets replaced by x'

7.2 Search and Amplitude Amplification

where $[\tilde{x}]$, $[x^m]$ indicates that the encoding of the new input and the training inputs is arbitrary. To find the nearest neighbour of the new input $\tilde{x}$ amongst the training inputs x^m, one has to compute the distance between the new input and all training points (see also Sect. 6.2.4) in superposition and write it into the last register,

$$|[\tilde{x}]\rangle \sum_{m=0}^{M-1} |m\rangle |[x^m]\rangle |[\mathrm{dist}(\tilde{x}, x^m)]\rangle.$$

This "distance register" stores the "cost" of a training point and can serve as a lookup table for the cost function. Iteratively reducing the subspace of possible solutions to the training points that have a lower cost than some current candidate x^l will eventually find the closest neighbour.

7.2.2 Adapting Grover's Search to Data Superpositions

In some contexts it can be useful to restrict Grover's search to a subspace of basis vectors. For example, the data superposition in basis encoding,

$$\frac{1}{\sqrt{M}} \sum_{m=1}^{M} |x^m\rangle,$$

corresponds to a sparse amplitude vector that has entries $\frac{1}{\sqrt{M}}$ for basis states that correspond to a training input, and zero else. Standard Grover search rotates this data superposition by the average, and thereby assigns a nonzero amplitude to training inputs that were not in the database. This can decrease the success probability of measuring the desired result significantly.

Example 7.1 (*Common Grover search.* [18])
We want to amplify the amplitude of search string 0110 in a sparse uniform superposition,

$$|\psi\rangle = \frac{1}{\sqrt{6}}(|0000\rangle + |0011\rangle + |0110\rangle + |1001\rangle + |1100\rangle + |1111\rangle),$$

which in vector notation corresponds to the amplitude vector

$$\frac{1}{\sqrt{6}}(1, 0, 0, 1, 0, 0, 1, 0, 0, 1, 0, 0, 1, 0, 0, 1)^T.$$

In conventional Grover search, the first step is to mark the target state by a negative phase,

$$\frac{1}{\sqrt{6}}(1,0,0,1,0,0,-1,0,0,1,0,0,1,0,0,1)^T,$$

after which every amplitude α_i is 'rotated' or transformed via $\alpha_i \to -\alpha_i + 2\bar{\alpha}$ (where $\bar{\alpha}$ is the average of all amplitudes) to get

$$\frac{1}{2\sqrt{6}}(-1,1,1,-1,1,1,3,1,1,-1,1,1,-1,1,1,-1)^T.$$

In the second iteration we mark again the target state

$$\frac{1}{2\sqrt{6}}(-1,1,1,-1,1,1,-3,1,1,-1,1,1,-1,1,1,-1)^T$$

and each amplitude gets updated as

$$\frac{1}{8\sqrt{6}}(5,-3,-3,5,-3,-3,13,-3,-3,5,-3,-3,5,-3,-3,5)^T.$$

As we see, the basis states that are not part of the data superposition end up having a non-negligible probability to be measured.

Ventura and Martinez [18] therefore introduce a simple adaptation to the amplitude amplification routine that maintains the zero amplitudes. After the desired state is marked and the Grover operator is applied for the first time rotating all amplitudes by the average, a new step is inserted which marks *all* states that were originally in the database superposition. The effect gives all states but the desired one the same phase and absolute value, so that the search can continue as if starting in a uniform superposition of all possible states.

Example 7.2 (*Ventura-Martinez version of Grover search* [18])
Getting back to the previous example and applying the Ventura-Martinez trick, starting with the same initial state, marking the target, and 'rotating' the amplitudes for the first time,

$$\frac{1}{2\sqrt{6}}(-1,1,1,-1,1,1,3,1,1,-1,1,1,-1,1,1,-1)^T,$$

the adapted routine 'marks' all amplitude that correspond to states in the data superposition,

$$\frac{1}{2\sqrt{6}}(1,1,1,1,1,1,-3,1,1,1,1,1,1,1,1,1)^T,$$

followed by another rotation,

$$\frac{1}{4\sqrt{6}}(1,1,1,1,1,1,9,1,1,1,1,1,1,1,1,1)^T.$$

7.2 Search and Amplitude Amplification

From now on we can proceed with Grover search as usual. The second iteration marks again the target,

$$\frac{1}{4\sqrt{6}}(1, 1, 1, 1, 1, 1, -9, 1, 1, 1, 1, 1, 1, 1, 1, 1)^T,$$

and the rotation leads to

$$\frac{1}{16\sqrt{6}}(-1, -1, -1, -1, -1, -1, 39, -1, -1, -1, -1, -1, -1, -1, -1, -1)^T.$$

It is obvious from this example that the amplitude amplification process is much faster, i.e. leads to a much larger probability of measuring the target state than without the adaptation.

7.2.3 Amplitude Amplification for Perceptron Training

Another straight forward application of amplitude amplification to machine learning is suggested in [20] and targets the training of perceptrons. Perceptron models have the advantage that rigorous bounds for training are known. Using the standard learning algorithm outlined in Sect. 2.4.2.1 and for training vectors of unit norm[1] assigned to two classes that are each separated by a positive margin γ, the training is guaranteed to converge to a zero classification error on the training set in $\mathcal{O}(\frac{1}{\gamma^2})$ iterations over the training inputs. A slightly different approach to training in combination with amplitude amplification allows us to improve this to $\mathcal{O}(\frac{1}{\sqrt{\gamma}})$.

The basic idea is to use a dual representation of the hyperplanes that separate the data correctly. In this representation, the hyperplanes or decision boundaries are depicted as points on a hypersphere (determined by the normalised weight vector) while training points define planes that cut through the hypersphere and define a "bad" subspace in which the desired solution is not allowed to lie, as well as a "good" subspace of allowed solutions (see Fig. 7.2). Assume we have K randomly sampled decision boundaries, represented by their weight vectors $w^1, \ldots, w^K$. One can show that a sample decision boundary perfectly separates the training set with probability $\mathcal{O}(\gamma)$. In turn, this means we have to start with $\mathcal{O}(\frac{1}{\gamma})$ samples to make sure the procedure does work on average. This is the first 'quadratic speedup' compared to $\mathcal{O}(\frac{1}{\gamma^2})$, and so far a purely classical one. The advantage of this step is that we only have K potential decision boundaries, with which we can now perform a Grover search of the best.

The second speedup comes from Grover search itself and is therefore a quantum speedup. We basis encode the weight vectors in a data superposition

---

[1] An assumption that allows us to omit that the bounds also depend on the norm.

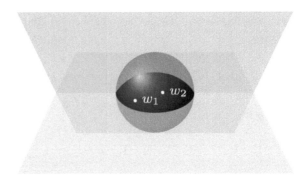

Fig. 7.2 In the dual representation, the normal vector of the separating hyperplanes w^1 and w^2 of a perceptron model are represented by points on a hypersphere, while the training set identifies a feasible region for the separating hyperplanes. Each training vector corresponds to a plane that 'cuts away' part of the hypersphere (illustrated by the grey planes)

$$\frac{1}{\sqrt{K}} \sum_{k=1}^{K} |w^k\rangle,$$

and gradually reinforce the amplitude of the weight vectors w^k that were found in the feasible subspace defined by the training data. For this purpose we need a quantum subroutine or 'oracle' that marks such 'desirable' weight vectors in the superposition. For example, considering only one training input, such an oracle would mark all weight vectors that classify the single input correctly. The number of iterations needed grows with $\mathcal{O}(\frac{1}{\sqrt{\gamma}})$. This convergence result can also be translated into an improvement of the mistake bound or the maximum number of misclassified data points in the test set after training [20].

7.3 Hybrid Training for Variational Algorithms

With full-blown fault-tolerant quantum computers still in the future, a class of hybrid classical-quantum algorithms has become popular to design near-term applications for quantum devices of the first generation. The idea of hybrid training of variational algorithms is to use a quantum device—possibly together with some classical processing—to compute the value of an objective function $C(\theta)$ for a given set of classical parameters θ. A classical algorithm is then used to optimise over the parameters by making queries to the quantum device [21].

As shown in Fig. 7.3, the quantum device implements a parametrised circuit $U(\theta)$ that prepares a state $U(\theta)|0\rangle = |\psi(\theta)\rangle$ which depends on a set of *circuit parameters* θ. A parametrised circuit can be thought of as a family of circuits where the parameters define one particular member of the family. For simplicity we will only

7.3 Hybrid Training for Variational Algorithms

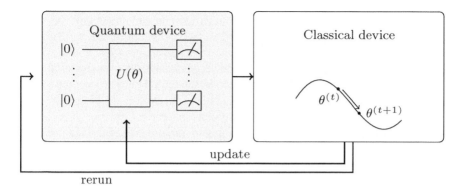

Fig. 7.3 The idea of a hybrid quantum-classical training algorithm for variational circuits is to use the quantum device to compute terms of an objective function or its derivatives, and subsequently use a classical device to compute better circuit parameters with respect to the objective. The entire routine gets iterated until the objective is minimised or maximised

consider unitary circuits here, but the concepts are easily extended to the evolution of subsystems and mixed states.

Measurements on the final state $|\psi(\theta)\rangle$ return estimates of expectation values, for example the energy expectation value, or the state of a certain qubit. These expectations depend on the circuit parameters θ. A cost function $C(\theta)$ uses the expectation values to define how good θ is in a given problem context. The goal of the algorithm is to find the circuit parameters θ of the variational circuit $U(\theta)$ that minimise $C(\theta)$.

To find the optimal circuit parameters, a classical algorithm iteratively queries the quantum device. These queries can either be the expectation values that define the cost, or—as we will see below—different expectation values that reveal the gradients of the variational circuit. Since training is a joint effort by a quantum and a classical algorithm, the training is called a "hybrid" scheme.

The variational circuit $U(\theta)$ is an ansatz that defines a set of all possible states $|\psi(\theta)\rangle$ it is able to prepare. It is typically much smaller than the space of all unitaries, since that would require a number of parameters that is quadratic in the Hilbert space dimension, which quickly becomes prohibitive with a growing number of qubits. Similar to the task of choosing a good model in machine learning, a fundamental challenge in variational algorithms lies in finding an ansatz that is rich enough to allow the parametrised state to approximate interesting solutions to the problem with as few parameters as possible. An interesting point is that for qubit-efficient circuits $U(\theta)$ that are not classically simulable, the overall training scheme exhibits an exponential quantum speedup, because the objective function could not be computed efficiently on a classical computer.

The great appeal of variational schemes is that they are suitable for near-term quantum technologies for the following reasons. Firstly, they do only require a fraction of the overall algorithm to run coherently (that is, as a quantum circuit), which leads to much smaller circuits. Second, there are many possible ansätze for the circuit, which means that it can be designed based on the strength of the device. Third, the fact that the circuit is learned can introduce robustness against systematic errors—if for example a certain quantum gate in the device systematically over-rotates the state, the parameters adjust themselves to correct for the error. Fourth, compared to other types of algorithms, optimisation as an iterative scheme can work with noise, for example in the estimations of the objective function terms.

In the context of this book, hybrid training of variational circuits is a promising candidate for quantum machine learning on near-term quantum devices. To show the working principle of variational circuits we will present two of the first variational algorithms from the quantum computing literature. We will then sketch two applications of these algorithms in quantum machine learning. The final section will have a closer look at how to perform the parameter updates on a classical computer, and explores different strategies of hybrid training that can be used in conjunction with these methods.

7.3.1 Variational Algorithms

The canon of quantum computing has been extended by a variety of variational algorithms. The first example, the variational eigensolver, stems from the physics-inspired problem of finding minimum energy eigenstates. The second example, the so called *quantum approximate optimisation algorithm* or *QAOA*, aims at solving combinatorial optimisation problems, and defines an interesting ansatz for the variational circuit itself.

7.3.1.1 Variational Eigensolvers

Variational algorithms were initially proposed as a prescription to find ground states—that is, lowest energy eigenstates—of quantum systems, where the cost is simply the energy expectation value. These schemes are also called *variational eigensolvers* [22]. The variational principle of quantum mechanics tells us that the ground state $|\psi\rangle$ minimises the expectation

$$\frac{\langle \psi | H | \psi \rangle}{\langle \psi | \psi \rangle}, \tag{7.3}$$

where H is the Hamiltonian of the system. The best approximation $|\psi(\theta^*)\rangle$ to the ground state given an ansatz $|\psi(\theta)\rangle$ minimises (7.3) over all sets of parameters θ. Assuming the kets are normalised, the cost function of the variational algorithm is therefore given by

$$C(\theta) = \langle \psi(\theta) | H | \psi(\theta) \rangle.$$

7.3 Hybrid Training for Variational Algorithms

The quantum device estimates $C(\theta)$ for an initial parameter set and hybrid training iteratively lowers the energy of the system by minimising the cost function.

In theory, we can perform measurements on the state $|\psi(\theta)\rangle$ to get an estimate of the expectation value of H. However, for general Hamiltonians this can involve a prohibitive number of measurements.

Example 7.3 (*(Estimation of energy expectation)*) Consider the Hamiltonian H of a $2^6 = 64$ dimensional Hilbert space, i.e., $|\psi(\theta)\rangle$ can be expressed by a 64-dimensional amplitude vector and describes a system of 6 qubits. Assume H has a -1 at the 13th diagonal entry and zeros else, in other words, the ground state of the Hamiltonian is the 13th basis state of the 6-qubit system. The expectation value $\langle \psi(\theta)|H|\psi(\theta)\rangle$ is then effectively the probability of measuring the 13th computational basis state, multiplied by (-1). Naively, to determine this probability we would have to measure the state $|\psi(\theta)\rangle$ repeatedly in the computational basis and divide the number of times we observe the 13th basis state by the total number of measurements. For a uniform superposition we need of the order of 2^n measurements to do this, which is infeasible for larger systems and defies the use of a quantum device altogether.

Luckily, in many practical cases H can be written as a weighted sum of local (i.e. 1- or 2-qubit) operators, $H = \sum_k h_k H_k$ with $h_k \in \mathbb{R}\ \forall k$. We have already discussed such a case for qubit-efficient Hamiltonian simulation in Sect. 5.4.1. The overall expectation is then given by a sum

$$C(\theta) = \sum_k h_k \langle \psi(\theta)|H_k|\psi(\theta)\rangle$$

of the estimates of 'local' expectation values $\langle \psi(\theta)|H_k|\psi(\theta)\rangle$. The local estimates are multiplied by the coefficients h_k and summed up on the classical device. These 'local' energy expectations are much easier to estimate, which reduces the number of required measurements dramatically. If the number of local terms in the objective function is small enough, i.e. it only grows polynomially with the number of qubits, estimating the energy expectation through measurements from the quantum device is qubit-efficient.

To give one example, remember that the Hamiltonian of a qubit system can always be written as a sum over Pauli operators,

$$H = \sum_{i,\alpha} h_\alpha^i \sigma_\alpha^i + \sum_{\substack{i,j \\ \alpha,\beta}} h_{\alpha,\beta}^{ij} \sigma_\alpha^i \sigma_\beta^j + \cdots, \tag{7.4}$$

and the expectation value becomes sum of expectations,

$$\langle H \rangle = \sum_{i,\alpha} h_\alpha^i \langle \sigma_\alpha^i \rangle + \sum_{\substack{i,j \\ \alpha,\beta}} h_{\alpha,\beta}^{ij} \langle \sigma_\alpha^i \sigma_\beta^j \rangle + \cdots,$$

where the superscripts i, j denote the qubit that the Pauli-operator acts on, while the subscripts define the Pauli operator $\alpha, \beta = 1, x, y, z$. From this representation we see that the energy expectation becomes qubit-efficient if the Hamiltonian can be written as a sum of only a few (i.e. tractably many) terms, each involving only a few Pauli operators. This is common in quantum chemistry, where Hamiltonians describe electronic systems under Born-Oppenheimer approximation, as well as in many-body physics and the famous Ising/Heisenberg model [22].

Generally speaking, variational quantum eigensolvers minimise the expectation value of an operator, here the Hamiltonian, using a quantum device to estimate the expectation. Eigensolvers are particularly interesting in cases where the estimation is qubit-efficient on a quantum device, while simulations on a classical computer are intractable.

7.3.1.2 Quantum Approximate Optimisation Algorithm

Another popular variational algorithm has been presented by Farhi and Goldstone, and in its original version solves a combinatorial optimisation problem [23]. For us, the most important aspect is the ansatz for $|\psi(\theta)\rangle$ for a given problem that will be adapted to prepare Gibbs states in the quantum machine learning algorithm below.

Consider an objective function that counts the number of statements from a predefined set of statements $\{C_k\}$ which are satisfied by a given bit string $z = \{0, 1\}^{\otimes n}$,

$$C(z) = \sum_k C_k(z).$$

A statement can for instance be "$(z_1 \wedge z_2) \vee z_3$", which is fulfilled for $z = z_1 z_2 z_3 = 000$ but violated by $z = 010$. If statement k is fulfilled by z, $C_k(z)$ is 1, and else it is 0. We can represent this objective function by the expectation values $\langle z|C|z\rangle$ of a quantum operator that we also call C. In matrix representation, the operator contains on its diagonal the number of statements satisfied by a bit string z, whereby the integer representation i of z defines the element $H_{i+1,i+1} = C(z)$ of the Hamiltonian (the '+1'is necessary to start the index count from 1). The matrix has only zero off-diagonal elements.

Example 7.4 Consider an objective function defined on bit strings of length $n = 2$ with statements $C_1(z_1, z_2) = (z_1 \wedge z_2)$ and $C_1(z_1, z_2) = (z_1 \vee z_2)$, the matrix expression of operator $C = C_1 + C_2$ would be given by

$$\begin{pmatrix} \langle 00|C|00\rangle & \langle 00|C|01\rangle & \langle 00|C|10\rangle & \langle 00|C|11\rangle \\ \langle 01|C|00\rangle & \langle 01|C|01\rangle & \langle 01|C|10\rangle & \langle 01|C|11\rangle \\ \langle 10|C|00\rangle & \langle 10|C|01\rangle & \langle 10|C|10\rangle & \langle 10|C|11\rangle \\ \langle 11|C|00\rangle & \langle 11|C|01\rangle & \langle 11|C|10\rangle & \langle 11|C|11\rangle \end{pmatrix} = \begin{pmatrix} 0 & 0 & 0 & 0 \\ 0 & 1 & 0 & 0 \\ 0 & 0 & 1 & 0 \\ 0 & 0 & 0 & 2 \end{pmatrix},$$

and contains the number of fulfilled statements on the diagonal.

7.3 Hybrid Training for Variational Algorithms

If we interpret a quantum state $|\psi\rangle = \sum_z \alpha_z |z\rangle$ as a qsample that defines a probability distribution over z-strings, then $\langle \psi | C | \psi \rangle$ is the expectation value of operator C under this distribution. Measuring $|\psi\rangle$ in the computational basis means to draw samples from the space of all bit strings.

We now define the ansatz $|\psi(\theta)\rangle$ of the variational algorithm. For this we need a second operator, a sum of Pauli σ_x or 'flip operators' on all bits,

$$B = \sum_{i=1}^{n} \sigma_x^i.$$

Together, B and C define the ansatz for the parametrised state preparation scheme. Consider the two parametrised unitaries

$$U(B, \beta) = \exp^{-i\beta B}, \quad U(C, \gamma) = \exp^{-i\gamma C}.$$

Starting with a uniform superposition $|s\rangle = \frac{1}{\sqrt{2^n}} \sum_z |z\rangle$ and alternately applying $U(B, \beta)\, U(C, \gamma)$ for short times t prepares the parametrised state

$$|\psi(\theta)\rangle = U(B, \beta_K) U(C, \gamma_K) \ldots U(B, \beta_1) U(C, \gamma_1) |s\rangle.$$

The set of parameters θ consists of the $2K$ parameters $\beta_1, \ldots, \beta_K, \gamma_1, \ldots, \gamma_K$.

Farhi and Goldstein have shown that for $K \to \infty$, the maximum expectation value over all θ is equal to the maximum of the objective function,

$$\lim_{K \to \infty} \max_{\theta} \langle \psi(\theta) | C | \psi(\theta) \rangle = \max_z C(z).$$

Let θ^* be the parameter set that maximises $\langle \psi(\theta) | C | \psi(\theta) \rangle$. The above suggests that sampling computational basis states from $|\psi(\theta^*)\rangle$ will reveal good candidates for z, since the state assigns more probability to basis states that correspond to z values which fulfil a lot of statements. For example, if in the extreme case there is only one z' that fulfills all statements and all other z fulfil none of them, we would get $|\psi(\theta^*)\rangle = |z'\rangle$ as the optimal solution, and sampling would always retrieve z'. Note that a "good candidate" in this context does not refer to a high probability to sample the optimal z, but to draw a reasonably good solution for which $C(z)$ is close to the global optimum. Farhi and Goldstone's investigations suggest that also for low K, even for $K = 1$, this algorithm can have useful solutions.

As a final side-note, the quantum approximate optimisation algorithm has its origins in the quantum adiabatic algorithm, in which a simple starting Hamiltonian (corresponding to B) gets slowly turned into the target Hamiltonian (here C) without leaving the ground state. Instead of a smooth continuous transition one uses two rapidly alternating evolutions defined by $U(B, \beta)$ and $U(C, \gamma)$. A subsequent paper by Farhi and Harrow [24] claims that sampling from $|\psi(\theta)\rangle$ even for small circuit depths K is classically intractable, giving this model a potential exponential quantum advantage.

7.3.2 Variational Quantum Machine Learning Algorithms

Based on these two variational algorithms, we present two suggestions of how to use variational algorithms in the context of quantum machine learning. First, we discuss how the variational eigensolvers can be extended to implement a binary classifier in which the expectation value of a quantum observable is interpreted as the output of a machine learning model. Second, we show how to use quantum approximate optimisation to prepare approximations of Gibbs states. These Gibbs states can be used as qsamples to train Boltzmann machines.

7.3.2.1 Variational Classifiers

The idea of a variational eigensolver can be extended for supervised learning tasks by interpreting the expectation value of an observable O as the output of a classifier,

$$f(x; \theta) = \langle \psi(x; \theta) | O | \psi(x; \theta) \rangle. \tag{7.5}$$

The resulting model is what we call a *variational classifier*. Note that the variational circuit also has to depend on the model inputs x.

One way to associate an expectation value with a binary model output $f(x; \theta)$ is to define

$$f(x; \theta) = \langle \psi(x; \theta) | \sigma_z^j | \psi(x; \theta) \rangle,$$

where the right side is the expectation value of the σ_z operator applied to the jth qubit. The σ_z operator measures whether a qubit is in state 0 or 1, and the expectation value is equivalent (up to a simple transformation) to measuring the probability of this qubit being in state 1, as we have discussed several times before.

Example 7.5 The expectation value of σ_z with respect to the first qubit of the quantum state $|\psi\rangle = a_0|00\rangle + a_1|01\rangle + a_2|10\rangle + a_3|11\rangle$ is given by

$$\langle \psi | \sigma_z^1 | \psi \rangle = |a_0|^2 + |a_1|^2 - |a_2|^2 - |a_3|^2,$$

while with respect to the second qubit the expectation value is

$$\langle \psi | \sigma_z^2 | \psi \rangle = |a_0|^2 - |a_1|^2 + |a_2|^2 - |a_3|^2.$$

The expectation value hence sums up the probabilities corresponding to basis states where the qubit is in state 1, and subtracts those where the qubit is in state 0. We can translate the expectation value into a probability of measuring the qubit in state 1 by performing the scaling-shift

$$p(q_j = 1) = 0.5 \langle \langle \psi | \sigma_z^j | \psi \rangle \rangle + 1.$$

7.3 Hybrid Training for Variational Algorithms

When defining the model output as an expectation value, the standard squared loss cost function, which measures the difference between the model outputs and the targets y^m for each training sample $m = 1, \ldots, M$, becomes

$$C(\theta) = \sum_{m=1}^{M} (\langle \psi(x^m; \theta) | \sigma_z^j | \psi(x^m; \theta) \rangle - y^m)^2.$$

Compared to variational eigensolvers, the cost function is no longer the expectation value itself, but a function of the expectation value. The model outputs $\langle \psi(x^m; \theta) | \sigma_z^j | \psi(x^m; \theta) \rangle$ are evaluated by the quantum device, while the classical device helps to update the parameters θ with respect the cost $C(\theta)$.

For models in which the outputs are not classical but quantum states of a single qubit, we can extend the above cost function idea to consider the expectation value of all three single-qubit Pauli operators and their target values [25],

$$C(\theta) = \sum_{m=1}^{M} \sum_{k=x,y,z} (\langle \psi(x^m, \theta) | \sigma_k | \psi(x^m, \theta) \rangle - \langle \sigma_k \rangle_{\text{target}}^m)^2.$$

Here $\langle \sigma_\alpha \rangle_{\text{target}}^m$ is the target value of the mth training input. Note that measuring a single-qubit σ_x [σ_y] observable can be realised by applying a σ_x [σ_y] gate to the qubit, thereby rotating its basis, and subsequently measuring the σ_z observable. Hence, measuring any Pauli operator is comparably simple.

7.3.2.2 Variational Preparation of Boltzmann Qsamples

The quantum approximate optimisation algorithm (QAOA) suggests an interesting ansatz for the quantum state prepared by a variational circuit. Here we will review [26] how to use the ansatz to prepare approximate qsamples of Boltzmann distributions

$$\rho_{\text{BM}} = \frac{e^{-\delta H}}{\text{tr}\{e^{-\delta H}\}}, \tag{7.6}$$

where

$$H = \sum_{i,j \in s} w_{ij} \sigma_i^z \sigma_j^z + \sum_{i \in s} b_i \sigma_i^z$$

is the Boltzmann Hamiltonian which corresponds to the energy function of a Boltzmann machine. The expression $i, j \in s$ means that the index runs over all qubits $q_1, \ldots, q_G$ that represent the units of the model, i.e. the visible *and* the hidden units. The density matrix ρ_{BM} corresponds to the probability distribution of Eq. (2.26) written in the language of quantum mechanics. In quantum theory it is also called a *Gibbs state* or *thermal state* with respect to the Hamiltonian H, since the evolution of the system does not change the state. Samples from the Gibbs state can help to compute the weight update in a gradient descent algorithm for the training of (quantum) Boltzmann machines [26]. The technique to prepare ρ_{BM} with the QAOA has been called *quantum approximate thermalisation* [26].

In quantum approximate thermalisation, the operator B of the QAOA is replaced by the *mixer Hamiltonian*

$$H_M = \sum_{i \in s} \sigma_x^i,$$

while the operator C is replaced by H from above. As in the QAOA, the Hamiltonians are applied for K iterations in an alternating fashion,

$$\prod_{k=1}^{K} e^{-i\beta_k H_M} e^{-i\gamma_k H},$$

with the real, classical parameters $\{\beta_1, \ldots, \beta_K, \gamma_1, \ldots, \gamma_K\}$. The goal is to find the parameters for which this sequence of operators approximately transforms the initial state[2]

$$\rho_0 = \frac{e^{-\delta H_M}}{\text{tr}\{e^{-\delta H_M}\}},$$

to the desired Boltzmann qsample from Eq. (7.6).

The iterative parameter optimization minimises the cost

$$C(\beta_1, \ldots, \gamma_K) = \text{tr}\{\rho_{\text{BM}}(\beta_1, \ldots, \gamma_K) H\}$$

by a numerical method such as Nelder-Mead. For this one has to estimate C with a quantum device or a simulation thereof. The cost is the expectation of H with respect to its approximate thermal state ρ_{BM}.

Recall from Sect. 2.4.2.4 that for maximum likelihood optimisation in the training of Boltzmann machines one has to not only sample the values of 'neurons' from the model distribution (which is the Boltzmann distribution we just looked at), but also from a "data" distribution where the visible units are 'clamped' to a training input. We therefore also need to prepare 'clamped' Boltzmann qsamples. For this one can use the same procedure as before, but has to use slightly different operators,

$$\tilde{H}_M = \sum_{i \in h} \sigma_x^i, \quad \tilde{H} = \sum_{i,j \in s \mid (i \in h) \vee (j \in h))} w_{ij} \sigma_i^z \sigma_j^z + \sum_{i \in h} b_i \sigma_i^z,$$

---

[2] The initial state plays the role of the uniform superposition that we started with in the original quantum approximate optimisation algorithm. It can be prepared by starting with the state $\bigotimes_j \sqrt{2\cosh(\delta)} \sum_{\pm} e^{\mp\frac{\beta}{2}|\pm\rangle_j |\pm\rangle_{E_j}}$ and tracing out the E_j ('environment') register (which has as many qubits as the visible register) [26].

where h is the set of qubits associated with hidden units. These clamped or 'partial' Hamiltonians exclude terms from the full Hamiltonians which are acting on visible units only. The initial state is now the thermal state of $\tilde{H}_M$ with the visible units prepared in a randomly sampled training input $|x^m\rangle$, in other words, they are 'clamped' to a specific value. An interesting variation uses either a superposition or a statistical mixture of training inputs for the visible qubits, called *quantum randomised clamping*, which was reported to improve the results of numerical simulations significantly [26].

7.3.3 Numerical Optimisation Methods

We have so far ignored the issue of how to optimise the classical parameters $\theta = \theta_1, \ldots, \theta_D$ in the hybrid training scheme. Since the quantum algorithm provides measurement samples for the expectation of an operator, we only have an estimation of the output, whose precision can be increased by repeating the algorithm and measurement. We distinguish three classes of methods to use these measurement estimates for optimisation, *derivative-free*, *numerical gradient-based* and *analytical gradient-based* methods.

7.3.3.1 Derivative-Free Methods

As the name suggests, derivative-free methods do not use gradients for optimisation. Amongst derivative-free methods one can again distinguish between *one-shot* and *iterative* methods. One-shot methods compute the optimal parameters on a classical computer and use the quantum device solely for state preparation. An example has been suggested in the context of the quantum approximate optimisation algorithm and the so-called *MaxCut problem for graphs with bounded degree* [23]. However, such classical solutions are very problem-specific, can be computationally expensive and may be difficult to construct.

Iterative derivative-free methods use successive evaluations of the objective function (and thereby of the expectation values estimated from the quantum computer)

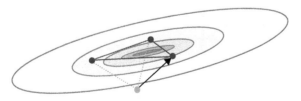

Fig. 7.4 The Nelder-Mead or simplex algorithm finds the minimum by successively updating the node of a simplex that has the highest cost value. It is gradient-free and simple to implement, but not very robust in high-dimensional landscapes

for optimisation. A prominent example is the Nelder-Mead method [27], in which the vertices of a $D+1$-simplex are iteratively updated, incrementally shrinking the area of the simplex around a minimum (Fig. 7.4). The algorithm terminates when either the distance between the points or the differences of the cost between two updates fall under a certain threshold. Other well-known candidates are genetic and particle swarm optimisation algorithms, but little is known about their performance in the context of variational quantum computing (for an exception see [28]).

7.3.3.2 Numerical Gradient-Based Methods

Gradient-based methods also have two sub-categories, *numerical* and *analytical methods*. If only black-box access to the cost function is provided, one can use finite-differences to compute the gradient numerically,

$$\frac{\partial C(\theta)}{\partial \theta_l} = \frac{C(\theta_1,..,\theta_l,..,\theta_D) - C(\theta_1,..,\theta_l + \Delta\theta_l,..,\theta_D)}{\Delta\theta_l} + \mathcal{O}(\Delta\theta_l^2) + \mathcal{O}(\frac{\epsilon}{\Delta\theta_l}). \tag{7.7}$$

The last term stems from the error of the estimation of the cost function $C(\theta)$, and ϵ is the error of the estimation.

Guerreschi et al. [29] provide some useful rules when considering the finite-differences method to compute gradients. Most importantly, we want the error intervals of the two cost function evaluations not to overlap. In other words, the difference between $C(\theta_1,\ldots,\theta_l,\ldots,\theta_D)$ and $C(\theta_1,\ldots,\theta_l+\Delta\theta_l,\ldots,\theta_D)$ has to be larger than twice the error (see Fig. 7.5), which means by Eq. (7.7) that

$$2\epsilon \leq \Delta\theta_l \frac{\partial C(\theta)}{\partial \theta_l}.$$

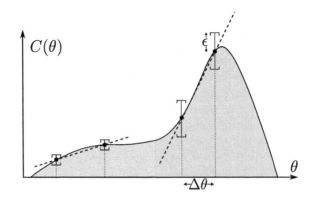

Fig. 7.5 The finite-difference method approximates a gradient with the linear function determined by function evaluations at two points with distance $\Delta\theta$ (here for a one-dimensional parameter space). The precision ϵ for each function evaluation $C(\theta)$ has to be smaller for small gradients (left) than for large ones (right)

7.3 Hybrid Training for Variational Algorithms

As a consequence, the smaller the gradient, the more precision we need in estimating the cost function, and the more repetitions of the algorithm are required. Numerical finite-differences methods are therefore particularly difficult in situations where the minimum has to be approximated closely, or when the optimisation landscape has many saddle points, and where the algorithm produces measurements with a high variance.

7.3.3.3 Analytical Gradient-Based Methods

The second type of gradient-based methods uses analytical gradients [29–31]. Consider a quantum circuit

$$U(\theta) = G(\theta_L) \ldots G(\theta_l) \ldots G(\theta_1),$$

that consists of L parametrised "elementary unitary blocks" G. These blocks are each defined by a set of parameters $\theta_l, l = 1, \ldots, L$. Let μ be an arbitrary parameter in θ. The analytical gradient $\nabla_\mu C(U(\theta))$ of a cost function $C(U(\theta))$ that depends on the variational circuit usually consists of the 'derivative of the circuit', $\partial_\mu U(\theta)$. We define the partial derivative of a matrix here as the matrix that results from taking the partial derivative of each element.

The problem with the expression $\partial_\mu U(\theta)$ is that these derivatives of circuits are themselves not necessarily unitaries, and therefore no quantum circuits. That means we cannot estimate any expectations from these circuits with the quantum device. However, one can often use tricks to estimate $\partial_\mu U(\theta)$ with the device, and thereby use the analytical gradients in a hybrid training scheme. If possible, this is much preferred over numerical gradients that are less robust.

Assume that μ only appears in the set of parameters θ_l (although the calculation becomes only slightly more complex when several parameters are tied to each other). Then

$$\partial_\mu U(\theta) = G(\theta_L) \ldots \partial_\mu G(\theta_l) \ldots G(\theta_1). \tag{7.8}$$

This is significant: Due to its linearity, the 'derivative of the circuit' is exactly the same as the original circuit except from one block.

A useful observation arises for cases in which

$$\partial_\mu G(\theta_l) = \sum_i a_i G_i(\theta_l) \tag{7.9}$$

holds. This means that the derivative of the unitary block can be computed as the weighed sum of other blocks. Below we will discuss two cases, one where $G_i(\theta_l) = G(r_i(\theta_l))$ (which means that only the parameters are transformed by a set of known and fixed functions r_i), and the other where the G_i form a decomposition of the generator O of the gate $G = \exp^{i\theta_l O}$ for a single parameter θ_l.

Inserted back into expression (7.8), one finds

$$\partial_\mu U(\theta) = \sum_i a_i G(\theta_L) \ldots G_i(\theta_l) \ldots G(\theta_1) = \sum_i a_i U_i(\theta). \tag{7.10}$$

The 'derivative of a circuit' is a linear combination of circuits. The gradient can therefore often be estimated by running the original variational circuit but with slightly different parameters or gates, and by combining the different outputs on the classical device. This has been termed a *classical linear combination of unitaries* [31]. The classical linear combination of unitaries trick allows us to use exactly the same quantum hardware to estimate gradients. Of course, the particulars of the computation depend on the parametrisation of the blocks G, the cost function and the expectation value that is evaluated from the circuit.

7.3.4 Analytical Gradients of a Variational Classifier

7.3.4.1 Derivatives of Gates

To give a concrete example of this still rather abstract notion of estimating gradients of circuits, consider a circuit constructed from elementary blocks that are general single qubit gates (see also [31] and Sect. 8.2.1),

$$G(\alpha, \beta, \gamma) = \begin{pmatrix} e^{i\beta} \cos\alpha & e^{i\gamma} \sin\alpha \\ -e^{-i\gamma} \sin\alpha & e^{-i\beta} \cos\alpha \end{pmatrix}. \tag{7.11}$$

"General" in this context means that any single qubit gate (up to an unmeasurable global phase factor) can be represented in this form [32]. The derivative of such a gate with respect to the parameters $\mu = \alpha, \beta, \gamma$ are as follows:

$$\partial_\alpha G(\alpha, \beta, \gamma) = G(\alpha + \frac{\pi}{2}, \beta, \gamma), \tag{7.12}$$

$$\partial_\beta G(\alpha, \beta, \gamma) = \frac{1}{2}G(\alpha, \beta + \frac{\pi}{2}, 0) + \frac{1}{2}G(\alpha, \beta + \frac{\pi}{2}, \pi), \tag{7.13}$$

$$\partial_\gamma G(\alpha, \beta, \gamma) = \frac{1}{2}G(\alpha, 0, \gamma + \frac{\pi}{2}) + \frac{1}{2}G(\alpha, \pi, \gamma + \frac{\pi}{2}). \tag{7.14}$$

Comparing this to Eq. (7.9), one sees that the transformations $r_i(\theta)$ of the parameters consists of shifting some parameters by $\frac{\pi}{2}$ and setting others to a constant. This is hence an illustration of the case where the linear combination of gates uses the same gate, but with transformed parameters, $G_i(\theta_l) = G(r_i(\theta_l))$. Controlled single qubit gates can be decomposed into linear combinations of unitaries in a similar fashion.

As a second example, consider a variational circuit that is parametrised as Pauli matrices, so that the building blocks read

$$G(\mu) = e^{i\mu O}$$

with O being a tensor product of Pauli operators σ_x, σ_y and σ_z including the identity $\mathbb{1}$, each acting on one of the n qubits. The derivative of a gate $e^{i\mu O}$ is formally given by

$$\partial_\mu e^{i\mu O} = iOe^{i\mu O},$$

and since we can apply the Pauli gates as unitaries, we can simply add O as a further gate to construct the derivative circuit [30]. If O was a non-unitary Hermitian operator, we could try to decompose it into a linear combination of unitaries.

7.3.4.2 Computing the Gradient

Now assume a variational classifier as introduced in Sect. 7.3.2.1, where the model is represented by a quantum expectation value $\langle \psi(x;\theta)|\sigma_z|\psi(x;\theta)\rangle$, with σ_z being applied to a single specified qubit. Formally, derivatives of this function are computed as

$$\partial_\mu \langle \psi(x,\theta)|\sigma_z|\psi(x,\theta)\rangle = \langle \partial_\mu \psi(x,\theta)|\sigma_z|\psi(x,\theta)\rangle + \langle \psi(x,\theta)|\sigma_z|\partial_\mu \psi(x,\theta)\rangle$$
$$= \text{Re}\{\langle \partial_\mu \psi(x,\theta)|\sigma_z|\psi(x,\theta)\rangle\}, \quad (7.15)$$

with $|\psi(x,\theta)\rangle = U(x,\theta)|0\rangle$. The derivative of the (quantum) model is nothing but the real part of the inner product between two states: The original variational circuit together with a final σ_z gate on a predefined output qubit, $\sigma_z|\psi(x,\theta)\rangle$, as well as the state produced by applying the derivative of the circuit, $|\partial_\mu \psi(x,\theta)\rangle = \partial_\mu U(x,\theta)|0\rangle$.

Using the classical linear combination of unitaries in Eq. (7.10) to compute the derivative $\partial_\mu U(x,\theta)$, we can express the right side of Eq. (7.15) by the real part of a linear combination of inner products, $\text{Re}\{\sum_i a_i \langle A_i|B\rangle\}$, between

$$|A_i\rangle = U_i(x,\theta)|0\rangle$$

and

$$|B\rangle = \sigma_z U(x,\theta)|0\rangle.$$

We showed in Sect. 6.1 how to compute the real part of the inner product of two quantum states. As a reminder, one has to prepare the two states $|A_i\rangle$ and $|B\rangle$ conditioned on the state of an ancilla qubit,

$$|0\rangle|A_i\rangle + |1\rangle|B\rangle,$$

and apply a Hadamard to the ancilla after which it gets measured. The advantage we have in this example is that $|A_i\rangle$ and $|B\rangle$ only differ in one elementary block, or in one qubit gate, as well as one σ_z operator. In other words, only two gates have to be applied conditioned to the ancilla qubit, and the rest of the circuit can be executed in its original version.

When we are dealing with the Pauli-matrix-generator-form of the second example from above, the evaluation of gradients becomes surprisingly simple [33]. Let us go back to the derivative of the expectation value in Eq. (7.15). If the gate that we want to derive is of the form $e^{-i\mu\sigma}$, where σ is an arbitrary Pauli matrix, we get

$$\begin{aligned}\partial_\mu \langle \psi(x,\theta)|\sigma_z|\psi(x,\theta)\rangle &= \langle 0|\ldots \partial_\mu e^{-i\mu\sigma}\ldots \sigma_z \ldots e^{i\mu\sigma}\ldots |0\rangle \\ &+ \langle 0|\ldots e^{-i\mu\sigma}\ldots \sigma_z \ldots \partial_\mu e^{i\mu\sigma}\ldots |0\rangle \\ &= \langle 0|\ldots (-i\sigma)e^{-i\mu\sigma}\ldots \sigma_z \ldots e^{i\mu\sigma}\ldots |0\rangle \\ &+ \langle 0|\ldots e^{-i\mu\sigma}\ldots \sigma_z \ldots (i\sigma)e^{i\mu\sigma}\ldots |0\rangle \\ &= \langle 0|\ldots (1-i\sigma)e^{-i\mu\sigma}\ldots \sigma_z \ldots (1+i\sigma)e^{i\mu\sigma}\ldots |0\rangle \\ &+ \langle 0|\ldots (1+i\sigma)e^{-i\mu\sigma}\ldots \sigma_z \ldots (1-i\sigma)e^{i\mu\sigma}\ldots |0\rangle\end{aligned}$$

In the last step we used the well-known relation

$$\exp^{i\mu\sigma} = \cos(\mu)\mathbb{1} + i\sin(\mu)\sigma.$$

Noting that $(1 \pm i\sigma) = e^{\pm i\frac{\pi}{2}\sigma}$, we find that we can estimate the analytical gradient of the expectation value by evaluating the original expectation once with an additional $e^{i\frac{\pi}{2}\sigma}$ gate, and once with an additional gate $e^{-i\frac{\pi}{2}\sigma}$ gate. The additional gate is inserted before (or after) the gate that we derive for. (Of course, this is equivalent to shifting the μ parameter by $\pm \frac{\pi}{2}$ for the two evaluations.) One can show that we can write an expression like $\mathrm{Re}\{\sum_i a_i \langle A_i|B\rangle\}$ as the difference of two expectation values for any generator that is unitary and Hermitian.

In summary, for certain cases of variational quantum machine learning protocols—namely those including derivatives of circuits that can be estimated from a classical linear combination of unitaries—we can use slightly different circuits to estimate analytical gradients. This is a very useful alternative to numerical methods.

7.4 Quantum Adiabatic Machine Learning

The last type of quantum algorithms for training we discuss in this chapter is based on the ideas of adiabatic quantum computing and annealing (see Sect. 3.3.3). The goal is to prepare the ground state of a quantum system, the quantum state of n qubits with the lowest energy. The ground state typically encodes the solution to a binary optimisation problem, or represents a qsample from which we want to draw samples. The dataset typically defines the Hamiltonian (and thereby the dynamics) of the system, which can be understood as a type of *Hamiltonian* or *dynamic encoding*.

Quantum annealing appears prominently in the quantum machine learning literature, which is partly due to the fact that it was one of the first technologies for which experiments beyond proof-of-principle setups were possible. An early generation of studies looked at how to formulate machine learning tasks as an optimisation problem that could be solved by quantum annealing (see following Sect. 7.4.1) and tested the ideas with the D-Wave device. However, experiments confirmed the ambiguous results of other quantum annealing studies, where there is still no conclusive answer as to which speedups are possible with this technique [34, 35]. It is therefore not clear where the advantage compared to classical annealing and sampling techniques lies. Noise and connectivity issues in this early-stage technology further complicated the picture.

Quantum machine learning research therefore focused on another twist of the story. Instead of employing annealing devices as an analogue solver of optimisation problems, they are used as an analogue sampler (see Sect. 7.4.2). The annealing procedure prepares a quantum state that can be interpreted as a qsample from which measurements can draw samples. One important observation was for example that the qsample distribution prepared by D-Wave may be sufficiently close to a classical Gibbs distribution to use the samples for the training of a Boltzmann machine.

Beyond annealing, other proposals, most of them experimental in nature, have sporadically been brought forward and some of them are summarised in Sect. 7.4.3.

7.4.1 Quadratic Unconstrained Optimisation

The first generations of D-Wave's commercial quantum annealing devices consisted of up to $n = 1024$ physical qubits with sparse programmable interaction strengths of an Ising-type model, and annealed objective functions in the format of *quadratic unconstrained binary optimisation*. The ground state or solution is the binary sequence $x_1 \ldots x_n$ that minimises the 'energy function'

$$\sum_{i \leq j=1}^{n} w_{ij} x_i x_j, \qquad (7.16)$$

with the coefficients or weights w_{ij}. A central task for this approach of quantum machine learning is therefore to cast a learning task into such an optimisation problem.

There have been a number of proposals of how to translate machine learning problems into an unconstrained binary optimisation problem (see also Sect. 8.3.3). Machine learning problems that lend themselves especially well for combinatorial optimisation are structure learning problems for graphs, for example of Bayesian nets [36]. Each edge of the graph is associated to a qubit in state 1, while a missing edge corresponds to a state 0 qubit, so that the graph connectivity is represented by a binary string. For example, a fully connected graph of K nodes is represented by a register of $\frac{K(K-1)}{2}$ qubits in state $|11\ldots1\rangle$, Under the assumption for Bayesian nets that no node has more than $|\pi|_{\max}$ parents, a 'score Hamiltonian' can be defined that assigns an energy value to every permitted graph architecture/bit string, and the string with the lowest energy is found via quantum annealing. Other examples have been in the area of image matching (recognising that two images show the same content but in different light conditions or camera perspectives) [37], as well as in software verification and validation [38].

A machine learning task that naturally comes in the shape of a quadratic unconstrained optimisation problem is the memory recall of a Hopfield neural network (see also Sect. 2.4.2.3). In Eq. (7.16) the weights w_{ij} are determined by a learning rule such as the Hebb rule, and define the solution to the corresponding minimisation problem. The x_i, x_j are bits of the new input pattern with binary features. Quantum annealing then retrieves the 'closest' pattern—the nearest minimum energy state—in the 'memory' of the Hopfield energy function.

7.4.2 Annealing Devices as Samplers

Much of the more recent work on quantum annealing for training uses annealing devices in a less obvious way, namely to sample from an approximation to the Gibbs distribution in order to estimate the expectations

$$\langle v_i h_j \rangle_{\text{model}} = \sum_{v,h} \frac{e^{-E(v,h)}}{Z} v_i h_j$$

in Eq. (2.30) for the training of Boltzmann machines. We have already seen a variation of this idea in the previous Sect. based on the quantum approximate optimisation algorithm.

It turns out that the natural distributions generated by quantum annealers can be understood as approximations to the Boltzmann distribution [39]. Strictly speaking these would be *quantum* Boltzmann distributions, where the (Heisenberg) energy function includes a transverse field term for the spins, but under certain conditions quantum dynamics might not play a major role. Experiments on the D-Wave device showed that the distributions it prepares as qsamples can in fact significantly deviate from a classical Boltzmann distribution, and the deviations are difficult to model due to out-of-equilibrium effects of fast annealing schedules [40]. However, the distributions obtained still seem to work well for the training of (possibly pre-trained) Boltzmann machines, and improvements in the number of training steps compared to contrastive divergence have been reported both for experimental applications and numerical simulations [40–44]. As an application, samples from the D-Wave device have been shown to successfully reconstruct and generate handwritten digits [41, 44]. A great advantage compared to contrastive divergence is also that fully connected Boltzmann machines can be used.

While conceptually, the idea of using physical hardware to generically 'simulate' a distribution seems very fruitful, the details of the implementation with the relatively new hardware are at this stage rather challenging [44]. For example, researchers report Gaussian noise perturbing the parameters fed into the system by an unknown function

7.4 Quantum Adiabatic Machine Learning

which prohibits precise calibration. A constant challenge is also the translation of the coupling of visible and hidden units (or even an all-to-all coupling for unrestricted Boltzmann machines) into the sparse architecture of the connections between qubits in the D-Wave quantum annealer. Fully connected models need to be encoded in a clever way to suit the typical *chimera graph* structure.

Figure 7.6a shows the connectivity of qubits in the early D-Wave devices. Only those qubits that are connected by an edge can communicate with each other. We cannot natively represent a fully connected graph by the qubits and their interaction, but have only limited connectivity between nodes (see Fig. 7.6b). One therefore needs clever embedding strategies in which one variable in the model or one node in the graph is represented by multiple qubits. Another solution to deal with connectivity is to consider only problems with a structure that suits the limitations of the chimera graph connectivity. In the example of Fig. 7.6c the task is to embed an image represented by a matrix of pixels into the chimera graph. Assuming that only pixels that are close to each other are correlated, we associate blocks of neighbouring pixels with blocks of densely interconnected qubits (where the connection runs through some hidden units). For instance, the blocks of pixel $1a$ and $4d$ are not directly connected in the chimera graph structure, but also lie in opposite corners of the image and are therefore not expected to be correlated as much as neighbouring pixels.

Another serious problem when using the D-Wave device as a sampler is the finite 'effective inverse temperature' of the device. The distribution implemented by the machine contains an inverse temperature parameter β in $\frac{e^{-\beta E(s,h)}}{Z}$, which can fluctuate

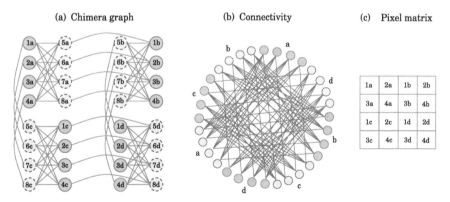

Fig. 7.6 In the original D-Wave annealing device, the qubits are connected by a so called *chimera graph* structure shown in **a** for a 32 qubit system. Some qubits in the chimera graph (green-yellow vertices) are used as visible units, while others (blue vertices with dotted outline) are used as hidden units. If we want the interactions between qubits to represent the weights between variables in a graphical model, we can therefore only encode sparse models with a connectivity shown in (**b**). However, in some applications the limited connectivity is not a problem. For example, when embedding the pixels of an image **c** into the chimera graph, we can use the assumption that not all pixels are strongly correlated, and associate pixels that are far away from each other with qubits in blocks that are not directly connected. Figure adapted from [42], with thanks to Marcello Benedetti

during the annealing process. Estimating this effective inverse temperature allows us to correct the samples taken from the annealer and use them to approximate the desired Boltzmann distribution [42].

The D-Wave device is only one example of an early-stage quantum technology, and we use it here to point out that another myriad of challenges appear when using real quantum devices for quantum machine learning. Connectivity and noise are two issues that occur in other platforms as well. These experimental challenges are constantly met with new engineering solutions, which may open another research branch to quantum machine learning altogether. We will discuss this point further in the conclusion when we consider quantum machine learning with intermediate-term quantum devices.

7.4.3 Beyond Annealing

Physical implementations other than quantum annealers or gate-based quantum computers have been used to investigate the idea of encoding the solution of a machine learning problem into the ground state of a quantum system. For example, an early contribution of a rather different kind was made by Horn and Gottlieb [46]. Their idea is to define a potential V for a Hamiltonian (which is traditionally decomposed into a kinetic and potential energy, $H = T + V$) such that a wave function with high probabilities in regions of high data density is the ground state of the Hamiltonian. The wave function serves as a kernel density estimator function introduced in Sect. 2.4.4.1. This idea has been subsequently used for clustering [47].

Another contribution is a demonstration that shows how an Hopfield-network-like associative memory recall can be executed with a nuclear magnetic resonance setup [45]. The proof-of-principle experiment only looks at binary strings of length 2. The idea is to encode the two binary values into the spectrum of nuclear spins in H and C atoms (see Fig. 7.7). A positive peak at the left region of the spectrum indicates a 1 while a negative peak in the right part of the spectrum indicates a -1. A 'no signal' encodes a 0, which means that the bit is unknown. Setting a weight parameter to $w = 1$ stores the two memory patterns $(-1, 1), (1, -1)$ into the spectrum of the nuclear magnetic resonance system, while the configuration $w = -1$ stores the patterns $(-1, -1), (1, 1)$. Starting the system with inputs x^{in} that have blanks at different positions leads to the desired memory recall. For example, for $w = 1$, the input $x^{\text{in}} = (-1, 0)$ recalls the memory pattern $x^* = (-1, 1)$. This is a minimal example of a quantum associative memory implemented by a quantum physical system. Interesting enough, the quantum model can lead to ground states that are superpositions of patterns.

These two examples illustrate that even though the focus of quantum machine learning lies on quantum computing devices, there are other quantum systems and experimental setups which can implement quantum machine learning algorithms or models.

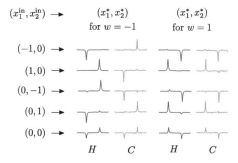

Fig. 7.7 Associated memory recall with atomic spectra of H and C atoms. Each spectrum encodes a -1 by a negative peak towards the left and a 1 by a positive peak towards the right. If both positive and negative peaks are present, the signal represents a superposition of -1 and 1. Depending on the 'weight' $w = 1, -1$, starting with an input signal x_1^{in}, x_2^{in} with a blank (0 or no signal) at different positions retrieves different solutions (x_1^*, x_2^*) from the memory. Image modified from [45]

References

1. Rebentrost, P., Bromley, T.R., Weedbrook, C., Lloyd, S.: A Quantum Hopfield Neural Network (2017). arXiv:1710.03599
2. Harrow, A.W., Hassidim, A., Lloyd, S.: Quantum algorithm for linear systems of equations. Phys. Rev. Lett. **103**(15), 150502 (2009)
3. Wiebe, N., Braun, D., Lloyd, S.: Quantum algorithm for data fitting. Phys. Rev. Lett. **109**(5), 050505 (2012)
4. Schuld, M., Sinayskiy, I., Petruccione, F.: Prediction by linear regression on a quantum computer. Phys. Rev. A **94**(2), 022342 (2016)
5. Rebentrost, P., Mohseni, M., Lloyd, S.: Quantum support vector machine for big data classification. Phys. Rev. Lett. **113**, 130503 (2014)
6. Zhao, Z., Fitzsimons, J.K., Fitzsimons, J.F.: Quantum Assisted Gaussian Process Regression (2015). arXiv:1512.03929
7. Berry, D.W., Childs, A.M., Kothari, R.: Hamiltonian simulation with nearly optimal dependence on all parameters. In: IEEE 56th Annual Symposium on Foundations of Computer Science (FOCS), pp. 792–809. IEEE (2015)
8. Wang, G.: Quantum algorithm for Linear Regression. Phys. Rev. A **96**, 012335 (2017)
9. Suykens, J.A.K., Vandewalle, J.: Least squares support vector machine classifiers. Neural Process. Lett. **9**(3), 293–300 (1999)
10. Hebb, D.O.: The Organization of Behavior: a Neuropsychological Theory. Wiley, New York (1949)
11. Aaronson, S.: Read the fine print. Nat. Phys. **11**(4), 291–293 (2015)
12. Clader, B.D., Jacobs, B.C., Sprouse, C.R.: Preconditioned quantum linear system algorithm. Phys. Rev. Lett. **110**(25), 250504 (2013)
13. Lloyd, S., Mohseni, M., Rebentrost, P.: Quantum principal component analysis. Nat. Phys. **10**, 631–633 (2014)
14. Cong, I., Duan, L.: Quantum discriminant analysis for dimensionality reduction and classification. New J. Phys. **18**, 073011 (2016)
15. Kerenedis, I., Prakash, A.: Quantum recommendation systems. In: Kerenidis, I., Prakash, A. (eds.), Quantum Recommendation Systems. LIPIcs-Leibniz International Proceedings in Informatics, vol. 67 (2017)
16. Wiebe, N., Kapoor, A., Svore, K.: Quantum nearest-neighbor algorithms for machine learning. Quantum Inf. Comput. **15**, 0318–0358 (2015)

17. Aïmeur, E., Brassard, G., Gambs, S.: Quantum speed-up for unsupervised learning. Mach. Learn. **90**(2), 261–287 (2013)
18. Ventura, D., Martinez, T.: Quantum associative memory. Inf. Sci. **124**(1), 273–296 (2000)
19. Dürr, C., Hoyer, P.: A Quantum Algorithm for Finding the Minimum (1996). arXiv:9607014v2
20. Kapoor, A., Wiebe, N., Svore, K.: Quantum perceptron models. In: Advances in Neural Information Processing Systems, pp. 3999–4007 (2016)
21. McClean, J.R., Romero, J., Babbush, R., Aspuru-Guzik, A.: The theory of variational hybrid quantum-classical algorithms. New J. Phys. **18**(2), 023023 (2016)
22. Peruzzo, A., McClean, J., Shadbolt, P., Yung, M.-H., Zhou, X.-Q., Love, P.J., Aspuru-Guzik, A., Obrien, J.L.: A variational eigenvalue solver on a photonic quantum processor. Nat. Commun. **5** (2014)
23. Farhi, E., Goldstone, J., Gutmann, S.: A Quantum Approximate Optimization Algorithm (2014). arXiv:1411.4028
24. Farhi, E., Harrow, A.W.: Quantum Supremacy Through the Quantum Approximate Optimization Algorithm (2016). arXiv:1602.07674
25. Wan, K.H., Dahlsten, O., Kristjánsson, H., Gardner, R., Kim, M.S.: Quantum generalisation of feed forward neural networks. npj Quantum. Inf. **3**(1), 36 (2017)
26. Verdon, G., Broughton, M., Biamonte, J.: A Quantum Algorithm to Train Neural Networks using Low-Depth Circuits (2017). arXiv:1712.05304
27. Nelder, J.A., Mead, R.: A simplex method for function minimization. Comput. J. **7**(4), 308–313 (1965)
28. Spagnolo, N., Maiorino, E., Vitelli, C., Bentivegna, M., Crespi, A., Ramponi, R., Mataloni, P., Osellame, R., Sciarrino, F.: Learning an unknown transformation via a genetic approach. Sci. Rep. **7**(1), 14316 (2017)
29. Guerreschi, G.G., Smelyanskiy, M.: Practical Optimization for Hybrid Quantum-Classical Algorithms (2017). arXiv:1701.01450
30. Farhi, E., Neven, H.: Classification with Quantum Neural Networks on Near Term Processors (2018). arXiv:1802.06002
31. Schuld, M., Bocharov, A., Wiebe, N., Svore, K.: A Circuit-Centric Variational Quantum Classifier (2018). arXiv:1804.00633
32. Barenco, A., Bennett, C.H., Cleve, R., DiVincenzo, D.P., Margolus, N., Shor, P., Sleator, T., Smolin, J.A., Weinfurter, H.: Elementary gates for quantum computation. Phys. Rev. A 52(5):3457 (1995)
33. Mitarai, K., Negoro, M., Kitagawa, M., Fujii, K.: Quantum Circuit Learning (2018). arXiv:1803.00745
34. Heim, B., Rønnow, T.F., Isakov, S.V., Troyer, M.: Quantum versus classical annealing of Ising spin glasses. Science **348**(6231), 215–217 (2015)
35. Troels, F.: Rnnow, Zhihui Wang, Joshua Job, Sergio Boixo, Sergei V. Isakov, David Wecker, John M. Martinis, Daniel A. Lidar, and Matthias Troyer. Defining and detecting quantum speedup. Science **345**, 420–424 (2014)
36. O'Gorman, B., Babbush, R., Perdomo-Ortiz, A., Aspuru-Guzik, A., Smelyanskiy, V.: Bayesian network structure learning using quantum annealing. Eur. Phys. J. Spec. Top. 224(1):163–188 (2015)
37. Neven, H., Rose, G., Macready, W.G.: Image recognition with an adiabatic quantum computer i. Mapping to Quadratic Unconstrained Binary Optimization (2008). arXiv:0804.4457
38. Pudenz, K.L., Lidar, D.A.: Quantum adiabatic machine learning. Quantum Inf. Process. **12**(5), 2027–2070 (2013)
39. Denil, M., De Freitas, N.: Toward the implementation of a quantum RBM. In: NIPS 2011 Deep Learning and Unsupervised Feature Learning Workshop (2011)
40. Korenkevych, D., Xue, Y., Bian, Z., Chudak, F., Macready, W.G., Rolfe, J., Andriyash, E.: Benchmarking Quantum Hardware for Training of Fully Visible Boltzmann Machines (2016). arXiv:1611.04528
41. Adachi, S.H., Henderson, M.P.: Application of Quantum Annealing to Training of Deep Neural Networks (2015). arXiv:1510.06356

References

42. Benedetti, M., Realpe-Gómez, J., Biswas, R., Perdomo-Ortiz, A.: Estimation of effective temperatures in quantum annealers for sampling applications: a case study with possible applications in deep learning. Phys. Rev. A **94**(2), 022308 (2016)
43. Amin, M.H., Andriyash, E., Rolfe, J., Kulchytskyy, B., Melko, R.: Quantum Boltzmann machine. Phys. Rev. X 8:021050 (2018)
44. Benedetti, M., Realpe-Gómez, J., Biswas, R., Perdomo-Ortiz, A.: Quantum-assisted learning of hardware-embedded probabilistic graphical models. Phys. Rev. X **7**, 041052 (2017)
45. Neigovzen, R., Neves, J.L., Sollacher, R., Glaser, S.J.: Quantum pattern recognition with liquid-state nuclear magnetic resonance. Phys. Rev. A 79(4):042321 (2009)
46. Horn, D., Gottlieb, A.: Algorithm for data clustering in pattern recognition problems based on quantum mechanics. Phys. Rev. Lett. **88**(1), 018702 (2002)
47. Cui, Y., Shi, J., Wang, Z.: Lazy quantum clustering induced radial basis function networks (LQC-RBFN) with effective centers selection and radii determination. Neurocomputing **175**, 797–807 (2016)

Chapter 8
Learning with Quantum Models

The last two chapters were mainly concerned with the translation of known machine learning models and optimisation techniques into quantum algorithms in order to harvest potential runtime speedups known from quantum computing. This chapter will look into 'genuine' quantum models for machine learning which either have no direct equivalent in classical machine learning, or which are quantum extensions of classical models with a new quality of dynamics. A quantum model as we understand it here is a model function or distribution that is based on the mathematical formalism of quantum theory, or naturally implemented by a quantum device. For example, it has been obvious from the last chapters that Gibbs distributions play a prominent role in some areas of machine learning. At the same time, quantum systems can be in a 'Gibbs state'. Previously, we described a number of attempts to use the quantum Gibbs states in order to sample from a (classical) Gibbs distribution. But what happens if we just use the 'quantum Gibbs distribution'? What properties would such models or training schemes exhibit? What if we use other distributions that are easy to prepare on a quantum device but difficult on a classical one, and construct machine learning algorithms from them? How powerful are the classifiers constructed from variational circuits in Sect. 7.3.2.1, that is if we use the input-output relation of a quantum circuit as a core machine learning model $f(x)$ and train the circuit to generalise from data?

This chapter sheds some light onto what has been earlier called the *exploratory* approach of quantum machine learning. It focuses less on speedups, but is interested in creating and analysing new models based on quantum mechanics, that serve as an addition to the machine learning literature. Although still in its infancy, the great appeal of this approach is that it is suited for near-term or noisy as well as non-universal quantum devices, because of the change in perspective: We start with a quantum system and ask whether it can be used for supervised learning, rather than translating models that were tailor-made for classical computers.

In the usual demeanour of this book we will present a couple of illustrative examples rather than an exhaustive literature review. Amongst them are quantum extensions of Ising models (Sect. 8.1) as well as variational circuits (Sect. 8.2). The last

Sect. 8.3 summarises further ideas, namely to replace Markov chains by quantum Markov chains, to use quantum superposition for ensemble methods, and to create special cost functions to be optimised by quantum annealing.

8.1 Quantum Extensions of Ising-Type Models

Hopfield networks and Boltzmann machines are both important machine learning methods that are based on one of the most successful family of models in physics, so called *Ising models*. The original Ising model describes ferromagnetic properties in statistical mechanics, namely the behaviour of a set of two-dimensional spin systems that interact with each other via certain nearest-neighbour interactions. We use the term in a wider sense here, to denote models of interacting units or particles of any connectivity and weight.

The Ising model is also no stranger to machine learning. Hopfield networks—which have a historical significance in reconnecting artificial neural networks with brain research—have in fact been introduced by a physicist, James Hopfield, who transferred the results from statistical physics to recurrent neural networks. Boltzmann machines can be understood as a probabilistic version of Hopfield networks. In both cases, the spins or particles are associated with binary variables or neurons, while the interactions correspond to the weights that connect the units. The physical equivalence gives rise to a well-defined quantum extension of these two methods which we will review in the following.

Let us first briefly summarise some central equations which have been explained in more detail in Sects. 2.4.2.3 and 2.4.2.4. Hopfield networks and Boltzmann machines are both a special case of recurrent neural networks consisting of G binary variables $s = s_1...s_G$ with $s_i \in \{-1, 1\}$ or $s_i \in \{0, 1\}$ for all $i = 1, \ldots, G$. The units can be distinguished into a set of visible and hidden units $s = vh = v_1...v_{N_v} h_1...h_{N_h}$ with $N_v + N_h = N$.[1] We can define an Ising-type energy function that assigns a real value to each configuration of the system,

$$E(s) = -\sum_{i,j} w_{ij} s_i s_j - \sum_i b_i s_i, \qquad (8.1)$$

with parameters $\{w_{ij}, b_i\}$ and $i, j = 1, \ldots, G$.

In Hopfield models the weights are symmetric ($w_{ij}=w_{ji}$) and non-self-connecting ($w_{ii} = 0$), and there is no constant field $b_i = 0$. With this architecture, the state of each unit is successively (synchronously or chronologically) updated according to the perceptron rule,

---

[1] In earlier chapters we referred to the number of visible units as $N + K$ to stress that in a supervised learning setting, the visible units can be divided into N input and K output units. For the sake of simplicity we will not make this distinction here.

8.1 Quantum Extensions of Ising-Type Models

$$s_i^{(t+1)} = \text{sgn}(\sum_j w_{ij} s_j^{(t)}).$$

The updates can be shown to lower or maintain the energy of the state in every time step $t \to t+1$, until it converges to the closest local minimum. These dynamics can be interpreted as an associative memory recall of patterns saved in minima.

The recall algorithm is similar to a Monte Carlo algorithm at zero temperature. In every step a random neuron s_i is picked and the state is flipped to $\bar{s}_i$ if does not increase the energy, that means if

$$E(s_1 \ldots \bar{s}_i \ldots s_N) \leq E(s_1 \ldots s_i \ldots s_N).$$

A generalised version of the Hopfield model [1] allows for a finite temperature and probabilistic updates of units in the associative memory recall. If one defines the probability of updating unit s_i to go from $s = s_1 \ldots s_i \ldots s_N$ to $\bar{s} = s_1 \ldots \bar{s}_i \ldots s_N$ as

$$p(\bar{s}) = \frac{1}{1 + e^{-2E(\bar{s})}},$$

we arrive at Boltzmann machines (with the temperature parameter set to 1). Here the binary units are random variables and the energy function defines the probability distribution over states $\{s\}$ via

$$p(s) = \frac{1}{Z} e^{-E(s)}. \tag{8.2}$$

As before, Z is the partition function

$$Z = \sum_s e^{-E(s)}. \tag{8.3}$$

For *restricted* Boltzmann machines, the parameters w_{ij} connecting two hidden or two visible units are set to zero.

8.1.1 The Quantum Ising Model

In the transition to quantum mechanics, the interacting units of the Ising model are replaced by qubits, the Hamiltonian energy function becomes an operator

$$H = -\frac{1}{2} \sum_{ij} w_{ij} \sigma_z^i \sigma_z^j - \sum_i b_i \sigma_z^i,$$

by replacing the values s_i by Pauli-σ_z operators acting on the ith or jth qubit. When we express the Hamiltonian as a matrix in the computational basis, it is diagonal. The reason for this is that the computational basis $\{|k\rangle\}$ is an eigenbasis to the σ_z operator, which means that the off-diagonal elements disappear, or $\langle k|\sigma_z^i|k'\rangle = 0$ and $\langle k|\sigma_z^i\sigma_z^j|k'\rangle = 0$ for all i, j if $k \neq k'$. The eigenvalues on the diagonal are the values of the energy function (8.1). For example, the Ising Hamiltonian for a 2-qubit system takes the following shape:

$$H = \begin{pmatrix} E(s=00) & 0 & 0 & 0 \\ 0 & E(s=01) & 0 & 0 \\ 0 & 0 & E(s=10) & 0 \\ 0 & 0 & 0 & E(s=11) \end{pmatrix},$$

where we omitted the dependence on θ from Eq. (8.1).

The quantum state of the system (here expressed as a density matrix) defines a probability distribution over the computational basis states. For an Ising model it is given by

$$\rho = \frac{1}{Z} e^{-H}, \qquad (8.4)$$

with the partition function being elegantly expressed by a trace,

$$Z = \mathrm{tr}\{e^{-H}\}.$$

For diagonal matrices H, the matrix exponential e^{-H} is equal to a matrix carrying exponentials of the diagonal elements on its diagonal. Consequently, ρ is diagonal with

$$\mathrm{diag}\{\rho\} = (\frac{1}{Z} e^{-E(s=0...0)}, \ ..., \ \frac{1}{Z} e^{-E(s=1...1)}).$$

The diagonal elements of the density matrix define the Boltzmann probability distribution of Eqs. (8.2) and (8.3) for all different possible states of the system's spins or neurons, and the operator formalism is fully equivalent to the classical formalism.

Having formulated the (still classical) Ising model in the language of quantum mechanics, it is not difficult to introduce quantum effects by adding terms with the other two Pauli operators that do not commute with σ_z. Non-commuting operators do not share an eigenbasis, and the resulting operator is not diagonal in the computational basis. It is therefore not any more a different mathematical formalism for the classical Ising model, but the non-zero off-diagonal elements give rise to genuine quantum dynamics.

It is common to only introduce a few 'quantum terms', such as the *transverse* or *x-field*,

$$H = -\frac{1}{2} \sum_{ij} w_{ij} \sigma_z^i \sigma_z^j - \sum_i b_i \sigma_z^i - \sum_i c_i \sigma_x^i, \qquad (8.5)$$

8.1 Quantum Extensions of Ising-Type Models

where the Pauli operator σ_x^i is applied to the ith qubit with a strength c_i. Since the Hamiltonian for the transverse Ising model is no longer diagonal in the computational basis, superpositions of basis states are eigenstates of H, and the system exhibits different dynamics compared to the classical system.

The strategy of extending the Hamiltonian illustrates a more general recipe to design quantum versions of machine learning models based on statistical physics:

1. Formulate the classical model in the language of quantum mechanics, for example using quantum state ρ to describe the probability distribution, and using operators to describe the 'evolution' of the system
2. Add terms that account for quantum effects, such as a transverse field in the Hamiltonian or, as we will see below, a coherent term in a master equation
3. Analyse the resulting quantum model for learning and characterise effects deriving from the extension.

We will now show some examples from the literature of how this has been done for Boltzmann machines and Hopfield networks. Point 3, the analysis of the quantum extension for machine learning purposes, is still mainly an open issue in the literature.

8.1.2 Training Quantum Boltzmann Machines

The above strategy of adding a transverse field to the Hamiltonian has been used to propose a quantum version of Boltzmann machines by Amin et al. [2]. The quantum Boltzmann machine defines the model as the quantum state ρ from Eq. (8.4) with the 'tansverse' Hamiltonian (8.5). One can understand ρ as a qsample, from which we can draw samples of computational basis states that correspond to states of the visible units. A crucial question is how to train such a model to learn the weights w_{ij}, b_i and c_i. As it turns out, the strategy from classical training does not carry over to the quantum model.

In order to write down the gradient from Eq. (2.29) in the language of quantum mechanics, we need to define the probability of observing the visible state v' by means of the model ρ,

$$p(v') = \text{tr}\{\Lambda_{v'}\rho\}.$$

The operator

$$\Lambda_{v'} = |v'\rangle\langle v'| \otimes \mathbb{1}_h,$$

is a projector that in matrix notation has a single 1 entry at the diagonal position $\langle v'| \cdot |v'\rangle$ and zeros else. In other words, the operator 'picks out' the diagonal element $\rho_{v'v'}$, which is the probability to measure the computational basis state $|v'\rangle$ or to sample v' from the quantum model. With this definition and inserting the expression for ρ from Eq. (8.4), the log likelihood objective becomes

$$C(\theta) = -\sum_{m=1}^{M} \log \frac{\text{tr}\{\Lambda_{v^m} e^{-H(\theta)}\}}{\text{tr}\{e^{-H(\theta)}\}}, \tag{8.6}$$

where we state the dependence of H on the parameters $\theta = \{w_{ij}, b_i, c_i\}$ now explicitly.[2] Compared to *maximum* likelihood estimation we consider the negative log-likelihood to formulate a cost function, which is *minimised*.

For gradient descent optimisation we need the derivative of the objective function for each parameter $\mu \in \theta$, which, applying logarithm derivation rules, is given by

$$\partial_\mu C(\theta) = \sum_{m=1}^{M} \left(\frac{\text{tr}\{\Lambda_v \partial_\mu e^{-H(\theta)}\}}{\text{tr}\{\Lambda_v e^{-H(\theta)}\}} - \frac{\text{tr}\{\partial_\mu e^{-H(\theta)}\}}{\text{tr}\{e^{-H(\theta)}\}} \right). \tag{8.7}$$

In the classical version of Eq. (2.9) we were able to execute the partial derivation via the chain rule, which for the interaction parameters $\mu = w_{ij}$ ended up to be the difference of two expectation values over units from the data versus the model distribution from Eq. (2.30),

$$\partial_{w_{ij}} C(\theta) = \langle s_i s_j \rangle_{\text{data}} - \langle s_i s_j \rangle_{\text{model}}.$$

These expectation values were approximated via sampling, for example with the contrastive divergence approach.

Using the quantum model, the operators $\partial_{w_{ij}} H$ and H do not commute, which means that the chain rule of basic derivation cannot be applied. However, due to the properties of the trace, one still finds a similar relation for the right term in Eq. (8.7),

$$\text{tr}\{\partial_{w_{ij}} e^{-H(\theta)}\} = -\text{tr}\{e^{-H(\theta)} \partial_{w_{ij}} H(\theta)\}.$$

The trace over 'operator' $\partial_{w_{ij}} H(\theta)$ for a Gibbs state $\rho = e^{-H(\theta)}$ divided by the partition function $\text{tr}\{e^{-H(\theta)}\}$ is nothing other than the expectation value over the Boltzmann distribution $\langle \partial_{w_{ij}} H(\theta) \rangle_{\text{model}} = \langle \sigma_z^i, \sigma_z^j \rangle_{\text{model}}$. For the left term of Eq. (8.7),

$$\frac{\text{tr}\{\Lambda_v \partial_{w_{ij}} e^{-H(\theta)}\}}{\text{tr}\{e^{-H(\theta)}\}}, \tag{8.8}$$

the additional operator inside the trace prohibits this trick. Without a strategy to approximate this term, common training methods are therefore not applicable, even if we only consider restricted Boltzmann machines. This is a good example of how quantum extensions can require different approaches for training.

A strategy proposed in [2] is to replace the Hamiltonian H in this troublesome term in Eq. (8.8) with a "clamped" Hamiltonian $H_m(\theta) = H(\theta) - \ln \Lambda_{v^m}$, which penalises any states other than $|v^m\rangle$. One can show that the new cost function is an

---

[2]Note that we simplified the original expression in [2] to be consistent with the more common definition of the log-likelihood introduced in Sect. 2.9.

8.1 Quantum Extensions of Ising-Type Models

upper bound for the original objective from Eq. (8.6). We can hence write the left term in Eq. (8.7) as

$$\sum_{m=1}^{M} p(v^m; \theta) \frac{\text{tr}\{e^{-H_m(\theta)} \partial_{w_{ij}} H_m(\theta)\}}{\text{tr}\{e^{-H_m(\theta)}\}} = \langle \partial_{w_{ij}} H_m(\theta) \rangle_{\text{data}} = \langle \sigma_z^i, \sigma_z^j \rangle_{\text{data}}.$$

One can show that this corresponds to minimising an upper bound of the objective [2]. This works well for parameter updates w_{ij}, b_i, but fails for the updating rule of the c_i, the parameters of the transverse field strength, which in the worst case have to be simply guessed. However, training with samples from a (clamped) quantum Boltzmann machine can have better convergence properties than a classical Boltzmann machine [2].

Other versions of quantum Boltzmann machines have been proposed, for example using a fermionic rather than a stochastic Hamiltonian, more general (i.e. quantum) training data sets, as well as relative entropy rather than a log-likelihood objective [3].

8.1.3 Quantum Hopfield Models

A number of studies on 'quantum Hopfield models' follow the same approach as in the previous section and investigate how the introduction of an transverse field term as in Eq. (8.5) changes the dynamics of the associative memory recall. For example, qualitative differences regarding noise and the capacity (ratio of stored patterns to system size) of the network have been reported [4–7]. But, as we will establish in this section, one can also look at quantum extensions of Ising-type models starting from the differential equations of the system. As an example, we follow Rotondo et al. [8] who formulate a quantum version of the Hopfield model as an open quantum system governed by a *quantum master equation*. We have not covered this topic much in the foundations of Chap. 3, and will therefore only sketch the general idea here. Interested readers can refer to [9] for more details on the theory and description of open quantum systems.

Unitary evolutions in quantum mechanics – which we have discussed at length— are the solution of the *Schrödinger equation* (3.1.3.3), which applies to closed quantum systems. In closed quantum systems, no information about the state is lost and the evolution is fully reversible. But the associative memory dynamics of a memory recall is a *dissipative* evolution in which the energy of the system is not preserved, which means that we do not have reversible dynamics. This requires an open quantum systems approach, where the equivalence of the *Schrödinger equation* is a *quantum master equation*.[3] A popular formulation of a quantum master equation

---

[3] Master equations appear in many other sub-disciplines of physics.

describing the evolution of a dissipative quantum system is the *Gorini-Kossakowski-Sudarshan-Lindblad equation* that we mentioned in Sect. 3.1.3.6. As a reminder, the GKSL equation defines the time evolution of a density operator, in this case

$$\frac{d}{dt}\rho = -i[H, \rho] + \sum_{k=1}^{K} \sum_{\gamma=\pm} \left(L_{k\gamma} \rho L_{k\gamma}^\dagger - \frac{1}{2} L_{k\gamma}^\dagger L_{k\gamma} \rho - \frac{1}{2} \rho L_{k\gamma}^\dagger L_{k\gamma} \} \right). \quad (8.9)$$

The first part, $-i[H, \rho]$, describes the coherent part of the evolution, namely the quantum effects. The second part, containing the Lindblad operators $L_{k\gamma}$, describes a stochastic evolution, which can be roughly thought of as transitions between states of the system represented by the $L_{k\gamma}$, according to a probability defined by ρ.

For the quantum version of the Hopfield model's dynamic, the *jump operators* $L_{k\gamma}$ are defined as

$$L_{k\pm} = \frac{e^{\mp 1/2 \Delta E_k}}{\sqrt{2 \cosh(\Delta E_k)}} \sigma_\pm^k.$$

The value of ΔE_k is the energy difference of a system in state s and a system in state $\bar{s}$ resulting from flipping the state of the kth unit. Furthermore, the plus/minus Pauli operators are defined as $\sigma_\pm^i = (\sigma_x^i \pm i\sigma_y^i)/2$. For the coherent term, the Hamiltonian can for example be chosen as the transverse field term from before,

$$H = c \sum_i \sigma_x^i,$$

but with constant coefficients $c_i = c$. If the coherent term is not included, the Lindblad equation becomes a classical equation. Rotondo et al. [8] show this in detail in the Heisenberg picture, where instead of the dynamics of the state ρ in Eq. (8.9), one looks at the same (but purely classical) dynamics for the σ_z^k operators,

$$\frac{d}{dt} \sigma_z^k = \sum_{k=1}^{K} \sum_{\gamma=\pm} \left(L_{k\gamma} \sigma_z^k L_{k\gamma}^\dagger - \frac{1}{2} L_{k\gamma}^\dagger L_{k\gamma} \sigma_z^k - \frac{1}{2} \sigma_z^k L_{k\gamma}^\dagger L_{k\gamma} \right)$$

and finds that the operator expectation value $\langle \sigma_z^k \rangle$ has the same expression as the well-known mean field approximation for the value of each spin in the classical spin glass model,

$$\langle s_k \rangle = \tanh \left(\sum_j w_{ij} \langle s_j \rangle \right).$$

The dynamics of the quantum model can be analysed with mean-field techniques that lead to a phase diagram [8]. Phase diagrams are important tools in statistical physics to study the behaviour of a system in different parameter regimes. In this case we are interested in the phase diagram of field strength parameter c and the temperature T. For any temperature, if c is sufficiently low, the general structure of the classical associative memory remains valid. For large temperatures and large c there is a parametric phase where the model converges to a zero pattern. A qualitatively new dynamics enters for small temperatures and large field strengths and is a limit cycle phase. This means that the state of the system periodically goes through a fixed sequence of states, a limit cycle. Such limit cycles are known from classical Hopfield models with asymmetric connections $w_{ij} \neq w_{ji}$. For example, an asymmetric Hopfield network with two nodes s_1, s_2 and connections $w_{12} = -0.5$, $w_{21} = 0.5$ will update the state of the two nodes according to

$$s_1^{(t+1)} = \varphi(-0.5 s_2^{(t)}), \qquad s_2^{(t+1)} = \varphi(0.5 s_1^{(t)}),$$

which, starting with $s_1 = 1$, $s_2 = 1$ and with φ being a step function with a threshold at 0, sends the state into the limit cycle

$$\begin{pmatrix} 1 \\ 1 \end{pmatrix} \to \begin{pmatrix} -1 \\ 1 \end{pmatrix} \to \begin{pmatrix} -1 \\ -1 \end{pmatrix} \to \begin{pmatrix} 1 \\ -1 \end{pmatrix} \to \begin{pmatrix} 1 \\ 1 \end{pmatrix}.$$

Hopfield remarked that Hopfield networks with asymmetric connections show a similar behaviour to dynamics of the Ising model at finite temperature [10]. Rotondo et al. argue that the limit-cycle phase of the quantum Hopfield model is instead a product of competition between coherent and dissipative dynamics, not a result of asymmetry or self-connections. The quantum model therefore promises to exhibit another quality of dynamics compared to the classical one.

8.1.4 Other Probabilistic Models

Not only Ising-type models have natural quantum extensions. Another probabilistic model that lends itself easily to this purpose are hidden Markov models, where a system undergoes a sequence of state transitions under observations, and either the transition/observation probabilities, or the next transition is to be learned (see Sect. 2.4.3.2). State transitions and observation probabilities are central to quantum physics as well, where they can be elegantly described in the formalism of density matrices and open quantum systems.

Monras, Beige and Wiesner [11] introduce a quantum formulation of hidden Markov models, in which the state of the system is a density matrix ρ and state transitions are formally represented by so called *Kraus operators* $\mathcal{K}$ with associated probabilities $\text{tr}\{\mathcal{K}\rho\}$ [9]. Although the primary idea is to use these quantum models to learn about partially observable quantum systems, Monras et al. observe that the number of internal states necessary for generating some stochastic processes is smaller than of a comparable classical model.[4] The 'reinforcement learning' version of hidden Markov models are *partially observable Markov decision processes* and have been generalised to a quantum setting by Barry, Barry and Aaronson [13]. They show that a certain type of problem—the existence of a sequence of actions that can reach a certain goal state—is undecidable for quantum models. Another variation of a hidden Markov model is proposed in [14] and is based on a certain formulation of quantum Markov processes.

A further branch of 'quantum-extendable' probabilistic models are graphical models. Leifer et al. [15] give a quantum mechanical version of belief propagation, which is an algorithm for inference in graphical models. Directed graphical models in which a directed edge stands for a causal influence are called *causal models*. A goal that gains increasing attention in the classical machine learning community is to discover causal structure from data, which is possible only to some extent [16] in the classical case. Quantum systems exhibit very different causal properties, and it is sometimes possible to infer the full causal structure of a model from (quantum) data [17, 18]. Costa and Shrapnel [19] introduce quantum analogues of causal models for the discovery of causal structure, and propose a translation for central concepts such as the 'Markov condition' or 'faithfulness'. Whether such quantum models can have any use for classical data (or even for quantum data) is a largely open question.

8.2 Variational Classifiers and Neural Networks

The second approach presented in this chapter is based on variational quantum circuits which have been introduced in Sect. 7.3 in the context of hybrid training. As a reminder, a variational circuit is a quantum circuit that consists of parametrised gates, and where the parameters (and thereby the circuit) can be optimised or learnt according to a certain objective. If we use some circuit parameters to feed inputs into the circuit, and understand the result computed by the circuit as an output, we can interpret a variational circuit as a mathematical model for supervised learning.

Learning models based on variational circuits are a prime example for the exploratory approach to quantum machine learning. The quantum device is used as a computational (sub-)routine in a mathematical model. It is treated as a black-box that produces—or assists in producing—predictions when fed with (possibly pre-processed) inputs. Hence, one works with a model that can be naturally implemented by a quantum device.

In the following we will revisit the ansatz of Sect. 7.3.3.2, where we considered a decomposition of a variational circuit into parametrised single-qubit gates and controlled versions thereof. We want to show now how to interpret such a quantum gate as a linear layer in the language of neural networks. This may help to tap into

---

[4] In a later paper, Monras and Winter [12] also perform a mathematical investigation into the task of learning a model from example data in the quantum setting, i.e. to ask whether there is there a quantum process that realises a set of measurements or observations.

8.2 Variational Classifiers and Neural Networks

the rich theory for neural nets in order to develop and analyse quantum models based on variational circuits. It also raises interesting questions about the number of parameters needed to build a sufficiently powerful classifier. We will finally look at some proposals from classical machine learning to decompose $N \times N$-dimensional unitaries into matrices which can be computed in time $\mathcal{O}(N)$ or below.

8.2.1 Gates as Linear Layers

Let us consider a variational circuit that consists of parametrised gates as defined in Eq. (7.11) of the previous section. Instead of trigonometric functions, we can use two complex numbers $z, u \in \mathbb{C}$ with each a real and an imaginary value as parameters,

$$G(z, v) = \begin{pmatrix} z & u \\ -u^* & z^* \end{pmatrix}. \tag{8.10}$$

To neglect the global phase we can include the condition that $|z|^2 + |v|^2 = 1$. An advantage of this parametrisation is that it does not introduce nonlinear dependencies of the circuit parameters on the model output, while a disadvantage in practice is that during training the normalisation condition has to be enforced. We will use this parametrisation here simply for ease of notation.

Let us have a look at the structure of the overall matrix representation of a (controlled) single qubit gate. As a reminder, a single qubit gate applied to the ith qubit of an n qubit register has the following structure:

$$G_{q_i} = \mathbb{1}_1 \otimes \cdots \otimes \underbrace{G(z, v)}_{\text{position } i} \otimes \cdots \otimes \mathbb{1}_n.$$

The $2^n \times 2^n$ matrix G_{q_i} is sparse and its structure depends on which qubit i the gate acts on.

Example 8.1 (*Parametrised single qubit gate*). For $n = 3$ and $i = 1$:

$$G_{q_1} = \begin{pmatrix} z & v & 0 & 0 & 0 & 0 & 0 & 0 \\ -v^* & z & 0 & 0 & 0 & 0 & 0 & 0 \\ 0 & 0 & z & v & 0 & 0 & 0 & 0 \\ 0 & 0 & -v^* & z^* & 0 & 0 & 0 & 0 \\ 0 & 0 & 0 & 0 & z & v & 0 & 0 \\ 0 & 0 & 0 & 0 & -v^* & z^* & 0 & 0 \\ 0 & 0 & 0 & 0 & 0 & 0 & z & v \\ 0 & 0 & 0 & 0 & 0 & 0 & -v^* & z^* \end{pmatrix}, \tag{8.11}$$

while for $n = 3$ and $i = 2$:

$$G_{q_2} = \begin{pmatrix} z & 0 & v & 0 & 0 & 0 & 0 & 0 \\ 0 & z & 0 & v & 0 & 0 & 0 & 0 \\ -v^* & 0 & z^* & 0 & 0 & 0 & 0 & 0 \\ 0 & -v^* & 0 & z^* & 0 & 0 & 0 & 0 \\ 0 & 0 & 0 & 0 & z & 0 & v & 0 \\ 0 & 0 & 0 & 0 & 0 & z & 0 & v \\ 0 & 0 & 0 & 0 & -v^* & 0 & z^* & 0 \\ 0 & 0 & 0 & 0 & 0 & -v^* & 0 & z^* \end{pmatrix}, \qquad (8.12)$$

and lastly $n = 3$ and $i = 3$ yields

$$G_{q_3} = \begin{pmatrix} z & 0 & 0 & 0 & v & 0 & 0 & 0 \\ 0 & z & 0 & 0 & 0 & v & 0 & 0 \\ 0 & 0 & z & 0 & 0 & 0 & v & 0 \\ 0 & 0 & 0 & z & 0 & 0 & 0 & v \\ -v^* & 0 & 0 & 0 & z^* & 0 & 0 & 0 \\ 0 & -v^* & 0 & 0 & 0 & z^* & 0 & 0 \\ 0 & 0 & -v^* & 0 & 0 & 0 & z^* & 0 \\ 0 & 0 & 0 & -v^* & 0 & 0 & 0 & z^* \end{pmatrix}. \qquad (8.13)$$

To make the elementary gate set universal, we also consider the application of the single qubit gate G_{q_i} controlled by another qubit j, which we denote by $c_{q_j} G_{q_i}$. A single qubit gate applied to qubit i and controlled by qubit j carries a 'diagonal unit entry' in some columns and rows as opposed to the single qubit gate.

Example 8.2 (*Controlled parametrised single qubit gate*). G_{q_1} from Eq. (8.11) can be controlled by q_3 and becomes

$$c_{q_3} G_{q_2} = \begin{pmatrix} 1 & 0 & 0 & 0 & 0 & 0 & 0 & 0 \\ 0 & z & 0 & u & 0 & 0 & 0 & 0 \\ 0 & 0 & 1 & 0 & 0 & 0 & 0 & 0 \\ 0 & -u^* & 0 & z^* & 0 & 0 & 0 & 0 \\ 0 & 0 & 0 & 0 & 1 & 0 & 0 & 0 \\ 0 & 0 & 0 & 0 & 0 & z & 0 & u \\ 0 & 0 & 0 & 0 & 0 & 0 & 1 & 0 \\ 0 & 0 & 0 & 0 & 0 & -u^* & 0 & z^* \end{pmatrix}. \qquad (8.14)$$

As a product of matrices, a variational circuit

$$U(\theta) = G(\theta_L) \cdots G(\theta_1)$$

with $\theta_1, \ldots, \theta_L \subset \theta$ can be understood as a sequence of linear layers of a neural network where the layers have a constant size. This perspective allows us to use graphical representations to visualise the connectivity power of a single qubit gate and helps to understand the mechanics of the unitary circuit as a classifier. The position of the qubit as well as the control qubit determine the architecture of each layer, i.e. which units are connected and which weights are tied (i.e., have the same value).

8.2 Variational Classifiers and Neural Networks

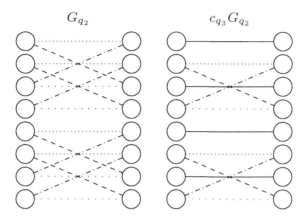

Fig. 8.1 The single qubit gate G_{q_2} (left) and the controlled single qubit gate $c_{q_3}G_{q_2}$ (right) from the examples applied to a system of 3 qubits drawn in the graphical representation of neural networks. The gates take a quantum state with amplitude vector $(\alpha_1, \ldots, \alpha_8)^T$ to a quantum state $(\alpha'_1, \ldots, \alpha'_8)^T$. The solid line is an 'identity' connection of weight 1, while all other line styles indicate varying weight parameters. Two lines of the same style means that the corresponding weight parameters are tied, i.e. they have the same value at all times

To illustrate this, consider the matrix G_{q_2} from Eq. (8.11), a single qubit gate acting on the second of 3 qubits. If we interpret the $2^3 = 8$-dimensional amplitude vector of the quantum state that the gate acts on as an input layer, the 8-dimensional output state as the output layer, and the gate as a connection matrix between the two, this would yield the graphical representation from the left side of Fig. 8.1. The control breaks this highly symmetric structure by replacing some connections by identities which carry over the value of the previous layer (see right side of Fig. 8.1). It becomes obvious that a single qubit gate connects four sets of two variables with the same weights, in other words, it ties the parameters of these connections. The control removes half of the ties and replaces them with identities.

An interesting viewpoint is that stacking many such unitary layers creates a deep architecture that does not change the length of vectors, which can have useful properties for training [20].

8.2.2 Considering the Model Parameter Count

The number of parameters or weights trained in a conventional neural network is at least as large as the input, since each input unit has to be connected to the following layer. In a variational quantum circuit, we have typically much fewer parameters than inputs. Mathematically speaking, a single qubit gate implements a matrix multiplication that transforms *all* N amplitudes, however large N is, controlled by only three parameters. Of course, this matrix multiplication has a lot of structure or 'strongly tied weights'.

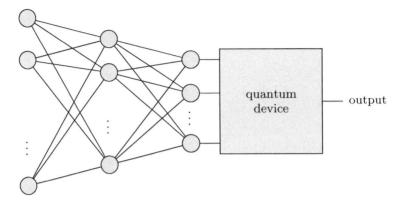

Fig. 8.2 In hybrid architectures for neural networks, the quantum device computes part of the model. We demonstrate this for a feed-forward neural network. The classical neural network (layers of green and blue circles) can be understood as a feature selection procedure that learns to feed useful and highly compressed features into the quantum device, which then 'calculates' the output of the model

It is an interesting question what representational power a quantum classifier based on a variational circuit has if we keep the number of parameters poly-logarithmic in the input dimension of the data. For example, imagine a 10 qubit system where the gate matrices act on 2^{10}-dimensional vectors, while the circuit has only a depth of 100 gates and hence of the order of 300 parameters. This is reminiscent of compact representations of weight matrices with tensor networks, to which some rich connections can be made [21]. First investigations on quantum classifiers based on variational circuits with a poly-logarithmic number of parameters in the size of the Hilbert space are promising [22], but there are yet many open questions.

Keeping the parameter count low is especially important if we use the feature maps from Sect. 6.2 to introduce nonlinearities into the model. If the feature map maps an input to an infinite-dimensional input layer which the circuit is applied to, we necessarily have much fewer parameters than 'hidden neurons' in the layer.

On the other end of the spectrum, there is a situation in which a parameter count of variational circuits can be of the order of the input. While so far we considered the variational circuit to define the quantum model that is trained by a classical algorithm, one can also hybridise inference—in other words the model—by using *quantum-assisted* or *hybrid architectures*. A hybrid architecture of a neural network has a 'quantum layer', while all remaining layers are standard classical neural network layers. Typically, the quantum layer would be of low-width and in the 'deep end' of a deep neural network (see Fig. 8.2), where it only has to deal with low-dimensional features. This circumvents the need for a resource-intense classical-quantum interface to feed the data into the quantum device. If the input dimension to the quantum layer is low enough, we can also easily allow for a number of parameters that is polynomial in this dimension. The idea of hybrid architectures has been originally proposed for probabilistic models, where one samples the states of some hidden units

8.2 Variational Classifiers and Neural Networks

from a quantum device and computes classical samples for the next hidden and visible layer from the quantum samples [23]. Whether this approach shows significant advantages in real applications is still mostly an open research question.

8.2.3 Circuits with a Linear Number of Parameters

An alternative connection between unitary circuits and neural networks comes interestingly enough from the classical machine learning literature. In the context of recurrent neural networks, unitary weight matrices that update the neurons in every time step are used to avoid the exploding/vanishing gradients problem [24–26]. A unitary matrix has the property of preserving the length of the vector it is applied to and therefore maintains the scale of the signal fed forward (in time), and the errors propagated backwards.

The proposal in Arjovsky et al. [24] aims at a decomposition of an $N \times N$ unitary matrix U into several unitary matrices which all have the property that they can be computed in time $O(N)$,

$$U = D_3 T_2 F^{-1} D_2 \Pi T_1 F D_1.$$

The decomposition includes the following types of matrices:

- The three matrices D_1, D_2, D_3 are diagonal with $\text{diag}\{D\} = (e^{i\omega_1}, ..., e^{i\omega_N})$ and real parameters $\omega_1, ..., \omega_N$.
- The two matrices T_1, T_2 are reflection matrices defined by a N-dimensional reflection vector v via

$$T_i = \mathbb{1} - 2\frac{v_i v_i^\dagger}{||v_i||^2},$$

for $i = 1, 2$.
- The fixed matrices F, F^{-1} implement a Fourier transform and its inverse.
- The fixed matrix Π implements a constant permutation.

The learnable parameters are in this case the $\{\omega\}$ of each diagonal matrices, as well as the entries of the reflection vectors that define T_1 and T_2. In total, these are $5N$ parameters, and all computations can be performed linear in N. To a quantum computing expert, the building blocks of the unitary layer are all very familiar from quantum computing, and such a layer—excluding the final nonlinear activation used in the paper [24], could be implemented as a variational circuit. Most notably, the Fourier transform can be implemented in poly-logarithmic time relative to N, a well-known exponential speedup (see Sect. 3.5.2).

Another unitary decomposition ansatz from the literature on unitary recurrent neural nets allows us to tune through architectures from $\mathcal{O}(\frac{N}{2})$ parameters to a full representation of a general unitary with $\mathcal{O}(N^2)$ parameters. It is inspired by an influential physics paper by Reck, Zeilinger, Bernstein and Bertani [27], and its later

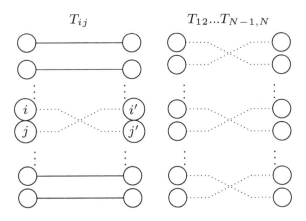

Fig. 8.3 Right: A 2-level rotation T_{ij} can be interpreted as a linear layer that connects only two dimensions i and j. Summarising several 2-level rotations to a single *2-level rotation layer* connects pairs of adjacent dimensions with each other

extension [28]. The original paper showed how to experimentally realise any unitary transformation through a network of certain building blocks (in their case, between a set of optical channels via beam splitters). On a more abstract level, these building blocks can be written as 2-level rotation operators, each of which has the form

$$T_{ij}(\alpha, \beta) = \begin{pmatrix} 1 & 0 & \cdots & & \cdots & & \cdots & \cdots & 0 \\ 0 & 1 & & & & & & & \vdots \\ \vdots & & \ddots & & & & & & \vdots \\ \vdots & & & e^{i\alpha}\cos\beta & -\sin\beta & & & & \vdots \\ \vdots & & & e^{i\alpha}\sin\beta & \cos\beta & & & & \vdots \\ & & & & & \ddots & & & \vdots \\ & & & & & & & 1 & 0 \\ 0 & \cdots & & \cdots & & \cdots & & 0 & 1 \end{pmatrix}. \tag{8.15}$$

Applied to an amplitude vector, this operator only touches the ith and jth amplitude and rotates them similar to a single qubit rotation controlled by all other qubits. In the graphical representation of layers we introduced above, this operation correlates two input nodes as depicted in Fig. 8.3.

Any unitary matrix can be decomposed into the network of 2-level-rotations shown in Fig. 8.4 (see [28] for a proof). Formally the network can be written as

$$U = (T_{1,2}...T_{N-1,N})(T_{2,3}...T_{N-2,N-1})(T_{1,2}...T_{N-1,N})...(T_{1,2}...T_{N-1,N}).$$

In other words, the network is a concatenation of even and odd 2-level rotation layers (summarised in the round brackets). Odd layers connect neighbours [1, 2], [3, 4]..., while even layers connect neighbours [2, 3], [4, 5].... The full parametrisation

8.2 Variational Classifiers and Neural Networks

Fig. 8.4 A square grid of 8 rotation layers (where the 28 parametrised T_{ij} rotations from Eq. (8.15) sit at intersections of the dotted lines) can represent any unitary transformation of an 8-dimensional input vector, represented by the circles

requires $N(N-1)/2$ rotation gates. Reducing the number of layers can be a useful strategy to limit the number of model parameters, as shown in [25].

Not surprisingly, also this building block of a unitary layer is common in quantum computing. The study of unitary layers in neural networks therefore offers a fruitful connection between quantum machine learning models based on variational circuits and classical machine learning.

8.3 Other Approaches to Build Quantum Models

In this last section, three other ideas to create quantum models for machine learning are summarised in order to demonstrate the variety of the explorative approach. The first idea is again to replace a classical formalism, Markov chains or random walks, by the quantum version [29]. The second idea looks at quantum superposition and how it can be used to create ensembles of classifiers in parallel to get a "training-free" predictor [30]. The final idea was one of the first demonstrations of quantum machine learning and constructs a cost function for binary classification with ensembles, which is tailor-made for the D-Wave device [31].

8.3.1 Quantum Walk Models

Any graph of N vertices—such as the graph in Fig. 8.5—gives rise to a Markov chain. A Markov chain is a sequence of states that is governed by a stochastic process (see also Sect. 3.1.2.2). Markov chains are described by a stochastic matrix $\mathbb{R}^{N \times N}$ with $\sum_{j=1}^{N} m_{ij} = 1$ and entries m_{ij} representing the weight of the directed edge going

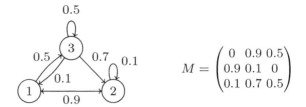

Fig. 8.5 A weighed directed graph of three nodes and the corresponding Markov chain represented by stochastic matrix M

from vertex i to j. These weights can be interpreted as a transition probability to go from site i to j. Repeatedly applying M to a n-dimensional stochastic vector π with $\sum_{l=1}^{N} \pi_l = 1$ evolves an initial probability distribution through discrete time steps, whereby $\pi(t+1)$ only depends on $\pi(t)$ and the graph architecture (hence the name 'Markov chain'). With this formalism, the probability of being at vertex i changes according to

$$\frac{d\pi_i}{dt} = -\sum_j M_{ij} \pi_j(t). \quad (8.16)$$

Markov chains with equal probability to jump from site i to any of the d sites adjacent to i are also known as *random walks* on a graph. Random walks are based on the idea of an abstract walker who in each step tosses a d-dimensional coin to choose one of d possible directions at random (or, in the language of graphs, chooses a random vertex that is connected to the vertex the walker is currently at). Random walks have been proven to be powerful tools in constructing efficient algorithms in computer science (see references in [32]).

In the most common quantum equivalent of random walks [33–37], a quantum walker walks between sites by changing its position state $|i\rangle \in \{|1\rangle, ..., |N\rangle\}$. To decide the next position, a 'quantum coin' is tossed, which is a qubit register with N degrees of freedom. The coin toss is performed by applying a unitary operator, and the next position is chosen conditioned on the state of the 'coin register', which can of course be in a superposition. The difference to classical walks is twofold: First, the various paths interfere with one another, and a measurement samples from the final distribution. Secondly, the unitarity of quantum walks implies that the evolution is reversible.

There is another, equivalent definition [38] which derives quantum walks via the quantisation of Markov chains [39], and which featured more prominently in quantum machine learning. We follow [29, 40] and their concise presentation to briefly outline the formalism of these *Szegedy quantum walks*. Given a Markov chain M on a graph of N nodes. We represent an edge between the ith and jth node by the quantum state $|i\rangle|j\rangle$. Define the operator

$$\Pi = \sum_{j=1}^{N} |\psi_j\rangle\langle\psi_j|,$$

8.3 Other Approaches to Build Quantum Models

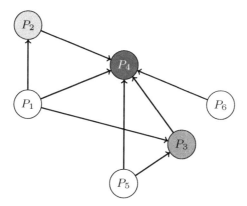

Fig. 8.6 In Google's PageRank algorithm, a graph of links (edges) between homepages (nodes) is created and the Markov process or random walk on the graph will show a high probability for nodes with lots of neighbours linking to them, which is represented by the darkness of the node. Here, P_4 is the most strongly ranked page, followed by P_3 and P_2. Quantum walks show evidence of weighing 'second hubs' such as node P_3 and P_2 stronger than classical PageRank

with[5]

$$|\psi_j\rangle = \sum_{k=1}^{N} \sqrt{M_{k,j}} |j\rangle |k\rangle,$$

as well as the swap operator

$$S = \sum_{j,k=1}^{N} |j,k\rangle \langle k,j|.$$

One step of the quantum walk corresponds to applying the operator $U = S(2\Pi - \mathbb{1})$. While the the term in the brackets is a reflection around the subspace spanned by the $|\psi_j\rangle$, the swap operator can be interpreted as taking the step to the next node.

Random walks are similar to Grover search and the Ising model in the sense that where ever they appear one can make the attempt of replacing them by their quantum version and hope for an improvement. The goal of the exercise is either to obtain a (usually quadratic) speedup in the time of reaching a desired node, or to obtain a different quality of the Markov chain's dynamics. Paparo and Martin-Delgado's [29] approach to Google's PageRank algorithm with tools of Szegedy quantum walks is an example of the latter. PageRank attributes weights to homepages and was the initial core routine of the famous Google search routine, now replaced by more advanced methods. The ranking is based on a rule that gives high weights to sites that are linked by many other pages, while not linking to many other pages themselves.

---

[5]Note that this is something in between an amplitude encoded matrix and a qsample according to the terminology developed in Sect. 3.4.

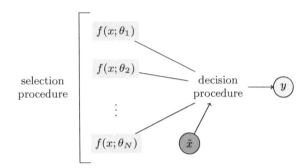

Fig. 8.7 Ensemble methods consult the predictions of varios classifiers to increase the overall predictive power of a model. Adapted from [30]

Homepages can therefore be understood as a graph where directed edges represent links between pages (see Fig. 8.6). A Markov chain on the graph will after many iterations lead to a probability distribution over nodes, where nodes that have many directed edges leading to them have a high probability of being visited, while those with few connections or many outgoing edges get a much smaller 'weight'.

Paparo and Martin-Delgado quantise the Markov chain associated with the PageRank algorithm using a Szegedy quantum walk and analyse the properties of the resulting quantum Markov chain via its spectral properties [29]. Their simulations show that the quantum version weighs pages differently than the classical PageRank algorithm. In a follow-up study Paparo et al. find that the "quantum PageRank algorithm is capable of unveiling the structure of the graph to a finer degree", for example by giving more weight to "secondary hubs" of less importance than the most referenced pages [41]. Another study [42] confirms these findings and notes that the quantum version "resolves ranking degeneracy" of pages with a similar weight. Even though being not directly related to supervised learning, this line of research is a beautiful example for the qualitative difference a quantum extension of a model can give rise to, as we already saw in Sect. 8.1.

Quantum walks have been used in the context of reinforcement learning [41] based on the *projective simulation* model [43]. Projective simulation is a formulation of supervised learning which considers a graph of interconnected 'memory clips', and upon a stimulus (a new input or percept) the agent performs an internal random walk on the graph until one of the memory clips is connected to an action (output). Replacing the random walk by a quantum walk can yield a quadratic speedup in the transversing time of the graph from inputs to outputs. For certain graph types, exponential speedups in hitting times have been established [44], and it remains to be seen if this can find application in machine learning as well.

8.3.2 Superposition and Quantum Ensembles

In the preceding chapter we saw that coherent or 'quantum' training is difficult, especially if the model leads to a nonconvex objective function for the selection of

8.3 Other Approaches to Build Quantum Models

model parameters θ. We will now demonstrate how a change in perspective, namely to consider integration instead of optimisation, could potentially be a solution to this problem. This idea is derived from classical *ensemble methods* and their links to Bayesian learning.

Consider a parametrised deterministic model

$$y = f(x; \theta), \tag{8.17}$$

for a supervised learning task with input x and parameters θ. The model also depends on a set of hyperparameters, for example defining the architecture of a neural network. Now, training the model can lead to very different results depending on the initial choice of parameters. Even if we pick the best set of parameters by consulting the test set, we will most likely neglect other candidates that recover a local structure in input space much better than the best overall candidate. For example, one model might have learned how to deal with outliers very well, but at the expense of being slightly less accurate with the rest of the inputs.

Ensemble methods try to construct better classifiers by considering not only one trained model but an entire committee of them. An ensemble of trained models (also called 'experts', 'ensemble members', 'committee members' or 'classifiers') takes the decision for a new prediction together, thereby combining the advantages of its members (see Fig. 8.7). Considering how familiar the principle of shared decision making is in most societies, it is surprising that this thought only gained widespread attention in machine learning as late as the 1990s. It is by now standard practice to use ensemble methods on top of a core machine learning routine.

There are numerous proposals for ensemble methods, and they can be distinguished by the selection procedure of choosing the members of the ensemble, as well as the procedure of decision making. For example, *mixtures of experts* train a number of classifiers using a specific error function and in the next step train a 'gating network' that weighs the predictions of each expert to produce a final answer [45]. Bagging [46] and Boosting [47] train classifiers on different partitions of the training set and decide by a majority voting rule. Any ensemble method that selects different versions of the model family (8.17) can be formally written as

$$y = g\left(\int_\theta w(\theta) f(x; \theta) \right), \tag{8.18}$$

where g defines the map to the chosen output space (i.e., a sign function). The argument of g is the expectation value of all models over a weight distribution $w(\theta)$. Note that this principle is similar to a perceptron, where the incoming nodes are the model predictions instead of inputs. The weighing distribution $w(\theta)$ defines which models are selected and which are not, and of course it has to ensure that the integral actually exists. In case that a finite number of models are considered, the integral is replaced by a sum.

The expression (8.18) closely resembles the Bayesian learning model in Sect. 2.2.3 if we associate $f(x;\theta)$ with the likelihood $p(x|\theta)$, and $w(\theta)$ with the conditional probability of a parameter given the data $\mathcal{D}$, $p(\theta|\mathcal{D})$. This opens up an important link between ensembles and Bayesian learning (studied in the theory of *Bayesian Model Averaging* [48]): Instead of training a few classifiers and combining their decisions, we can weigh and add *all possible* classifiers with a clever rule.

Thinking about ensembles from a quantum perspective, quantum parallelism immediately comes to mind. We want to review a very general framework of how, given a quantum classifier, one can use superposition to construct an ensemble of classifiers according to (8.18). The details to this idea can be found in [30], together with a more concrete example for application.

We need two basic ingredients. First, we need a *quantum classifier*, by which we mean a quantum routine $\mathcal{A}$ which maps a quantum state that 'encodes a model' $|[\theta]\rangle$ via its parameters θ, as well as a quantum state $|[x]\rangle$ encoding an input x, to an output state $|f(x;\theta)\rangle$,

$$\mathcal{A} \, |[x]\rangle|[\theta]\rangle|0\rangle \to |[x]\rangle|[\theta]\rangle|f(x;\theta)\rangle.$$

To keep things as abstract as possible, we indicate by the brackets "[·]" that the information encoding strategy can be arbitrary, except from the output which is basis encoded. Assuming the task is binary classification, we can understand the a single output qubit to encode the final result. We have discussed in Chap. 6 how to construct such models. Note that $\mathcal{A}$ can be implemented in parallel to a superposition of states in the 'model register'.

$$\mathcal{A} \, |[x]\rangle \sum_\theta \alpha_\theta |[\theta]\rangle \, |0\rangle \to |[x]\rangle \sum_\theta \alpha_\theta |[\theta]\rangle \, |f(x;\theta)\rangle.$$

Second, we need a state preparation routine that defines the weighing distribution and thereby the *quantum ensemble* (given the base quantum classifier). Consider a uniform superposition $\frac{1}{\sqrt{2^n}} \sum_\theta |[\theta]\rangle$ of all 2^n possible models encoded into the n qubits of register $|[\theta]\rangle$. We define $\mathcal{W}$ as a routine that prepares a qsample from a uniform distribution over the classifiers,

$$\mathcal{W} \, \frac{1}{\sqrt{2^n}} \sum_\theta |[\theta]\rangle|0\rangle \to \sum_\theta \sqrt{w_\theta}|\theta\rangle|0\rangle.$$

Note that the distribution is here discrete because we are thinking of qubit systems, but a continuous variable quantum system may be used as well.

Applying the two routines one after the other yields

$$\mathcal{W}\mathcal{A} \, |[x]\rangle \frac{1}{\sqrt{2^n}} \sum_\theta |[\theta]\rangle|0\rangle \to |[x]\rangle \sum_\theta \sqrt{w_\theta}|[\theta]\rangle|f(x;\theta)\rangle.$$

The state on the right hand side is a weighed superposition of all predictions $f(x;\theta)$.

8.3 Other Approaches to Build Quantum Models

The expectation value of the σ_z operator, which, as remarked in Sect. 7.3.2.1, is equivalent to the probability of measuring the output qubit in state 1, reveals the average prediction of the ensemble. In short, model selection is shifted from optimisation to find the best set of parameters θ^*, to preparing a qsample of a distribution that strengthens the influence of good models towards a collective decision. The distribution could for instance favour models with a good performance on the training set, which can be evaluated in quantum parallel for each model [30]. Finally, the distributions can include complex and negative amplitudes, so that the prediction is the result of quantum interference. Once more, the question in which situations this can be useful is still investigated by current research.

8.3.3 QBoost

QBoost was developed by researchers at *Google* and *D-Wave* labs [31, 49, 50] before quantum machine learning became popular, and it is a beautiful illustration of how casting a problem into the format amenable for quantum annealing devices can lead to new methods for machine learning. The basic machine learning model is—again— an ensemble of K binary classifiers $f_k(x)$, $k = 1, \ldots, K$, that are combined by a weighted sum of the form $f(x) = \text{sgn}(\sum_{k=1}^{K} w^T f_k(x))$ with $x \in \mathbb{R}^N$ and $f(x) \in \{-1, 1\}$. We assume that the individual classifiers are trained, and therefore omit their model parameters.

Training the ensemble means to find the weights of the individual classifiers. We choose to minimise a least-squares loss function, and add a L_0 regularisation term to prevent overfitting,

$$\frac{1}{M} \sum_{m=1}^{M} (f(x^m) - y^m)^2 + \lambda ||w||_0,$$

where $|| \cdot ||_0$ counts the number of nonzero parameters (and is hence a sparsity-inducing norm), and λ tunes the strength of regularisation. The problem is already in the form of quadratic unconstrained binary optimisation as can be seen by evaluating the square brackets,

$$\sum_{k,k'=1}^{K} w_k w_{k'} \left(\sum_{m=1}^{M} f_k(x^m) f_{k'}(x^m) \right) + \sum_{k=1}^{K} w_k \left(\lambda - 2 f_k(x^m) y^m \right).$$

The known terms $\sum_m f_k(x^m) f_{k'}(x^m)$ and $\lambda - 2 f_k(x^m) y^m$ serve as the interaction and field strengths of the Ising model, while the weights take the role of the x_i, x_j from Eq. (7.16). This is sometimes called an *inverse Ising model*.

However, the formulation requires that the weights w_i, $i = 1, \ldots, K$ are binary variables. One can replace them by binary strings with a little more modelling effort,

Fig. 8.8 Illustration of QBoost. Several classifiers $h_k(x)$ are combined by a perceptron-like gating network that builds the weighed sum of the indiviual predictions and applies a step activation function on the result to retrieve the combined classification decision y

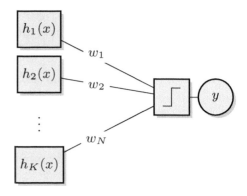

but current quantum hardware limits us to a very low bit depth τ if $w_i \in \{0, 1\}^\tau$. Luckily, estimations show that low-bit representations of parameters can still represent a sufficiently rich space of decision boundaries, or more precisely that "the required bit depth for weight variables only grows logarithmically with the ratio of the number of training examples to the number of features" [49]. This is consistent with some successful application of binary weights to deep neural networks [51] (Fig. 8.8).

Neven and his co-workers claim that compared to AdaBoost, the structure of the QBoost cost function shows various advantages even if executed on a classical computer (and with a predefined set of weak classifiers $f_k(x)$). It produces a strong classifier that in some cases is able to beat classical AdaBoost through a lower generalization error, employing fewer weak classifiers and requiring fewer boosting iterations. These advantages are suspected to be largely due to an intrinsic regularisation by the low-bit depth binary representation of the weights and the good performance of the least-square objective function (as opposed to AdaBoost's stepwise weighing of the training vectors).

Denchev et al. [52] add to this work by introducing a special loss function (*q-loss*) for perceptron models of the form used for QBoost. This loss function can be generically formulated as a quadratic unconstrained optimisation problem, and promises better robustness against large outliers than the square loss. Less optimistic results are obtained by Dulny and Kim [53], who find that in a realistic setting using a *Kaggle*[6] competition dataset, classical state-of-the-art methods perform better than QBoost. Even though research in cost functions tailor-made to quantum annealers has since been less in the focus of quantum machine learning, QBoost illustrates once more the exploratory approach, but this time with a cost function adapted to a quantum device, rather than the model itself.

---

[6]See https://www.kaggle.com/.

References

1. Amit, D.J., Gutfreund, H., Sompolinsky, H.: Spin-glass models of neural networks. Phys. Rev. A **32**(2), 1007–1018 (1985)
2. Amin, M.H., Andriyash, E. Rolfe, J., Kulchytskyy, B., Melko, R.: Quantum Boltzmann machine. Phys. Rev. X **8**, 021050 (2018)
3. Kieferova, M., Wiebe, N.: Tomography and generative data modeling via quantum Boltzmann training. Phys. Rev. A **96**, 062327 (2017)
4. Inoue, J.: Application of the quantum spin glass theory to image restoration. Phys. Rev. E **63**(4), 046114 (2001)
5. Shcherbina, M., Tirozzi, B.: Quantum Hopfield model. arXiv:1201.5024v1 (2012)
6. Nishimori, H., Nonomura, Y.: Quantum effects in neural networks. J. Phys. Soc. Jpn. **65**(12), 3780–3796 (1996)
7. Inoue, J-I.: Pattern-recalling processes in quantum Hopfield networks far from saturation. In *Journal of Physics: Conference Series*, vol. 297, p. 012012. IOP Publishing (2011)
8. Rotondo, P., Marcuzzi, M., Garrahan, J.P., Lesanovsky, I., Müller, M.: Open quantum generalisation of Hopfield neural networks. J. Phys. A: Math. Theor. **51**(11), 115301 (2018)
9. Breuer, H-P., Petruccione, F.: The Theory of Open Quantum Systems. Oxford University Press (2002)
10. Hopfield, J.J.: Neural networks and physical systems with emergent collective computational abilities. Proc. Natl. Acad. Sci. **79**(8), 2554–2558 (1982)
11. Monras, A., Beige, A., Wiesner, K.: Hidden quantum Markov models and non-adaptive read-out of many-body states. Appl. Math. Comput. Sci. **3** (2011)
12. Monras, A., Winter, A.: Quantum learning of classical stochastic processes: the completely positive realization problem. J. Math. Phys. **57**(1), 015219 (2016)
13. Barry, J., Barry, D.T., Aaronson, S.: Quantum partially observable Markov decision processes. Phys. Rev. A **90**, 032311 (2014)
14. Cholewa, M., Gawron, P., Głomb, P., Kurzyk, D.: Quantum hidden Markov models based on transition operation matrices. Quantum Inf. Process. **16**, 101 (2015)
15. MS Leifer and David Poulin: Quantum graphical models and belief propagation. Ann. Phys. **323**(8), 1899–1946 (2008)
16. Pearl, J.: Causality. Cambridge University Press (2009)
17. Ried, K., Agnew, M., Vermeyden, L., Janzing, D., Spekkens, R.W., Resch, K.J.: A quantum advantage for inferring causal structure. Nat. Phys. **11**(5), 414–420 (2015)
18. Brukner, Č.: Quantum causality. Nat. Phys. **10**(4) (2014)
19. Costa, F., Shrapnel, S.: Quantum causal modelling. New J. Phys. **18**, 063032 (2016)
20. Saxe, A.M., McClelland, J.L., Ganguli, S.: Exact solutions to the nonlinear dynamics of learning in deep linear neural networks. arXiv preprint arXiv:1312.6120 (2013)
21. Stoudenmire, E., Schwab, D.J.: Supervised learning with tensor networks. In: Advances in Neural Information Processing Systems, pp. 4799–4807 (2016)
22. Schuld, M., Bocharov, A., Wiebe, N., Svore, K.: A circuit-centric variational quantum classifier. arXiv preprint arXiv:1804.00633 (2018)
23. Benedetti, M., Realpe-Gómez, J., Perdomo-Ortiz, A.: Quantum-assisted Helmholtz machines: a quantum-classical deep learning framework for industrial datasets in near-term devices. Quantum Sci. Technol. **3**, 034007 (2018)
24. Arjovsky, M., Shah, A., Bengio, Y.: Unitary evolution recurrent neural networks. J. Mach. Learn. Res. **48** (2016)
25. Jing, L., Shen, Y., Dubček, T., Peurifoy, J., Skirlo, S., LeCun, Y., Tegmark, M., Soljačić, M.: Tunable efficient unitary neural networks (EUNN) and their application to RNN. In: International Conference on Machine Learning, pp. 1733–1741 (2017)
26. Wisdom, S., Powers, T., Hershey, J., Le Roux, J., Atlas, L.: Full-capacity unitary recurrent neural networks. In: Advances in Neural Information Processing Systems, pp. 4880–4888 (2016)

27. Reck, M., Zeilinger, A., Bernstein, H.J., Bertani, P.: Experimental realization of any discrete unitary operator. Phys. Rev. Lett. **73**(1), 58 (1994)
28. Clements, W.R., Humphreys, P.C., Metcalf, B.J., Kolthammer, W.S., Walmsley, I.A.: Optimal design for universal multiport interferometers. Optica **3**(12), 1460–1465 (2016)
29. Paparo, G.D., Martin-Delgado, M.A.: Google in a quantum network. Sci. Rep **2** (2012)
30. Schuld, M., Petruccione, F.: Quantum ensembles of quantum classifiers. Sci. Rep. **8**(1), 2772 (2018)
31. Neven, H., Denchev, V.S., Rose, G., Macready, W.G.: Training a large scale classifier with the quantum adiabatic algorithm. arXiv preprint arXiv:0912.0779 (2009)
32. Moore, C., Russell, A.: Quantum walks on the hypercube. In: Randomization and Approximation Techniques in Computer Science, pp. 164–178. Springer (2002)
33. Kendon, V.: Decoherence in quantum walks. A review. Math. Struct. Comput. Sci. **17**, 1169–1220 (2007). (11)
34. Kempe, J.: Quantum random walks: an introductory overview. Contemp. Phys. **44**(4), 307–327 (2003)
35. Venegas-Andraca, S.E.: Quantum walks: A comprehensive review. Quantum Inf. Process. **11**(5), 1015–1106 (2012)
36. Wang, J., Manouchehri, K.: Physical Implementation of Quantum Walks. Springer (2013)
37. Travaglione, B.C., Milburn, G.J.: Implementing the quantum random walk. Phys. Rev. A **65**, 032310 (2002)
38. Wong, T.G.: Equivalence of Szegedys and coined quantum walks. Quantum Inf. Process. **16**(9), 215 (2017)
39. Szegedy, M.: Quantum speed-up of Markov chain based algorithms. In: Proceedings of 45th Annual IEEE Symposium on Foundations of Computer Science, 2004, pp. 32–41. IEEE (2004)
40. Loke, T., Wang, J.B.: Efficient quantum circuits for Szegedy quantum walks. Ann. Phys. **382**, 64–84 (2017)
41. Paparo, G.D., Dunjko, V., Makmal, A., Martin-Delgado, M.A., Briegel, H.J.: Quantum speedup for active learning agents. Phys. Rev. X **4**(3), 031002 (2014)
42. Loke, T., Tang, J.W., Rodriguez, J., Small, M., Wang, J.B.: Comparing classical and quantum pageranks. Quantum Inf. Process. **16**(1), 25 (2017)
43. Briegel, H.J., De las Cuevas, G.: Projective simulation for artificial intelligence. Sci. Rep. **2**, 1–16 (2012)
44. Kempe, J.: Quantum random walks hit exponentially faster. Probab. Theory Relat. Fields **133**, 215–235 (2005)
45. Jacobs, R.A., Jordan, M.I., Nowlan, S.J., Hinton, G.E.: Adaptive mixtures of local experts. Neural Comput. **3**(1), 79–87 (1991)
46. Breiman, L.: Random forests. Mach. Learn. **45**(1), 5–32 (2001)
47. Schapire, R.E.: The strength of weak learnability. Mach. Learn. **5**(2), 197–227 (1990)
48. Minka, T.P.: Bayesian model averaging is not model combination. http://www.stat.cmu.edu/minka/papers/bma.html (2000). Comment available electronically
49. Neven, H., Denchev, V.S., Rose, G., Macready, W.G.: Qboost: large scale classifier training with adiabatic quantum optimization. In: Asian Conference on Machine Learning (ACML), pp. 333–348 (2012)
50. Neven, H., Denchev, V.S., Rose, G.. Macready, W.G.: Training a binary classifier with the quantum adiabatic algorithm. arXiv preprint arXiv:0811.0416 (2008)
51. Courbariaux, M., Hubara, I., Soudry, D., El-Yaniv, R., Bengio, Y.: Binarized neural networks: training deep neural networks with weights and activations constrained to $+1$ or -1. arXiv preprint arXiv:1602.02830 (2016)
52. Denchev, V.S., Ding, N., Vishwanathan, S.V.N., Neven, H.: Robust classification with adiabatic quantum optimization. In: Proceedings of the 29th International Conference on Machine Learning (ICML-12), pp. 863–870 (2012)
53. Dulny III, J., Kim, M.: Developing quantum annealer driven data discovery. arXiv preprint arXiv:1603.07980 (2016)

Chapter 9
Prospects for Near-Term Quantum Machine Learning

In order to run the quantum machine learning algorithms presented in this book we often assumed to have a universal, large-scale, error-corrected quantum computer available. *Universal* means that the computer can implement any unitary operation for the quantum system it is based on, and therefore any quantum algorithm we can think of. *Large-scale* refers to the fact that we have a reasonably high number of qubits (or alternative elementary quantum systems) at hand. *Error-corrected* means that the outcomes of the algorithm are exactly described by the theoretical equations of quantum theory, in other words, the computer does not make any errors in the computation besides those stemming from numerical instability.

But as we discussed in the introduction, quantum computing is an emerging technology, and the first generation of quantum computers, which have been called *noisy intermediate-term devices* [1], does not fulfil these conditions. Firstly, intermediate-term devices are not necessarily universal. Sometimes they do not aim at universality, for example in the case of quantum annealers that are specialised to solve a specific problem. But even quantum computers that in principle are designed as universal devices may not have fully connected architectures, or only implement a subset of gates reliably. Secondly, we usually have a rather small number of qubits available, which is why intermediate-term devices are also called *small-scale devices*. Thirdly, early-generation quantum computers are noisy. Not only do the gates have a limited fidelity or precision, they also do not have mechanisms to detect and correct errors that decohere the qubits and disturb the calculation. Hence, we can only apply a small number of gates before the result of the computation is too noisy to be useful.

While for non-universality, every technology has different challenges to which solutions will be increasingly available, the small scale and noisiness of most early technologies impose general limitations on the 'size' of the quantum algorithms they can implement: We only have a small number of qubits (i.e., a low *circuit width*) and have to keep the number of gates (the *circuit depth*) likewise small enough to contain error propagation. As a rule of thumb, the quantum community currently speaks of intermediate-term algorithms if we use of the order of 100 qubits and of the order of 1,000 gates. An important question is therefore what approaches to quantum machine learning are actually feasible in noisy intermediate-term devices. In other words, *are there quantum machine learning algorithms for real-life problems that use only about* 100 *qubits, have a circuit depth of about* 1,000 *gates and are robust to some reasonable levels of noise?*

This last question is the subject of active research, and given its difficulty it might remain so for some time. In this final chapter we want to discuss some aspects of the question in the context of what has been presented in this book, both as a summary and as an outlook.

9.1 Small Versus Big Data

We have attributed a lot of space in this book to questions of data encoding, or how to feed the full information we have for a certain problem, most notably the dataset, into the quantum device. In many situations, it is the bottleneck of an algorithm. This was driven to the extreme in our introductory 'Titanic' classification example in Chap. 1, where besides data encoding, the algorithm only took one Hadamard gate and two measurements to produce a prediction. It is not surprising that in general, data encoding costs linear time in the data size, because every feature has to be addressed. Even fancy technologies like quantum Random Access Memories would still require the data to be written into the memory. The linear costs pose a dilemma for intermediate-term applications: A dataset of only 100 inputs, each of dimension 10, would already "use up" of the order of 1,000 gates. In addition to that, data representation in basis encoding has severe spatial limitations for the dimension of the data; if every feature is encoded in a τ-bit binary sequence, we can only encode $100/\tau$ features into 100 qubits.

The bottleneck of data encoding means that, besides some few special cases, quantum machine learning will most likely not offer intermediate-term solutions for big data processing. Of course, some algorithms—most notably hybrid training schemes—only process a subset of samples from the dataset at a time, thereby lifting the restrictions on the number M of data samples. For example, in Sect. 7.3.3.2 we looked at hybrid gradient descent training where the quantum device was used to estimate numerical or analytical gradients. Combining this with single-batch stochastic gradient descent, the quantum device only has to process one data point at a time, and the quantum subroutine is independent of the number of samples in the training set. Another example are hybrid kernel methods in which the quantum computer computes the 'quantum kernel' as a measure of distance between two data points (see Sect. 6.2.3). However, if the dimension N of the data samples is large, we still have to face the problem of extensive time and spatial resources needed to feed the samples into a quantum computer.

In order to deal with high-dimensional data we have two basic options. First, we can introduce structure that allows faster state preparation at the cost of accuracy. We call this largely unexplored strategy *approximate data encoding*. We have

9.1 Small Versus Big Data

indirectly touched upon this topic in the context of probabilistic models in Sect. 6.3. The task of preparing a qsample, which in principle is nothing else than arbitrary state preparation, was addressed by preparing a mean field approximation of the distribution instead, which was linear—rather than exponential—in the number of qubits. Another topic in which approximate data encoding played a role was in the quantum basic linear algebra routines in combination with density matrix exponentiation, where once the data was given as a density matrix, we could read out eigenvalues of its low-rank approximation qubit-efficiently (Sect. 5.4.3). Whether the 'sweet spot' in the trade-off between state preparation time/circuit length and accuracy of the desired state is useful for pattern recognition, and whether it suits the requirements of intermediate-term devices are both open questions.

Second, we have touched upon the suggestion of considering quantum-assisted model architectures, in which only a small part of the machine learning model is computed by the quantum device. It was suggested in the literature that the deep and compact layers in a deep learning model could play this role. Again, we introduce structure to reduce complexity. While approximate data encoding would try to be as faithful as possible to the original dataset, the idea of deep quantum layers is to use a classical part of the model to select a few powerful features that are fed into the quantum device. Hopes are that the quantum device can use the reduced features more easily, or explore a different class of feature reduction strategies. Again, a lot more research is required to determine the potential of quantum-assisted model architectures, and an important question is how to incorporate them into training algorithms such as gradient descent.

There is one setting in which high-dimensional (even exponentially large) inputs could be classified by quantum machine learning algorithms, and this is the idea of using 'quantum data'. This has been discussed in Sect. 1.1.3 of the introduction, but has not been the topic of later chapters because of a lack of research results at the time of writing. The idea was to use quantum machine learning algorithms to classify quantum states that are produced by a quantum simulator. Although machine learning for problems in quantum physics has shown fruitful applications, the combination of solving a quantum problem by a quantum machine learning algorithm is still a mostly untouched research agenda.

Although the promise of quantum methods for big data sounds appealing and the difficulties sobering, one should possibly not be too concerned. Besides the worldwide excitement about big data, problems where data collection is expensive or where data is naturally limited are plentiful, and there are many settings where predictors have to be built for extremely small datasets. An example is data generated by biological experiments, or specific data from the domain of reasoning. If quantum computing can show a qualitative advantage, for example in terms of what we discussed as 'model complexity' in Sect. 4.3, there will indeed be worthwhile applications in the area of *small data*.

9.2 Hybrid Versus Fully Coherent Approaches

In the course of this book we have seen quantum algorithms that range from the fully coherent training and classification strategies based on quantum basic linear algebra (blas) routines, to hybrid schemes where only a small part of the model is computed by a quantum device. Unsurprisingly, the latter is much more suitable for intermediate-term technologies. Hybrid quantum algorithms have the advantage of using quantum devices only for relatively short routines intermitted by classical computations. Another great advantage of hybrid techniques is that the parameters are available as classical information, which means they can be easily stored and used to predict multiple inputs. In contrast, many *quantum blas*-based algorithms produce a quantum state that encodes the trained parameters, and classification consumes this quantum state so that training has to be fully repeated for every prediction. Even if 'quantum memories' to store the quantum parameter state were developed, the no-cloning theorem in quantum physics prohibits the replication of the state.

Amongst hybrid algorithms, variational circuits are particularly promising (see Sects. 7.3 and 8.2). Here, an ansatz of a parametrised circuit is chosen and the parameters are fitted to optimise a certain objective. For example, one can define an input-output relation with regards to the quantum circuit and train it to generalise the input-output relations from the training data. While the idea is rather simple, the implementation opens a Pandora's box of questions, some of which we have previously mentioned. What is a good ansatz for such a circuit for a given problem? We would like the ansatz to be slim in the circuit depth, width and the number of parameters used, but also as expressive as possible. This is nothing other than one of the fundamental questions of machine learning, namely to find simple but powerful models. Every hyperparameter allows more flexibility, but also complicates model selection for practical applications. Training is another issue. How can we train a model that is not given as a mathematical equation, but as a physical quantum algorithm? Can we do better than numerical optimisation, for example by using techniques such as the classical linear combination of unitaries presented in Sect. 7.3.3.2? The parametrisation of the quantum circuit can play a significant role in the difficulty to train a model, as it defines the landscape of the objective function. What are good parametrisation strategies? Do we have to extend the tricks of classical iterative training, such as momentum and adaptive learning rates, by those fitted to quantum machine learning? In short, variational algorithms for machine learning open a rich research area with the potential of developing an entirely new subfield of classical machine learning. In the larger picture, quantum machine learning could even pioneer a more general class of approaches in which 'black-box' analogue physical devices are used as machine learning models and trained by classical computers.

9.3 Qualitative Versus Quantitative Advantages

Quantum computing is a discipline with strong roots in theory. While the mathematical foundations were basically laid in the 1930 s, we are still exploring the wealth of their practical applications today. This is also true for quantum computing, where algorithmic design had to revert to theoretical proofs to advertise the power of quantum algorithms, since without hardware numerical arguments were often out of reach. Quantum machine learning seems to follow in these footsteps, and a large share of the early literature tries to find speedups that quantum computing could contribute to machine learning (see for example the Section on quantum blas for optimisation 7.1). In other words, the role of quantum computing is to offer a quantitative advantage defined in terms of asymptotic computational complexity as discussed in Sect. 4.1. The methods of choice are likewise purely theoretical, involving proofs of upper and lower bounds for runtime guarantees, and the 'holy grail' is to show exponential speedups for quantum machine learning. Such speedups are typically proven by imposing very specific constraints on the data, and little is known on applications that fulfil these constraints, or whether the resulting algorithm is useful in practice. Needless to say, judging from the journey quantum computing has taken so far, it is highly unlikely that it provides solutions to NP-complete problems and can solve all issues in machine learning.

On the other hand, machine learning methods tend to be based on problems that are in general intractable, a fact that does not stop the same methods from being very successful for specific cases. Consider for example non-convex optimisation in high-dimensional spaces, which is generally intractable, but still solved to satisfaction by the machine learning algorithms in use. Methods employed in machine learning are largely of practical nature, and breakthroughs often the result of huge numerical experiments.[1] Computational complexity is one figure of merit among others such as the generalisation power of a model, its ease of use, mathematical and algorithmic simplicity and wide applicability, whether it allows the user to interpret the results, and if its power and limits are theoretically understood.

Quantum machine learning may therefore have to rethink its roots in quantum computing and develop into a truly interdisciplinary field. We have motivated this in our distinction between explorative versus translational approaches in Sect. 1.1.4. Especially in the first generation of papers on quantum machine learning, it was popular to choose the prototype of a classical machine learning model and try to reproduce it with a quantum algorithm that promises some asymptotic speedup, in other words to translate the model into the language of quantum computing. The explorative approach that we highlighted in Chap. 8 is interested in creating new models, new dynamics and new training strategies that extend the canon of machine learning. Instead of theoretical analysis, these contributions benchmark their models against standard algorithms in numeric experiments. Their paths may diverge

---

[1] A prominent joke attributes the success in machine learning to "graduate descent", describing a worldwide army of graduate students that manually search through the infinite space of possible models.

from what is popular in classical machine learning at the moment. For example, it was demonstrated in Sects. 6.2 and 6.3 where it was argued that the ideas of kernel methods and probabilistic models are much closer to quantum theory than the principle of huge feed-forward neural networks. The overall goal of the explorative approach is to identify a new quality that quantum theory can contribute to pattern recognition.

Of course, whether to focus on qualitative versus quantitative advantages is not necessarily an 'either-or' question. Constraints stemming from computational complexity are a major factor in shaping successful algorithms, and speedups, even quadratic ones, can prove hugely useful. It may not be possible to simulate a quantum model classically, which is in itself an exponential speedup. And the theoretical rigour of quantum researchers can prove a useful resource to develop the theory side of machine learning. But with intermediate-term quantum technologies paving the way to turn quantum computing into a numerical playground, quantum machine learning has good reasons to be at the forefront of these efforts.

9.4 What Machine Learning Can Do for Quantum Computing

This book tried to show ways of how quantum computing can help with machine learning, in particular with supervised learning. We want to conclude by turning the question around and ask what machine learning has to offer for quantum computing. One answer is trivial. Many researchers hope that machine learning can contribute a 'killer app' that makes quantum computing commercially viable. By connecting the emerging technology of quantum computing with a multi-billion dollar market, investments are much more prone to flow and help building large-scale quantum computers.

But another answer has been touched upon in the previous three sections of this conclusion. Quantum computing on the one hand is predominantly focused on theoretical quantum speedups, rather than hybrid algorithms that offer a new quality and which draw their motivation from successful numerical implementations rather than proofs. Machine learning on the other hand has to deal with uncertainty, noise, hard optimisation tasks and the ill-posed mathematical problem of generalisation. Maybe machine learning can inspire quantum computing in the era of intermediate-scale devices to add a number of methods to the toolbox, methods that are less rigorous and more practical, less quantitative and more qualitative.

Finally, machine learning is not only a field in computer science, but also based on philosophical questions of what it means to learn. Quantum machine learning carries the concept of learning into quantum information processing, and opens up a lot of questions on an abstract level—questions that aim at increasing our knowledge rather than finding commercial applications.

Reference

1. Preskill, J.: Quantum computing in the NISQ era and beyond. arXiv preprint arXiv:1801.00862 (2018)

Index

A
Accuracy, 32
Activation function, 49, 50, 53
　nonlinear, 50
　rectified linear units, 53, 185
　sigmoid, 53, 185
　step function, 185
　tanh, 53
AdaBoost, 3, 270
Adiabatic quantum computing, 107
Adiabatic state generation, 158
Amplitude, 79
Amplitude amplification, 114, 223
Amplitude-efficient, 128, 140, 154, 160
Amplitude encoding, 121, 148
Amplitude vector, 11, 80
Analytical gradient-based methods, 233
Ancilla register, 142
Approach
　exploratory, 7, 127, 136, 247
　translational, 7
Artificial intelligence, 21
Associative memory, 55, 218, 249, 253

B
Backpropagation, 49, 52
Backpropagation algorithm, 39
Backpropagation through time, 55
Bagging, 267
Bags of words, 26
Basis change, 85
Basis encoding, 141
Bayes formula, 34, 35, 65
Bayesian learning, 267, 268
Bayesian model averaging, 268
Bayesian network, 60, 62, 205, 239
Belief network, 60
Bell state, 100
Bernoulli distribution, 147, 208
Bernstein-Vazirani algorithm, 133, 135
Bias gate, 182
Big data, 274
Binary classification, 146
Black-body radiation, 77
Bloch sphere, 94
Board games, 25
Boltzmann distribution, 57
Boltzmann machine, 39, 45, 56, 205, 207, 239, 240, 248, 249
　quantum, 251
　restricted, 57, 249
Boltzmann qsamples, 231, 240
Boosting, 267
Born approximation, 90
Born-Oppenheimer approximation, 228
Born rule, 88
Bra, 85
Branch selection, 121, 156, 162, 212

C
Cascade, 151
Chimera graph, 241
Circuit depth, 273
Circuit parameter, 225
Circuit width, 273
Classical linear combination of unitaries, 236
Classical statistics, 83
Classification, 22, 219
　binary, 67

Classification task, 26
Classifier
    distance based, 196
    quantum squared-distance, 13
    squared-distance, 8, 9
Clifford group, 16
Clipping, 55
Cluster state, 108
Coherent state, 195
Completeness relation, 85
Complexity
    asymptotic computational, 127, 128
    computational, 127
    model, 127
    quantum query, 133
    sample, 127
Composite system, 89
Computational basis, 93, 96
Concentric circles, 38
Concept class, 131
Confidence interval, 147
Contrastive divergence, 59
Copies of quantum states, 194
Cost, 32
Cost function, 32, 40, 42
CQ case, 139
Cross entropy, 41

D
Data-efficient, 129
Data encoding, 139
Data fitting, 45
Data matrix, 47, 214
Data pre-processing, 13, 26
Data superposition, 141, 142
Decision, 23
Deep neural networks, 54
Density matrix, 83, 85, 89
Density matrix exponentiation, 167, 212, 216
Density operator, 85
Dephased quantum channel, 135
Derivative-free method, 233
Design matrix, 47
Deutsch algorithm, 103
Deutsch-Josza algorithm, 104, 133
Diagnosis, 23
Dimensionality reduction, 219
Dirac notation, 75, 84, 85
Discrete spectrum, 87
Distance-based classifier, 196
Distribution
    discriminative model, 29
    generative model, 29
DNA analysis, 63
Dual form of a model, 37
Dürr-Høyer algorithm, 219
D-Wave, 4, 239, 240, 263

E
Encoding
    amplitude, 13, 108, 110, 158, 179, 184, 194, 199
    angle, 186, 194
    approximate data, 275
    basis, 108, 109, 182, 184, 193, 196
    data, 13, 15
    dynamic, 108, 113, 238
    Hamiltonian, 113, 163, 238
    one-hot, 26
    qsample, 108, 112, 159, 204
Ensemble method, 267
Ensemble of classifier, 263
Entanglement, 89
Event, 78
Expectation value, 78, 230

F
False negatives, 32
False positives, 32
Feature, 8
Feature engineering, 26
Feature map, 37, 46, 67, 190
    canonical, 190
    nonlinear, 48
Feature scaling, 26
Feature selection, 26
Feature space, 37, 68, 189
    infinite dimensional, 38
Feature vector, 26
Fixed point arithmetic, 184
Fock states, 195
Forecast, 22
Fractional bit, 184
Function
    balanced, 103
    constant, 103

G
Gate
    CNOT, 99
    Hadamard, 98
    multi-qubit, 99

NOT, 97
parametrised single qubit, 257, 258
S, 98
SWAP, 99
Toffoli, 99, 145
X, 97, 98
Y, 98
Z, 98
Gating network, 267
Gaussian distribution, 35
Gaussian noise, 41
Gaussian process, 7, 45, 68
   kernel, 69
Generalisation, 32
Generalised coherent states, 195
Gibbs sampling, 59
Gibbs state, 247
Global property, 103
Gorini-Kossakowski-Sudarshan-Lindblad equation, 90, 254
Gradient-based method, 234
Gradient descent, 42, 51
Gram matrix, 36, 69, 163, 189
   density, 201
Graphical model, 45, 60, 256
Grover algorithm, 114
Grover operator, 115
Grover-Rudolph scheme, 160
Grover search, 116, 128, 159, 211, 219, 223, 265
   common, 221
   Ventura-Martinez version, 222
Guess, 23

H
Hadamard gate, 7
Hadamard transformation, 9, 11, 14, 15
Hamiltonian
   sparse, 166
   strictly local, 166
Hamiltonian evolution, 212
Hamiltonian simulation, 120, 164, 218
   qubit-efficient, 227
Handwriting recognition, 63
Harrow-Hassidim-Lloyd (HHL) algorithm, 211, 213
Hebbian learning rule, 217
Hermitian operator, 78, 80, 86
Hidden Markov model, 62
   quantum, 255
Hidden units, 55, 56
Hilbert space, 78, 84, 94

reproducing kernel, 190, 191
History of quantum mechanics, 76
Hopfield model, 217
   generalised, 249
   quantum, 253
Hopfield network, 55, 136, 217, 248
   asymmetric, 255
Hopfield neural network, 56, 240
Hybrid architecture, 260
Hybrid training, 225
Hyperparameters, 28
Hypothesis guessing, 24

I
Image recognition, 23
Index register, 96, 118
Inductive bias, 31
Inference, 17, 22, 31, 62, 173
Inner product, 174
Input, 23
Input-to-hidden layer, 54
Integer bit, 184
Integrate-and-fire principle, 48
Integration, 35
Interference
   constructive, 106
   destructive, 106
Interference circuit, 177, 199
Intermediate-term device, 3, 273
Intermediate-term quantum device, 242
Ising/Heisenberg model, 228
Ising model, 247, 265
   quantum, 249
Ising-type energy function, 55, 57
Iterative method, 233

K
Kaggle, 8, 270
Kernel, 36, 67, 68
   cosine, 194
   Gaussian, 38, 195
   homogenous polynomial, 194
   linear, 194
   positive semi-definite, 189
   quantum, 193
   radial basis function, 195
Kernel density estimation, 64
Kernelised binary classifier, 199
   quantum, 196
Kernel method, 45, 64, 184, 189
Kernel trick, 37, 189

Ket, 85
Killer app, 278
K-nearest neighbour, 65
Kraus formalism, 159
Kraus operator, 255

L
Lazy learners, 31
Learning
    Bayesian, 34, 35, 41, 65, 69
    impatient, 133
    Probably Approximately Correct (PAC), 132
    quantum reinforcement, 6
    reinforcement, 25
    supervised, 6, 22, 25
    unsupervised, 6, 25
Learning rate, 43
Least squares estimation, 47
Likelihood, 34
Lindblad operator, 91, 254
Linear model, 45
    kernelised, 192
Linear regression, 33, 46
Loading register, 142
Log-likelihood, 41
Look-up table, 150
Loss, 32, 33, 40
    12, 40
    hinge, 40
    logistic, 40
    squared, 40
Lüders postulate, 88

M
Machine learning, 3
    quantum, 1, 4
    quantum-assisted, 1
    supervised, 8
Margin, 67
Marginalisation, 28, 82, 160
Markov chain, 263
Markov Chain Monte Carlo method, 59
Markov condition, 61
Markov decision process, 256
Markov equivalence, 62
Markov logic network, 209
Markov process, 63, 81
Matrix completion, 219
Matrix inversion, 119, 121, 211, 217
Matrix multiplication algorithm, 119
MaxCut problem, 233

Maximum a posteriori estimate, 30
Maximum a posteriori estimation, 35
Maximum likelihood estimation, 35, 41, 147
Measurement, 17, 80
    computational basis, 97
Membership query, 131
Memory branch, 142
Message passing algorithm, 62
Mixed state, 83, 85, 97
Mixture of experts, 267
MNIST dataset, 23
Model, 28
    deterministic, 28
    probabilistic, 28
Model complexity, 135
Model family, 28
Model function, 28
Monte Carlo algorithm, 249
Monte Carlo sampling, 62
Multi-controlled rotation, 150

N
Nearest neighbour, 8
Nearest neighbour method, 7
Near-term quantum device, 226
Nelder-Mead method, 234
Neural network, 3, 7, 39, 45, 48
    deep, 21, 49, 60, 270
    feed-forward, 49, 51, 55, 184
    Hopfield, 56
    linear layer, 174
    probabilistic recurrent, 56
    quantum, 186
    recurrent, 55, 261
    single-hidden-layer, 50
Neuron, 54
Newtonian mechanics, 76
Noise, 135
Nonlinear regression, 45, 48
Non-locality, 91
Non-parametric model, 192
NP-hard, 159, 186
NP-hard problem, 62, 91, 115
Nuclear magnetic resonance, 242
Numerical gradient-based method, 233

O
Observable, 80, 86
One-hot encoding, 41
One-shot method, 233
One-versus-all scheme, 26

Open quantum system, 90, 253, 255
Optimisation, 22, 35, 39, 42, 185
  convex, 68
  convex quadratic, 42
  least-squares, 40, 42
  non-convex, 42
  quadratic, 67
  quadratic unconstrained binary, 239
Output, 23
Overfitting, 32, 33, 44, 136

P

PageRank algorithm, 265
  quantum, 266
Partial trace, 83, 89
Parzen window estimator, 65
Pattern classification, 64
Pattern matching, 55
Pattern recognition
  supervised, 25, 56
Pauli matrix, 98, 230
Pauli operators, 250
Perceptron, 49, 50, 66, 219, 223, 248, 267, 270
Phase estimation, 169
Planck's constant, 87
Positive Operator-Valued Measure (POVM), 88
Posterior, 34
Postselection, 121, 158, 159, 162, 176, 198
Prediction, 22, 23
Prediction task, 25
Preprocessing, 17
Primal form of a model, 37
Prior, 34
Probabilistic graphical model, 204
Probabilistic model, 28
Probability vector, 11
Processing branch, 142
Product state, 82
Programmes, 42
Projective measurement, 88
Projective simulation model, 266
Projector, 79
Pure state, 83, 87

Q

QBoost, 269, 270
QBoost algorithm, 4
Q-loss, 270
QQ case, 136
Qsample, 112, 134, 135

Quadratic Unconstrained Binary Optimisation (QUBO), 108
Quantum adiabatic algorithm, 229
Quantum advantage, 128
Quantum algorithm, 2, 17, 91
Quantum annealing, 107, 238, 239
Quantum approximate optimisation algorithm, 226, 231
Quantum approximate thermalisation, 232
Quantum-assisted architecture, 260
Quantum associative memory, 4, 242
Quantum blas, 212, 218, 276
Quantum circuits, 93
Quantum classifier, 260, 268
Quantum coherence, 2
Quantum coins, 13
Quantum complexity theory, 128
Quantum computer, 1, 2, 6
  fault-tolerant, 224
Quantum computing, 1
  continuous-variable, 108
  measurement-based, 108
Quantum counting, 116
Quantum data, 6, 275
Quantum enhancement, 128
Quantum ensemble, 268
Quantum example oracle, 134
Quantum extensions, 18
Quantum Fourier transform, 117, 169
  inverse, 118, 168
Quantum gate, 2, 93, 257
Quantum information, 75
Quantum information processing, 91, 128
Quantum interference, 7
Quantum kernel, 193
Quantum learning theory, 131
Quantum linear systems of equations routine, 119
Quantum logic gate, 97
Quantum machine learning, 17
Quantum Markov chain, 248
Quantum Markov process, 256
Quantum master equation, 90, 253
Quantum matrix inversion, 212
Quantum mechanics, 76
Quantum model, 136
Quantum oracle, 133
Quantum page rank, 266
Quantum parallelism, 101
  for data processing, 183
Quantum phase estimation, 117, 120, 182, 212, 214
Quantum programming language, 2

Quantum random access memory, 144, 156, 274
Quantum randomised clamping, 233
Quantum simulation, 6
Quantum speedup, 128
  common, 129
  limited, 129
  potential, 129
  provable, 129
  strong, 129
Quantum state, 80, 83
Quantum statistics, 83
Quantum supremacy, 128
Quantum system, 76
  composite, 82
  evolutions of, 81
  time evolution, 87
Quantum theory, 2, 75
Quantum tomography, 146
Quantum tunneling, 107
Quantum Turing machine, 91
Quantum walk, 166, 264
  Szegedy, 264, 265
  unitary, 264
Quantum walk model, 263
Qubit, 2, 3, 76, 92
Qubit-efficient, 128, 140, 149, 154, 156, 160, 169, 178
Qubit register, 96
Quine-McCluskey method, 185
Quron, 187

R
Random variable, 78
Random walk, 264
Regression, 27
  linear, 45
  nonlinear, 45
Regularisation, 32, 42, 136
Regulariser, 33, 40
Rejection sampling, 162
Repeat-until-success circuit, 187
Representer theorem, 36, 189, 191, 192, 196

S
Sample complexity, 131
Schrödinger equation, 87, 253
Self-adjoint matrix, 80
Self-adjoint operator, 80, 86
Separability, 85
Separable state, 89
Sequence-to-sequence modeling, 55

Set
  training, 33
  validation, 33
Singular value decomposition, 47, 48
Singular vectors, 48
Small data, 275
Small scale device, 3, 273
Softmax, 27
Spectral representation, 86
Spectral theorem, 80
Spooky action at a distance, 91
State estimation, 63
State preparation, 139
  amplitude encoding, 212
  linear time, 150
  oracle based, 156
  parallelism-based, 154
Statistical learning theory, 127
Stochastic gradient descent, 42, 44
Stochastic matrix, 11, 81
Stochastic process, 63
  doubly embedded, 63
Storage register, 142
Super-luminal communication, 159
Superposition principle, 93
Support vector, 67
Support vector machine, 3, 45, 66, 215
  least-squares, 215
Suzuki-Trotter formula, 165, 218
Swap test, 175

T
Tensor network, 260
Test set, 33
Text prediction, 63
Thermalisation, 59
Thermal state, 231
Thermodynamics, 159
Time-evolution operator, 87
Time series forecasting, 24
Training, 30, 39
Training error, 32
Training strategies, 22

U
Unitary matrix, 81, 82

V
Vapnik-Chervonenkis dimension, 132, 134, 136
Variational algorithm

    hybrid training, 224
Variational circuit, 195, 225, 256
Variational classifier, 230, 237
Variational eigensolver, 226
Visible units, 56
Von Neumann equation, 87, 90

W
Wald interval, 147

Weak-coupling, 90
Weierstrass approximation theorem, 46
Weight vector, 28
Wilson score interval, 147

X
XOR function, 37

Printed by Printforce, the Netherlands